Yanpei Liu
Combinatorial Functional Equations

Also of Interest

Algebraic Elements of Graphs
Yanpei Liu, 2017
ISBN 978-3-11-048073-3, e-ISBN (PDF) 978-3-11-048184-6,
e-ISBN (EPUB) 978-3-11-048075-7

Topological Theory of Graphs
Yanpei Liu, 2017
ISBN 978-3-11-047669-9, e-ISBN (PDF) 978-3-11-047949-2,
e-ISBN (EPUB) 978-3-11-047922-5

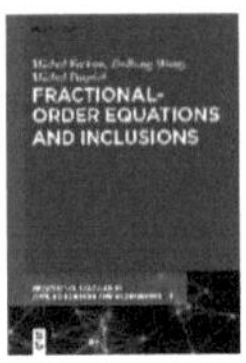

Fractional-Order Equations and Inclusions
Michal Fečkan, JinRong Wang, Michal Pospíšil, 2017
ISBN 978-3-11-052138-2, e-ISBN (PDF) 978-3-11-052207-5,
e-ISBN (EPUB) 978-3-11-052155-9

Elementary Functional Analysis
Marat V. Markin, 2018
ISBN 978-3-11-061391-9, e-ISBN (PDF) 978-3-11-061403-9,
e-ISBN (EPUB) 978-3-11-061409-1

De Gruyter Series in Nonlinear Analysis and Applications
Jürgen Appell et al. (Eds.)
ISSN 0941-813X

Yanpei Liu

Combinatorial Functional Equations

Volume 1: Basic Theory

DE GRUYTER

Author
Prof. Yanpei Liu
Dept. of Mathematics
School of Science
Beijing Jiaotong University
3 Shangyuancun
Haidian District
100044 Beijing
People's Republic of China
ypliu@bjtu.edu.cn

ISBN 978-3-11-062391-8
e-ISBN (PDF) 978-3-11-062583-7
e-ISBN (EPUB) 978-3-11-062422-9

Library of Congress Control Number: 2019945471

Bibliographic information published by the Deutsche Nationalbibliothek
The Deutsche Nationalbibliothek lists this publication in the Deutsche Nationalbibliografie; detailed bibliographic data are available on the Internet at http://dnb.dnb.de.

Cover image: DCrane08 / iStock / Getty Images Plus
Typesetting: VTeX UAB, Lithuania
Printing and binding: CPI books GmbH, Leck

www.degruyter.com

Preface

This monograph, consisting of two books, I and II, includes fresh approaches in the two branches of combinatorics and functional equations, concentrating on algebraic approaches to establishing a rigorous theory for discussing the property of being well-defined and solutions for which it is not necessary to care about convergence or non-convergence and suitability. Its central feature is in building up a theory for unifying the theories of counting distinct classes in classifications under a variety of isomorphisms on a variety of combinatorial configurations, particularly maps (rooted and un-rooted), embeddings of graphs on surfaces, even graphs themselves and so forth, with an infinite partition vector as given parameter.

This monograph is on the basis of my previous books: *Enumerative Theory of Maps* published by Kluwer (Springer) in 1999, *General Theory of Map Census* by Science in 2009 and *Theory of Combinatorial Functional Equations* (in Chinese) published by USTC Press in 2015, by introducing the extension of an integral domain which is a ring of obeying the cancelation law, a mathematical theory for a series of combinatorial functional equations, discovered mostly in the last 30 years or more in specific cases, relevant for counting non-isomorphic classes, via certain classifications of combinatorial configurations, particularly combinatorial maps, graphic embedding on surfaces, even graphs themselves, lattices, networks, hypergraphs, matroids, words, designs, cryptographics, to name only a few, with an infinite partition vector as given parameter.

This monograph might be seen as an advanced version of the previous one, reflecting a series of items of progress made since then.

First, almost all equations are generalized to have constant coefficients of certain terms for each equation. These constants are arbitrarily given mostly in $\mathbb{Z}_+$ because of their original usage.

Second, all functional equations related to plane trees and near-trees are shown to have their solutions in the form of an explicision (or explicit expression) summation-free obtained only by transformations on the extension of integral domain.

Third, all functional equations related to outer planar maps are shown to have their solutions in the form of an explicision obtained only by transformations on the extension of integral domain.

Fourth, all functional equations related to planar maps are indirectly shown to have their solutions in the form of explicisions implied from an investigation of a corresponding planar embedding by introducing a new extra parameter vector.

Fifth, all functional equations related to maps on surfaces are still indirectly shown to have their solutions in the form of explicisions looking not so complicated as in the third case via investigations of the embedding of underlying graphs with symmetry considered.

https://doi.org/10.1515/9783110625837-201

Sixth, all solutions of equations considered have a specific case as one of the applications done just from determining the number of certain equivalent classes of maps under a given parameter vector.

The whole book is concentrated on contemplating the constructibility and the realizability not only for systematization in theory but also for efficientization in running and for intelligentization in usage.

The constructibility enables us to emphasize on exploiting the inner constructions for the consistency theory established on each equation considered in the extension of integral domain. The realizability enables us to evaluate the solution via a number of operations on the extension of integral domain itself as well.

Although all equations have a combinatorial meaning as a special case or in one of their applications, the basic theoretical principles presented in this subject can be seen as pure mathematics, independent of combinatorics, particularly, from the point of view of maps related to graph theory.

For each equation with meson functional, from certain restrictions, a number of function equations, difference (straight and slope) equations, and differential (ordinary and partial) equations are also involved with as an application for classifying varieties of combinatorial maps. From these, explicit expressions of the solutions for the corresponding meson functional equations are indirectly extracted, as shown in each of the chapters from Chapter 3 through Chapter 21.

The monograph is divided into two books, I and II. In Book I, the central content is on basic theory of equations with or without a functional. And book II is on an advanced theory of meson functional equations because of its universality.

This volume is Book I titled *Combinatorial Functional equations* I—*Basic Theory*, which contains Introduction and Chapter 1 through Chapter 10.

Introduction provides an overall view of all the equations particularly mentioned as representatives.

Chapter 1 and Chapter 2 are for the main background from algebra, theory of functions and functionals, and only for meson functionals with the general equation.

From Chapter 3 on through Chapter 7, basic equations including function equations with one and more variables and functional equations with infinite variables under basic functionals such as straight and/or slope differences, ordinary or partial differentials and most simple meson equations under the meson functional are, to a certain extent, investigated. The origins of them are with certain enumerations of a variety of maps in a finite number of parameters on surfaces of smaller genera.

Chapter 3 and Chapter 4 are concerned with equations of functions of, respectively, one and at least two variables.

Chapter 5 is involved with basic functional equations under a straight difference, a slope difference, or both.

Chapter 6 and Chapter 7 are concentrated on, respectively, ordinary and partial differential equations.

From Chapter 8 through Chapter 10, we are only concerned with basic meson functional equations. They are all solved directly by extracting the solution with the coefficient of each term in the solution as an explicision (*i. e.*, explicit expression) in the form of a summation-free, or finite sum with all terms positive in the extension of integral domain.

Chapter 8 addresses two types of tree equations under a meson functional with constant coefficients.

Chapter 9 and Chapter 10 address two types of near-tree equations, or, as we may say, uni-cyclic equation and wintersweets equations, under the meson functional with constant coefficients on the extension of the integral domain.

An attempt has been made to keep the presentation of this book as self-contained as possible. It should not be necessary to read more specialized books beforehand whatever; the concepts of the extension of an integral domain (a ring obeying the cancelation law) and of the meson functional have to be clearly understood.

Because of the concentration only on algebraic qualitative and quantitative theories of all equations considered, many articles on map enumeration are omitted, for which I would like to apologize. However, for more details as regards the references one is referred to the bibliographies of my previous two books: *Enumerative Theory of Maps* [44] (1999, pp. 392–406) and *General Theory of Map Census* [51] (2009, pp. 451–470).

Many people, I'should mention, are, directly or indirectly, contributors to this book. However, I can only name a few because of the limited space. First of all, I have to express my appreciation to Professor W. T. Tutte for indicating to me the potential topic on map enumeration when I was in the University of Waterloo 30 years ago. This topic should now be seen as the origin of the present book. Many of my cooperators used to do, or are still doing, research on the topics around, such as Junliang Cai, Han Ren, Rongxia Hao, Zhaoxiang Li, Wenzhong Liu, Yan Xu, Yongli Zhang, and Liyan Pan, and they are presenting a number of new relevant results. Last but not least, I have also to express my heartiest thanks to Juniang Cai, Rongxia Hao, Zhaoxiang Li and Liangxia Wan for carefully reading the manuscripts to avoid too many errors and mistakes. Of course, any error or mistake remaining belongs to myself.

Daotiancun, Beijing
October, 2018

Y. P. Liu

Contents

Introduction

From Chapter 3 on through Chapter 10, all typical equations considered are listed for an overall picture of this book. The book is divided in four parts.

Part one consists of two chapters: Chapter 3 and Chapter 4 dealing with function equations.

In Chapter 3, six types of equations are covered.

First, we consider the quadratic equation for $f \in \mathcal{R}\{z\}$ with four independent constant parameters

$$\begin{cases} azf^2 - bf + c = 0; \\ f|_{z=0} = d, \end{cases} \tag{1}$$

where $a, b, c, d \in \mathbb{Z}_+$. In this equation, the coefficient az of f^2 involves the variable z.

Second, we consider the quadratic equation for $f \in \mathcal{R}\{z\}$ with four independent constant parameters

$$\begin{cases} af^2 - bf + cz = 0; \\ f|_{z=0} = d, \end{cases} \tag{2}$$

where $a, b, c, d \in \mathbb{Z}_+$. In this equation, the term independent of f involves the variable z.

Third, we consider the quadratic equation for $f \in \mathcal{R}\{z\}$ with four independent constant parameters

$$\begin{cases} a(1+z)f^2 - b(1+z)f + cz = 0; \\ f|_{z=0} = d, \end{cases} \tag{3}$$

where $a, b, c, d \in \mathbb{Z}_+$, $abc > 0$. Several coefficients (like $a(1+z)$, $b(1+z)$ and cz) involve the variable z.

Fourth, we consider the cubic equation for $f \in \mathcal{R}\{z\}$

$$\begin{cases} f^3 + \dfrac{a(1-z)}{z^2}f^2 + \dfrac{b(z-2)}{z}f + c = 0; \\ f|_{z=0} = d, \end{cases} \tag{4}$$

where $a, b, c, d \in \mathbb{Z}_+$.

Fifth, we consider the cubic equation for $f \in \mathcal{R}\{z\}$

$$\begin{cases} azf^3 - 3bzf^2 + (3z-1)f + c = 0; \\ f|_{z=0} = d, \end{cases} \tag{5}$$

where $a, b, c, d \in \mathbb{Z}_+$.

https://doi.org/10.1515/9783110625837-202

Sixth, we consider the quartic equation for $f \in \mathcal{R}\{z\}$

$$\begin{cases} zf^4 - a(1-z)f^3 + b(1-3z)f^2 + 3zf - cz = 0; \\ f|_{z=0} = d, \end{cases} \tag{6}$$

where $a, b, c, d \in \mathbb{Z}_+$.

In Chapter 4, five types of equations are covered.

First, we consider the quadratic equation for $f \in \mathcal{R}\{x, y\}$ with four independent constant parameters

$$\begin{cases} axy^2f^2 + (x-1)f + c(x-1) = 0; \\ f|_{x=0,y=0} = d, \end{cases} \tag{7}$$

where $a, c, d \in \mathbb{Z}_+$, $ac > 0$.

Second, we consider the quadratic equation for $f \in \mathcal{R}\{x, y\}$

$$\begin{cases} \dfrac{axyf}{x-f} - (1+y)f + cx^2y = 0; \\ f_{x=0,y=0} = d, \end{cases} \tag{8}$$

where $a, c, d \in \mathbb{Z}_+$.

Third, we consider the linear equation for $f \in \mathcal{R}\{x, y\}$

$$\begin{cases} f = cx^2y + \dfrac{axy}{1-xy}\left(\dfrac{x}{1-x}h - \dfrac{1}{1-x}f\right) \\ f|_{x=0,y=0} = d, \end{cases} \tag{9}$$

where $a, c, d \in \mathbb{Z}_+$ and $h = f(1, y)$.

Fourth, we consider the quadratic equation for $f \in \mathcal{R}\{x, y\}$

$$\begin{cases} ax^2y(1-x^2)f^2 - (1-x^2+x^2y)f + c(1-x^2) + bx^2yh = 0; \\ f|_{x=0,y=0} = d, \end{cases} \tag{10}$$

where $a, b, c, d \in \mathbb{Z}_+$ and $h = f(1, y)$.

Fifth, we consider the equation of higher degree for $f \in \mathcal{R}\{x, y\}$

$$\begin{cases} f = cx^2y + \dfrac{ax^2y(f-h)}{x^2(1+h)^2 - (1+f)^2}; \\ f|_{y=0} = d(\Rightarrow h_{y=0} = d), \end{cases} \tag{11}$$

where $a, c, d \in \mathbb{Z}_+$ and $h = f(1, y)$.

Part two consists of basic functional equations which involve fundamental functionals as differences (straight and slope) in Chapter 5.

In Chapter 5, five types of equations are covered.

First, we consider the equation in which only one straight difference occurs for $f \in \mathcal{R}\{x, y\}$

$$\begin{cases} f(1 - axy\delta_{1,x}(xf)) = c; \\ f|_{x=y=0} = d, \end{cases} \tag{12}$$

where $a, c, d \in \mathbb{Z}_+$ and $\delta_{1,x}(xf) = ((xf)_{x=1} - (xf))/(1 - x)$ is the straight difference of xf between 1 and x shown in Chapter 2.

Second, we consider the equation in which two straight differences occur for $f \in \mathcal{R}\{x, y\}$

$$\begin{cases} f(1 + ayzt(f_{t=1} + f_{z=1})) = c + byzt(\delta_{1,t}(tf) + \delta_{1,z}(zf)); \\ f|_{y=0 \Rightarrow z=t=0} = d, \end{cases} \tag{13}$$

in which, both straight differences of z and t occur for $a, b, c, d \in \mathbb{Z}_+$.

Third, we consider the equation in which only one slope difference occurs for $f \in \mathcal{R}\{x, y\}$

$$\begin{cases} f = c + ax^2yf^2 + by\partial_{1,x}(x^2f) - xyhf - (h-1)(f-1); \\ f|_{x=0 \Rightarrow y=0} = d \quad \text{(initial condition!)}, \end{cases} \tag{14}$$

where $a, b, c, d \in \mathbb{Z}_+$ and $h = f|_{x=1} \in \mathcal{R}\{y\}$ with only one slope difference of x shown in Chapter 2.

Fourth, we consider the equation in which two slope differences occur for $f \in \mathcal{R}\{y, z, t\}$

$$\begin{cases} f = 2cyz^2t + \dfrac{ayzt\partial_{1,z}f}{1 - \frac{\partial_{1,z}f_{t=1}}{2}} - \dfrac{byzt\partial_{1,t}f}{1 - \frac{\partial_{1,t}f_{z=1}}{2}} \\ f|_{y=0 \Rightarrow z=t=0} = d, \end{cases} \tag{15}$$

where $a, b, c, d \in \mathbb{Z}_+$.

Fifth, we consider the equation in which one straight difference and one slope difference occur for $f \in \mathcal{R}\{x, y\}$

$$\begin{cases} f = c + bxy\delta_{1,x}(xf) + \dfrac{ax^2y(\delta_{1,x}f)^2}{1 - (1 + \partial_{1,x}f)}; \\ f|_{y=0 \Rightarrow x=0} = d, \end{cases} \tag{16}$$

such that $f \in \mathcal{R}\{x, y\}$ for $a, b, c, d \in \mathbb{Z}_+$.

Part three consists of basic functional equations involving fundamental functionals: ordinary differential and partial differential in, respectively, Chapter 6 and Chapter 7.

In Chapter 6, six ordinary differential equations are covered.

First, we consider the equation in which one ordinary differentiation occurs with a variable parameter for $h \in \mathcal{R}\{y\}$

$$\begin{cases} y\dfrac{\mathrm{d}h}{\mathrm{d}y} = 2\tau\left(2ay\dfrac{\mathrm{d}h}{\mathrm{d}y} + bh\right); \\ h|_{y=0} = d, \end{cases} \tag{17}$$

where $a, b, d \in \mathbb{Z}_+$ and τ is known from

$$\partial_y^n \tau = \begin{cases} 0, & \text{when } n = 0; \\ 1, & \text{when } n = 1; \\ 3\sum_{i=1}^{n-1} \partial_y^i \tau \partial_y^{n-i} \tau, & \text{when } n \geq 2, \end{cases}$$

for $n \geq 0$.

Second, we consider the equation in which one ordinary differentiation occurs for $h \in \mathcal{R}\{x\}$

$$\begin{cases} 2x^2\dfrac{\mathrm{d}h}{\mathrm{d}x} = -c + a(1-x)h; \\ h_0 = h|_{x=0} = d, \end{cases} \tag{18}$$

where $a, c, d \in \mathbb{Z}_+$.

Third, we consider the quadratic equation in which one ordinary differentiation occurs for $f \in \mathcal{R}\{x\}$

$$\begin{cases} 2x^2\dfrac{\mathrm{d}f}{\mathrm{d}x} = -c + b(1-x)f - axf^2; \\ f_0 = f|_{x=0} = d, \end{cases} \tag{19}$$

where $a, b, c, d \in \mathbb{Z}_+$

Fourth, we consider the quadratic equation in which one ordinary differentiation occurs for $f \in \mathcal{R}\{x\}$ with two variable parameters

$$\begin{cases} 4x^2\dfrac{\mathrm{d}f}{\mathrm{d}x} = b\alpha(x)f - axf^2 - 2cx\beta(x); \\ \left.\dfrac{\mathrm{d}f}{\mathrm{d}x}\right|_{x=0} = d, \end{cases} \tag{20}$$

for $a, b, c, d \in \mathbb{Z}_+$ where

$$\begin{cases} \alpha(x) = 1 - 2x - 2xf_{\mathrm{Orien}}; \\ \beta(x) = f_{\mathrm{Orien}} + 2x\dfrac{\mathrm{d}f_{\mathrm{Orien}}}{\mathrm{d}x}, \end{cases}$$

and f_{Orien} the solution of equation (19).

Fifth, we consider the quadratic equation in which one ordinary differentiation occurs for $f \in \mathcal{R}\{x\}$ with three independent constant parameters

$$\begin{cases} ax^2\dfrac{\mathrm{d}f}{\mathrm{d}x} = -d + (1 - bx)f - cxf^2 \\ f|_{x=0} = d, \end{cases} \tag{21}$$

where $a, b, c, d \in \mathbb{Z}_+$.

Sixth, we consider the quadratic equation in which one ordinary differentiation of second order occurs for $f \in \mathcal{R}\{x\}$

$$\begin{cases} \left(2az + 5bf - 3cz\dfrac{\mathrm{d}f}{\mathrm{d}z}\right)\dfrac{\mathrm{d}^2f}{\mathrm{d}z^2} = 48z; \\ f|_{z=0} = d, \quad \dfrac{\mathrm{d}f}{\mathrm{d}z}\Big|_{z=0} = d, \end{cases} \tag{22}$$

where $a, b, c, d \in \mathbb{Z}_+$.

In Chapter 7, seven partial differential equations are covered.

First, we consider the quadratic function equation for $f \in \mathcal{R}\{x, y\}$

$$\begin{cases} x^4yf^2 + a(y - x^2)f - bx^2yf^* + c(x^2 - y) = 0; \\ f|_{x=y=0} = d, \end{cases} \tag{23}$$

where $a, b, c, d \in \mathbb{Z}_+$ and $f^* = \partial_x^2 f$ used in the partial differential equations appearing below.

Second, we consider the system of partial differential equations about $(g, f) \in \mathcal{R}^2\{x, y\}$

$$\begin{cases} g = \dfrac{x^4y(f + x\frac{\partial f}{\partial x}) - yx^2g^*}{x^2 - y - 2x^4yf}; \\ f = \dfrac{x^4yf^2 - x^2yf^* + x^2 - y}{x^2 - y}; \\ f|_{x=y=0} = 1, \quad g|_{x=y=0} = 0, \end{cases} \tag{24}$$

where $f^* = \partial_x^2 f$ and $g^* = \partial_x^2 g$.

Third, we consider the system of partial differential equations for $(g, f, h) \in \mathcal{R}^3\{x, y\}$

$$\begin{cases} x^4y\left(z\dfrac{\partial g}{\partial z}\right)\Big|_{z=x} = x^2(1 - 2x^2yh)f - yf_{i_x \geq 4}; \\ x^3zy\delta_{z,x}(uh|_{x=u}) = (x^2 - 2x^4yh)g - yg_{i_x \geq 3}; \\ x^4yh^2 + (y - x^2)h - x^2yh_{2_x} + x^2 - y = 0; \\ f|_{x=y=0} = 0; \quad g|_{x=z=y=0} = 1; \quad h|_{x=y=0} = 1, \end{cases} \tag{25}$$

where $f_{i_x \geq 4}$ and $g_{i_x \geq 3}$ are the results from functions f and g deleting the terms of x with degrees less than and equal to, respectively, 4 and 3.

Fourth, we consider the system of partial differential equations for $(g, f, h, p) \in \mathcal{R}^4\{x, y\}$

$$\begin{cases} x^4y\left(x\frac{\partial g}{\partial x} + \left[z\frac{\partial h}{\partial z}\right]_{z=x}\right) = x^2f - x^4y(g + g^2 + 2pf) - y(f - x^2\partial_x^2 f); \\ x^4y\left(p + x\frac{\partial p}{\partial x}\right) = x^2(x^2 - y - 2x^4yp)g + x^2y\partial_x^2 g; \\ \frac{x^3zy}{z - x}\delta_{z,x}(up|_{x=u}) = x^2(1 - 2x^2yp)h - y(h - x^2\partial_x^2 h); \\ x^4yp^2 + (y - x^2)p - x^2y\partial_x^2 p + x^2 - y = 0; \\ f|_{x=z=y=0} = g|_{x=z=y=0} = h|_{x=z=y=0} = 0; \quad p|_{x=z=y=0} = 1. \end{cases} \tag{26}$$

Fifth, we consider the equation in which two partial differentiations occur for $f \in \mathcal{R}\{x, y\}$ with three constant parameters

$$\begin{cases} axy\left(2y\frac{\partial f}{\partial y} - x\frac{\partial f}{\partial x}\right) = (1 - xyf|_{x=1})f - c; \\ f|_{x=0.y=0} = d, \end{cases} \tag{27}$$

where $a, c, d \in \mathbb{Z}_+$ and $a \neq 0$.

Sixth, we consider the equation in which one partial differentiation with a constant parameter occurs for $f \in \mathcal{R}\{x, y\}$ with four constant parameters

$$\begin{cases} ax^3y\frac{\partial f}{\partial x} = \left(1 - ax^2y + \frac{xy}{1 - x}\right)f - \frac{bx^2y}{1 - x}f|_{x=1} - c(xy + 1); \\ f|_{x=0,y=0} = d, \end{cases} \tag{28}$$

where $a, b, c, d \in \mathbb{Z}_+$ and $a \neq 0$.

Seventh, we consider the equation in which one partial differentiation occurs for $f \in \mathcal{R}\{x, y\}$ with four constant parameters

$$\begin{cases} 2ax^4y\frac{\partial f}{\partial x^2} = \left(1 - ax^2y + \frac{x^2y}{1 - x^2}\right)f - \frac{bx^2y}{1 - x^2}f|_{x=1} - c; \\ f|_{x=0,y=0} = d, \end{cases} \tag{29}$$

where $a, b, c, d \in \mathbb{Z}_+$ and $a \neq 0$.

Part four consists of basic equations involving the meson functional in Chapter 8 through Chapter 10.

In Chapter 8, two types of basic functional equations are covered.

First, we consider a most simple equation with the meson functional and three independent constant coefficients for $f \in \mathcal{R}\{x, \mathbf{y}\}$

$$\begin{cases} \displaystyle\int\!\!\int_y \frac{ay^2 f}{1-cyf} = f - by_1; \\ f|_{\mathbf{y}=\mathbf{0}} = d, \end{cases} \tag{30}$$

where $a, b, c, d \in \mathbb{Z}_+, f = f(\mathbf{y}) \in \mathcal{R}\{x, \mathbf{y}\}$ and $\mathbf{y} = (0, y_2, y_3, \ldots)$.

Second, we consider another most simple equation with the meson functional and three independent constant coefficients for $f \in \mathcal{R}\{x, \mathbf{y}\}$

$$\begin{cases} ax \displaystyle\int_y y\delta_{x,y}(uf|_{x=u}) = f - c; \\ f|_{x=0,\mathbf{y}=\mathbf{0}} = d, \end{cases} \tag{31}$$

where $a, c, d \in \mathbb{Z}_+$ and $f = f(x, \mathbf{y}) \in \mathcal{R}\{x, \mathbf{y}\}$ for $\mathbf{y} = (y_1, y_2, y_3, \ldots)$.

In Chapter 9, only one type of basic equation involving the meson functional is covered.

Consider the meson equation for $f \in \mathcal{R}\{x, \mathbf{y}\}$ with three independent constant coefficients,

$$\begin{cases} a_2 x \displaystyle\int_y y\partial_{x,y} f|_{x=u} = f - a_1 x^2 f_{\text{gtree}}; \\ f|_{x=0,\mathbf{y}=\mathbf{0}} = a_0, \end{cases} \tag{32}$$

where $a_0, a_1, a_2 \in \mathbb{Z}_+, f_{\text{gtree}} \in \mathcal{R}\{x, \mathbf{y}\}$ is the solution of equation (31) with $a = c = d = 1$. Because the solution of equation (32) when $a_0 = a_1 = a_2 = 1$ is the enufunction of general plane rooted trees with the root-vertex valency (x) and the vertex partition vector ($\mathbf{y}$) as parameters, equation (32) is called a type of *unicycle model*.

In Chapter 10, another type of basic equation involving with the meson functional is covered.

Consider the meson equation for $f \in \mathcal{R}\{x, \mathbf{y}\}$ with four independent constant coefficients as

$$\begin{cases} a_2 x \displaystyle\int_y y\delta_{x,y}(uf|_{x=u}) = \left(1 - \frac{a_3 x y_3}{1-y_2}\right) f - a_1; \\ f|_{x=0\Leftrightarrow\mathbf{y}=\mathbf{0}} = a_0, \end{cases} \tag{33}$$

where $a_0, a_1, a_2, a_3 \in \mathbb{Z}_+$.

Because of a solution of equation (33) for $a_0 = a_1 = a_2 = a_3 = 1$, meaningful in wintersweets as outer planar maps, this equation is called a *wintersweets model*.

1 Preliminaries

For the sake of brevity we adopt, throughout this book, the usual logical conventions: disjunction, conjunction, negation, implication, equivalence, universal quantification, belong to, and existential quantification, denoted by the familiar symbols: $\vee$, $\wedge$, $\neg$, $\Rightarrow$, $\Leftrightarrow$, $\forall$, $\in$ and $\exists$, respectively.

1.1 Sets and mappings

A *set* consists of objects considered to have a property in common. The objects are called *elements* of the set. If a set consists of all objects considered, then the set is said to be *universal*, and denoted by Ω. Usually, sets are represented by capital letters, as A, B, C, and elements by lower case ones, as a, b, c. The statement "a is an element of the set A" is denoted "$a \in A$". If any element of set A is an element of set B, then A is called a *subset* of B, denoted by $A \subseteq B$. A set without an element is the *empty set*, denoted by $\emptyset$.

Any set A is a subset of A itself and the empty set is always a subset of any set. We denoted by $\mathcal{O}$ the set, or *family* of all subsets, in the universal set, *i. e.*, $\mathcal{O} = \{A \mid A \subseteq \Omega\}$, or 2^{Ω}. Naturally, $\emptyset, \Omega \in \mathcal{O}$.

For two sets $A, B \in \mathcal{O}$, we denote by $\cup$ and $\cap$ the two operations called, respectively, *union* and *intersection*, *i. e.*,

$$A \cup B = \{x \mid x \in A \text{ or } x \in B\} \quad \text{and} \quad A \cap B = \{x \mid x \in A, x \in B\}.$$

Set A *minus* set B, denoted by $A \backslash B$, is the result of deleting all elements of B from A, or called the *difference* of A and B.

If $B \subseteq A$, the difference is denoted by $A - B$. If $A = \Omega$, the difference is denoted by $\bar{B} = \Omega - B$, called the *complement* of B.

Because each of $\cup$ and $\cap$ satisfies the *commutative law* and the *associative law*, we are allowed to adopt

$$\bigcup_{i=1}^{n} A_i \quad \text{and} \quad \bigcap_{i=1}^{n} A_i \tag{1.1.1}$$

where $A_i \in \Omega$ and both i and n, where $1 \leq i \leq n$ and $n \geq 1$, are positive integers. Moreover, they satisfy the distributive law. These laws are similar to what appeared for addition and multiplication in arithmetics. However, the idempotent law, the absorption law, the unitary law and the universal bound law are not available for similarity as regards arithmetics (see § 1.1 in [52] (Liu YP)). On the basis of what was mentioned above, the following results can be found.

https://doi.org/10.1515/9783110625837-001

Theorem 1.1.1. *For any $A, X \subseteq \Omega$, we have*

$$\begin{cases} (A \cap X = A) \vee (A \cup X = X) \Rightarrow A = \emptyset; \\ (A \cap X = X) \vee (A \cup X = A) \Rightarrow A = \Omega. \end{cases} \tag{1.1.2}$$

Theorem 1.1.2. *For any $A, B \subseteq \Omega$, we have*

$$A \cap B = A \Leftrightarrow A \cup B = B. \tag{1.1.3}$$

Theorem 1.1.3. *For any $A, B, C \subseteq \Omega$, we have*

$$(A \cap B = A \cap C) \wedge (A \cup B = A \cup C) \Leftrightarrow B = C. \tag{1.1.4}$$

Theorem 1.1.4. *For any $A \subseteq \Omega$, we*

$$\overline{\overline{A}} = A. \tag{1.1.5}$$

Theorem 1.1.5. *For any $A, B \subseteq \Omega$, we have*

$$\overline{A \cup B} = \overline{A} \cap \overline{B}; \quad \overline{A \cap B} = \overline{A} \cup \overline{B}. \tag{1.1.6}$$

Let $A, B \subseteq \Omega$. A *mapping* from A to B is a correspondence between A and S such that any element of A has a corresponding element of B. An element in A is said to be *co-image* (or *back image*, or *initial image*), and an element in B, *image* (or *forward image*, or *end image*).

For two sets A and B,

$$A \times B = \{(x, y) \mid \forall x \in A, \forall y \in B\},$$

is called their *Cartesian product*. The Cartesian product of a set X and itself is called the *power* of X. For example, $X \times X = X^2$. Generally, $X^{n-1} \times X = X^n$ where $n \geq 1$. Particularly, $X^0 = \emptyset$ and $X^1 = X$.

It is seen that $\cup$, $\cap$ and $\setminus$ are all mappings from $2^\Omega \times 2^\Omega$ to 2^Ω ($2^\Omega \times 2^\Omega \to 2^\Omega$), and $^-$ from 2^Ω to itself ($2^\Omega \to 2^\Omega$).

An *injection* from set A to set B is a mapping $\alpha : A \to B$ such that, $\forall a, b \in A$,

$$a \neq b \Rightarrow \alpha(a) \neq \alpha(b).$$

An injection is also called a 1–1 mapping. A *surjection* is a mapping $\beta : A \to B$ such that, $\forall b \in B$,

$$\exists a \in A, \quad \beta(a) = b.$$

For example, union $\cup$, intersection $\cap$, and difference $\setminus$ are all surjections, but not injections. A mapping with both injection and surjection is called a *bijection*, or 1–1 correspondence. For example, the complement $^-$ is a bijection.

If repetition of an element in a set is allowable, then the set is said to be *multiple*; otherwise, *nonmultiple*. All sets considered in this book are nonmultiple unless specifically indicated otherwise. The number of occurrences of an element in a multiple set is called the *multiplier* of the element.

Two sets A and B with a bijection are said to have their *cardinalities* equal, *i. e.*, $|A| = |B|$. For example, the set of all positive integers and the set of all positive even numbers have the same cardinality. A set with its cardinality a finite number is said to be a *finite*; otherwise, an *infinite set*. Two finite sets have the same number of elements if, and only if they have the same cardinality. For two finite sets X and T, we have

$$|X \times Y| = |X| \times |Y| = |X||Y|.$$

Let X^Y be the set of all mappings from X to Y, then $|X^Y| = |Y|^{|X|}$.

An *isomorphism*, denoted by $A \sim B$, of two multiple sets A and B is a bijection between A and B such that corresponding elements have the same multiplier. It is easily seen that $A \sim B$ implies that their cardinalities are equal.

Theorem 1.1.6. *Two multiple sets A and B have an isomorphism τ if, and only if,*

$$|A| = |B| \tag{1.1.7}$$

with $\forall a \in A$, $m(a) = m(\tau(a))$ where m is the multiplier.

Proof. Necessity is from the definition of isomorphism. Sufficiency is from the definition of bijection. □

As consequence, two sets A and B are isomorphic if, and only if, $|A| = |B|$.

However, the recognition of isomorphism between two systems $(2^A; \cup, \cap, ^-)$ and $(2^B; \cup, \cap, ^-)$ is not so easy in general because of the three operations involved.

On a set $A \neq \emptyset$, if there is an operation, denoted by $\diamond$, such that the following four axioms: Group 1–Group 4 are satisfied, then we call A a *group*, denoted by $(A; \diamond, 1_\diamond)$.

Group 1 (closed law) $\forall x, y \in A$, $x \diamond y \in A$.
Group 2 (associative law) $\forall x, y, z \in A$, $(x \diamond y) \diamond z = x \diamond (y \diamond z)$.
Group 3 (identity law) $\exists 1_{(\diamond)}$ (simply 1) $\in A$, such that $\forall x \in S$, $x \diamond 1_{(\diamond)} = x$.
Group 4 (inverse law) $\forall x \in S$, $\exists y \in S$, $x \diamond y = 1_{(\diamond)}$.

If $\diamond$ satisfies the commutative law in the group, then it is called a *commutative group*, or *Abelian group*. For an Abelian group, the operation $\diamond$ is always replaced by $+$ and the identity by 0.

If there is another operation, denoted by $\cdot$, on an Abelian group $(A; +, 0)$ such that $\cdot$ satisfies Group 1–Group 3 and $\{+, \cdot\}$ satisfies the *distributive law*: $\forall a, b, c \in A$,

$$(a + b) \cdot c = a \cdot c + b \cdot c \qquad c \cdot (a + b) = c \cdot a + c \cdot b, \tag{1.1.8}$$

then A is a *ring*, denoted by $(A; +, \cdot, 0, 1)$.

On the ring, if $\cdot$ satisfies the commutative law, then the ring is said to be *commutative*. If a commutative ring A satisfies the *cancelation law*: $\forall a, b, c \in A$, $c \neq 0$, $a \cdot c = b \cdot c \Rightarrow a = b$, then the ring is called an *integral domain*.

On a commutative ring $(A; +, \cdot, 0, 1)$, if for any $a \in A$, $a \neq 0$, there exists a^{-1} such that $a \cdot a^{-1} = a^{-1} \cdot a = 1$, then it is called a *field*.

A *space* (exactly, *vector space* or *linear space*) over a field F, denoted by $(\mathcal{X}, F; +, \bullet)$ (simply $\mathcal{X}$), is an Abelian group $(\mathcal{X}, +)$, or $\mathcal{X}$, in company with the field $(F, +, \bullet)$, or simply F, satisfying the following four axioms: Space 1–Space 4. The operation "+" is called *vector sum*, "•" a *scalar product*. In the Abelian group $\mathcal{X}$ and the field F, addition symbols are the same. For the scalar product $a \bullet A = aA$ for $a \in F$ and $A \in \mathcal{X}$, one adopts the same symbol as the multiplication in F. An element of $\mathcal{X}$ is called a *vector*. An element of F is called a *scalar*.

Space 1 $\forall a \in F, \forall A, B \in \mathcal{X}, a(A + B) = aA + aB$.
Space 2 $\forall a, b \in F, \forall A \in \mathcal{X}, (a + b)A = aA + bA$.
Space 3 $\forall a, b \in F, \forall A \in \mathcal{X}, (ab)A = a(bA)$.
Space 4 $\forall A \in \mathcal{X}, 1A = A$.

One might like to understand the distinctions in symbols between vectors and scalars. $0_{\mathcal{X}}$ and 0_F denote the elements zero of, respectively, $\mathcal{X}$ and F. From Space 1–Space 4, it is seen that $\forall A \in \mathcal{X}$, $0_F A = 0_{\mathcal{X}}$ and $\forall a \in F$, $a0_{\mathcal{X}} = 0_{\mathcal{X}}$. Hence, both $0_{\mathcal{X}}$ and 0_F are only denoted by 0. For $\mathcal{Y} \subseteq \mathcal{X}$, if $\mathcal{Y}$ is a space itself with the same operations as in $\mathcal{X}$, then $\mathcal{Y}$ is called a *subspace* of $\mathcal{X}$, denoted by $\mathcal{Y} \subseteq_{\text{vect}} \mathcal{X}$ (or simply, $\mathcal{Y} \subseteq \mathcal{X}$ if there arises no confusion). Because of 0 being itself a space, called the *zero space* or a *trivial space*, denoted by 0 as well, 0 is a subspace of any space. The subspace consists of all vectors of order 2 and 0 is denoted by $\mathcal{J}$.

1.2 Functions and transformations

Let $\mathcal{O} = 2^{\mathbb{X}}$, when $\mathbb{X}$ is a set of numbers; a mapping ϕ from $A \in \mathcal{O}$ to $B \in \mathcal{O}$ is a *function*. A and B are, respectively, called *domain* and *co-domain* of ϕ. The sets $D = \{x \mid \exists y \in B, y = \phi(x)\} \subseteq A$ and $Y = \{y \mid \exists x \in A, y = \phi(x)\} \subseteq B$ are, respectively, the *co-domain set* and *image set*(or *range*) of ϕ. If $D \subseteq X^n$, n is a positive integer, then the function ϕ is called a *function of n variables* (or *n-function*), denoted by $\phi = \phi(\mathbf{x})$, $\mathbf{x} = (x_1, x_2, \ldots, x_n)$ is a row vector. When $n = 1$, ϕ is a *function of one variable*; otherwise, a *function of several variables*. If $\phi(0) = 0$, then ϕ is a *homogeneous function*.

Although the set of all integer numbers $\mathbb{Z}$, the set of all rational numbers $\mathbb{Q}$, the set of real numbers $\mathbb{R}$ and the set of all complex numbers $\mathbb{C}$ are involved in this book, domains and ranges are always in $\mathbb{R}$ unless specifically indicated otherwise.

By two functions f and g being equal, *i. e.*, $f = g$, is meant that their domains are the same and their ranges are the same with the property that, for any x in the domain, $f(x) = g(x)$.

If a function ϕ has the property: for any $\mathbf{x}$ and $\mathbf{y}$,

$$\phi(\mathbf{x}+\mathbf{y}) = \phi(\mathbf{x}) + \phi(\mathbf{y}), \tag{1.2.1}$$

then ϕ is said to be a *linear function*; otherwise, a *nonlinear function*. Let $\mathbf{a} = (a_1, a_2, a_3, \ldots)$ and $\mathbf{b} = (b_1, b_2, b_3, \ldots)$, then $(\mathbf{a}, \mathbf{b}) = a_1 b_1 + a_2 b_2 + \cdots + a_n b_n = \mathbf{a}(\mathbf{b})'$ is called the *inner product* of $\mathbf{a}$ and $\mathbf{b}$ where $'^{=\mathrm{T}}$ represents the *transpose*.

Theorem 1.2.1. *A homogeneous function $\phi : \mathbb{R}^n \to \mathbb{R}$ is linear if, and only if, there exists a constant vector $\mathbf{a}$ such that*

$$\phi = (\mathbf{a}, \mathbf{x}). \tag{1.2.2}$$

Proof. Since ϕ is a homogeneous function,

$$\phi(\mathbf{0}) = \mathbf{0}. \tag{1.2.3}$$

From the linearity, *i. e.*, (1.2.1), by substituting $\mathbf{y}$ for $-\mathbf{x}$, (1.2.3) leads to $\phi(\mathbf{x}) + \phi(-\mathbf{x}) = \mathbf{0}$. Hence,

$$\phi(-\mathbf{x}) = -\phi(\mathbf{x}). \tag{1.2.4}$$

On the basis of (1.2.1), for any positive integer n, we have

$$\phi\left(\sum_{i=1}^{n} \mathbf{x}_i\right) = \sum_{i=1}^{n} \phi(\mathbf{x}_i) \tag{1.2.5}$$

where $\mathbf{x}_i = (0, \ldots, 0, x_i, 0, \ldots, 0)$, $i = 1, 2, \ldots, n$.

Let $\mathbf{x} = \mathbf{x}_i$, $1 \le i \le n$, then, for any positive integer n,

$$\phi(n\mathbf{x}) = n\phi(\mathbf{x}). \tag{1.2.6}$$

Because of $\frac{n}{m}\mathbf{x} = \mathbf{y}$, *i. e.*, $n\mathbf{x} = m\mathbf{y}$, from (1.2.6), we have $n\phi(\mathbf{x}) = m\phi(\mathbf{y})$, *i. e.*,

$$\phi\left(\frac{m}{n}\mathbf{y}\right) = \phi(\mathbf{x}) = \frac{m}{n}\phi(\mathbf{y}). \tag{1.2.7}$$

From (1.2.4), (1.2.6) is valid for any rational number n. By the density of rational numbers and the connectedness of function ϕ, for any $\mathbf{a} \in \mathbb{R}$, we have

$$\phi(a\mathbf{x}) = a\phi(\mathbf{x}). \tag{1.2.8}$$

Necessity. Because of the linearity of ϕ, for any $\mathbf{x}, \mathbf{y} \in \mathbb{R}^n$, $\phi(\mathbf{x}+\mathbf{y}) = \phi(\mathbf{x}) + \phi(\mathbf{y})$. Since, for any $\mathbb{R}$,

$$\mathbf{x} = \sum_{i=1}^{n} x_i \mathbf{1}_i,$$

where $\mathbf{1}_i$ is the vector of all components 0 but only the ith 1, by (1.2.5),

$$\begin{aligned}\phi(\mathbf{x}) &= \phi\left(\sum_{i=1}^{n} x_i \mathbf{1}_i\right), \quad \text{by (1.2.8)},\\ &= \sum_{i=1}^{n} x_i \phi(\mathbf{1}_i).\end{aligned}$$

Therefore, (1.2.2) is valid where $\mathbf{a} = (\phi(\mathbf{1}_1), \phi(\mathbf{1}_2), \ldots, \phi(\mathbf{1}_n))$. The necessity is done.

Sufficiency. By considering $\mathbf{a}(\mathbf{x}' + \mathbf{y}') = \mathbf{a}\mathbf{x}' + \mathbf{a}\mathbf{y}'$, from (1.2.1), the sufficiency is done. □

The operation $\mathbf{x} = \mathbf{t} + \mathbf{a}$ on the function $\phi(\mathbf{x})$ is called a *translation*. It is easily seen that any nonhomogeneous function can be obtained by translation.

Let $A = (a_{i,j})_{1\le i\le n, 1\le j\le m}$ be an $n \times m$ (*i. e.*, n rows and m columns) matrix. By employing $\mathbf{x} = A\mathbf{z}'$, the n variables in $\mathbf{x}$ replaced by m variables in $\mathbf{z}$, the procedure is called doing a *linear transformation* on a function $\phi(\mathbf{x})$.

Theorem 1.2.2. *On a ring, all $m \times n$ linear transformations form a linear space.*

Proof. On the basis of the four axioms on a space: Space 1–Space 4, because of all $m \times n$ matrices forming an Abelian group for matrix addition, all linear transformations form a linear space. □

Although Theorem 1.2.2 is not new in linear algebra, attention should be paid to the fact that the valid area much larger than $\mathbb{Z}$, $\mathbb{Q}$, $\mathbb{R}$ and $\mathbb{C}$ *etc.*, or we say that elements in the ring are not necessarily numbers.

Only a square matrix may to have an inverse. All matrices which have an inverse are called *invertible*. It is well known that a matrix is invertible if, and only if, its determinant is not 0.

Theorem 1.2.3. *On a ring (not necessarily commutative) R, all square matrices of determinants not 0 form a noncommutative ring.*

Proof. By the definition of a ring, we arrive at the conclusion. □

Given two functions f and g of one variable x. If the range of g is the range of f, we have the *composition*, denoted by $f \circ g = fg$, *i. e.*, for any x available, $f \circ g(x) = f(g(x))$. However, when $g = f$, $f \circ g \ne f^2$ in general. For example, let $f(x) = x + 1$, then $f \circ f(x) = f(f(x)) = f(x+1) = x + 2$. But $f^2 = (x+1)^2 = x^2 + 2x + 1$. For avoiding confusion, take $f \circ f = f^{\circ 2}$. Recurrently, $f \circ f^{\circ n-1} = f^{\circ n}$, $n \ge 3$. When $f(x) = x$, $f^{\circ n} = f$, $n \ge 2$. For convenience of usage, $f^{\circ 1} = f$ is always assumed. This function is called the *identity*, denoted by 1. The function $f(x) = x^n$, $n \ge 0$, is called a *power function* where n is its *degree*. When $n = 0$, the power function $f(x) = 1$ is confirmed.

Given three functions $h : A \to B$, $g : B \to C$ and $f : C \to D$ of one variable x with both f and g a surjection. Observe that $f \circ g \circ h = fgh$. Because of

$$(fg)h(x) = fg(h(x)) = f(gh(x)) = f(gh)(x),$$

for any $x \in A$, we have $(fg)h = f(gh)$. Thus it is shown that the composition satisfies the associative law. Hence, fgh is meaningful.

Any function $f : A \to B$ has a *left identity*, denoted by 1_A, and a *right identity*, denoted by 1_B, such that

$$1_A f = f = f 1_B. \tag{1.2.9}$$

This is called the *identity law*.

For a function $r : S \to T$ where $S \subseteq A$ and $T \subseteq B$, if for any $x \in S$, $r(x) = f(x)$, the r is called the *restriction* of f, *i. e.*, $r = f|_S$. Conversely, f is called a *extension* of r. For a function $i : S \to A$, if for any $x \in S$, $i(x) = x$, then i is called the *inclusion*. Thus, $i = 1_A|_S$.

A function can always be represented as a composition of two functions: one is an injection and the other is a surjection. Assume the image set of f is $U \subseteq B$, then $f = r \circ i$, where the restriction $r = f|_A : A \to U$ is a surjection and the inclusion $i : U \to B$ is an injection.

Given two functions $f : A \to B$ and $g : B \to A$. We address composition. If $f \circ g = 1_B$, then f is the *left inverse* of g and g, the *right inverse* of f. If a function has both left inverse and the right inverse, then from the identity law (1.2.9), the two inverses are the same, called the *inverse*. The inverse of f is denoted by f^{-1}.

Theorem 1.2.4. *A function of domain not empty has a left inverse if, and only if, it is an injection. A function of domain not empty has a right inverse if, and only if, it is a surjection. A function of domain not empty has an inverse if, and only if, it is a bijection*

Proof. Assume $g : B \to A$ with the left inverse $f : A \to B$, then $fg = 1_B$. Hence, $g(x_1) = g(x_2)$ implies $x_1 = f(g(x_1)) = f(g(x_2)) = x_2$. This shows that g is an injection.

Conversely, assume $g : B \to A$ is an injection. Because of $B \neq \emptyset$, let $x_0 \in B$. Since g is an injection, for any $x \in A$, there is at most one $y \in B$ such that $g(y) = x$. Let the function

$$f(x) = \begin{cases} y(g(y) = x), & \text{when } x \text{ is in the image set of } g; \\ x_0, & \text{otherwise.} \end{cases}$$

It is easily checked, for any $y \in B$, that $f(g(y)) = y$, *i. e.*, $fg = 1_B$. Hence, g has the left inverse. This is the first statement.

Symmetrically, the second statement is a result of the first. Then the third is deduced from the first two statements. □

In the theorem, the condition $B \neq \emptyset$ is of no importance, otherwise it is a degeneracy of the theorem. On the basis of the uniqueness of inverse, we have

$$(g^{-1})^{-1} = g. \tag{1.2.10}$$

Further, the inverse of the composition of two functions obeys the rule of *reversing the order*:

$$(fg)^{-1} = g^{-1}f^{-1}. \tag{1.2.11}$$

Because of there being no commutative law for composition of two functions, the order has to be considered.

Let x be an undeterminate (not necessary a number!), then x^n, integer $n \geq 0$, is called a *monomial* of *degree n*. Take $x^0 = 1$. A *polynomial* is a linear combination of monomials. For example,

$$a_0 + a_1x + a_2x^2 + \cdots + a_nx^n \left(= \mathbf{ax}'_{[n+1]} = \sum_{i=0}^{n} a_ix^i \right) \tag{1.2.12}$$

is the general form of a polynomial of degree n, $n \geq 1$, where

$$\mathbf{a} = (a_0, a_1, a_2, \ldots, a_n) \quad \text{and} \quad \mathbf{x}_{[n+1]} = (1, x, x^2, \ldots, x^n).$$

The components of $\mathbf{a}$ are called the *coefficients* of the polynomial and $\mathbf{a}$, the *coefficient vector* shown in (1.2.12). A monomial with its coefficient in a polynomial is called a *term*.

For a ring R (precisely $(R; +, \circ, 0, 1)$, the symbol $\circ$ is always omitted in expressions); let

$$P_R = \{\mathbf{ax}'_{[n+1]} \mid \forall \mathbf{a} \in R^{n+1}, n \geq 0\},$$

then on P_R, the addition (+) and multiplication ($\circ$, often omitted!) are

$$\sum_{i=0}^{s} a_ix^i + \sum_{i=0}^{t} b_ix^i = \sum_{i=0}^{\max\{s,t\}} (a_i + b_i)_ix^i$$

for $a_i = 0$ $(i > s \geq 0)$ and $b_i = 0$ $(i > t \geq 0)$;

$$\left(\sum_{i=0}^{s} a_ix^i\right)\left(\sum_{j=0}^{t} b_jx^j\right) = \sum_{k=0}^{s+t} c_kx^k$$

where

$$c_k = \sum_{\substack{0 \leq i \leq s, 0 \leq j \leq t \\ i+j=k}} a_ib_j = \sum_{i=0}^{k}\sum_{j=0}^{k-i} a_ib_{k-j}.$$

Theorem 1.2.5. *$(P_R; +, \circ, 0, 1)$ is a ring.*

Proof. First, for addition and multiplication, it is easy to check the axioms: Group 1 (closed law), Group 2 (associative law), commutative law, and distributive law, *i. e.*, (1.1.8).

From (1.2.12), when $\mathbf{a} = (0, 0, 0, \ldots) = \mathbf{0}$, the polynomial is 0 (Group 3 for +) and when $\mathbf{a} = (1, 0, 0, \ldots.) = \mathbf{0}$, the polynomial is 1 (Group 3 for $\circ$).

For +, the inverse of polynomial determined by $\mathbf{a}$ is the polynomial by $-\mathbf{a}$ (Group 4 for +). However, Group 4 is not valid for $\circ$.

In consequence, $(P_R; +, 0)$ is an Abelian group and $(P_R; \circ, 1)$ can only be checked to satisfy the axioms: Group 1–Group 3. By considering the distributive law, the theorem is proved. □

Let $\mathcal{R} = (P_R; +, \circ, 0, 1)$. Because of $\mathcal{R} \subseteq \mathcal{R}[x]$, the ring $\mathcal{R}[x]$ is called an *extension* of $\mathcal{R}$.

In general, because of a polynomial of m ($m \geq 2$) variables and n ($n \geq 2$) degree have the form

$$p_n(\mathbf{x}_m) = \sum_{i=0}^{n} p_i(\mathbf{x}_{m-1}) x_m^i \tag{1.2.13}$$

where $p_i(\mathbf{x}_{m-1})$ is a polynomial of $m - 1$ variables and degree i, by employing Theorem 1.2.5 recurrently, we find that $\mathcal{R}[\mathbf{x}]$ is an extension of ring $\mathcal{R}$.

Theorem 1.2.6. *A polynomial $p(x)$ has $(x - a)$ as a factor if, and only if, $p(a) = 0$.*

Proof. Because of $p(x)$ with factor $(x - a)$, there exists a polynomial $q(x)$ whose degree is at least 1 less than the degree of $p(x)$, so that $p(x) = (x - a)q(x)$. Therefore, $p(a) = 0$.

Conversely, on the basis of $p(a) = 0$, assume $p(x) = c_0 + c_1 x + c_2 x^2 + \cdots + c_n x^n$; we have

$$\begin{aligned} p(x) &= \sum_{i=0}^{n} c_i x^i - \sum_{i=0}^{n} c_i a^i = \sum_{i=1}^{n} c_i (x^i - a^i) \\ &= (x - a)\left(c_1 + \sum_{i=2}^{n} c_i \sum_{j=0}^{i-1} a^j x^{i-1-j} \right). \end{aligned}$$

Therefore, $p(x)$ has a factor $(x - a)$. □

Theorem 1.2.7. *Two polynomials over an infinite field are equal if, and only if, their coefficient vectors are the same.*

Proof. Since the sufficiency is easily seen to be true, only it is necessary only to prove the necessity. Because of two polynomials are equal with the same degree, assume $p(x) = a_0 + a_1 x + \cdots + a_n x^n$ and $q(x) = b_0 + b_1 x + \cdots + b_n x^n$ are equal. If $a_i = b_i$ not for all $0 \leq i \leq n$, then from Theorem 1.2.6, $d(x) = p(x) - q(x)$ has at most n points such that

$d(x) = 0$ and $d(x) \neq 0$ for other points. This is a contradiction to $p(x)$ and $q(x)$ being equal. □

In Theorem 1.2.7, an infinite field cannot be replaced by a finite field. For example, on a mod 3 integer field Z_3, the polynomials $p(x) = x^3 - x$ and $q(x) = 0$ are the same, but their coefficients are different.

Theorem 1.2.8. *There is only one polynomial of degree not greater than n, satisfying the values at n + 1 distinct points.*

Proof. Suppose a_i, $0 \leq i \leq n$, are $n+1$ distinct points. We have $p(a_i) = y_i$, $0 \leq i \leq n$. Now, construct a polynomial $p(x)$ of degree n such that $p(a_i) = y_i$, $0 \leq i \leq n$, *i. e.*,

$$p(x) = \sum_{i=0}^{n} y_i \left(\frac{q_i(x)}{q_i(a_i)} \right) \tag{1.2.14}$$

where

$$q_i(x) = \prod_{\substack{0 \leq j \leq n \\ j \neq i}} (x - a_j).$$

Attention: Because of the distinction among a_i, $0 \leq i \leq n$, $q_i(a_i) \neq 0$, $0 \leq i \leq n$.

In what follows, observe the uniqueness. Assume $r(x)$ is another polynomial of degree n satisfying $r(a_i) = y_i$, $0 \leq i \leq n$. Because of $p(a_i) - r(a_i) = 0$, $0 \leq i \leq n$, the polynomial $p(x) - r(x)$ of degree at most n has $n+1$ factors of degree 1. From Theorem 1.2.6, the only possibility is $p(x) - r(x) = 0$, *i. e.*, $p(x) = r(x)$. □

In proving the theorem, (1.2.14) provides an explicit formula for determining a polynomial of degree n via the values at $n + 1$ distinct points. This formula is called *Lagrange interpolation*. Usage of Lagrange interpolation is not only for determining a polynomial but also for approximating a general nonlinear function.

1.3 Extensions of integral domain on series

A polynomial of infinite degree is called a *series*. The general form of a series is

$$s(x) = a_0 + a_1 x + a_2 x^2 + \cdots = \sum_{i=0}^{\infty} a_i x^i, \tag{1.3.1}$$

where, for any i, $0 \leq i \leq \infty$, $a_i \in R$, R is a ring.

Occasionally, for generality, finite monomials of negative powers are allowed, *i. e.*, for a nonnegative integer L,

$$s_{[-]}(x) = \sum_{i=-L}^{\infty} a_i x^i = \sum_{i=1}^{L} a_{-i} x^{-i} + \sum_{i=0}^{\infty} a_i x^i, \tag{1.3.2}$$

is called the *Laurent series*.

For convenience, a series is represented by its coefficients in a vector. For example, $s(x)$ in (1.3.1) is $\mathbf{a} = (a_0, a_1, a_2, \ldots, \infty)$.

The addition of two series **a** and **b** is defined by

$$\mathbf{a} + \mathbf{b} = \mathbf{c}, \quad \mathbf{c} = (a_0 + b_0, a_1 + b_1, a_2 + b_2, \ldots, \infty).$$

Let $S_R = \{\mathbf{a} \mid \forall \mathbf{a} \in R^{\infty}\}$, *i. e.*, the set od all series such that each of its coefficients is in the ring R, then, by considering the closed law on S_R for addition deduced from that on R, we have $\mathbf{c} \in S$. Similarly, the commutative law and the associative law on S_R are from those on R. Let $\mathbf{0} = (0, 0, 0, \ldots, \infty)$, then $\mathbf{0} + \mathbf{a} = \mathbf{a} + \mathbf{0} = \mathbf{a}$, $\mathbf{0}$ is the zero on S_R. For $\mathbf{a} \in S_R$, the inverse of $-\mathbf{a} = (-a_0 - a_1, -a_2, \ldots, -\infty)$ because of $(-\mathbf{a}) + \mathbf{a} = \mathbf{a} + (-\mathbf{a}) = \mathbf{0}$. Hence, $(S_R; +, \mathbf{0})$ is an Abelian group.

For two series $s_1, s_2 \in S_R$ defined by, respectively, **a** and **b**, let

$$\mathbf{a} * \mathbf{b} = \mathbf{c}, \quad \mathbf{c} = (c_0, c_1, c_2, \ldots, \infty),$$

where, for $i = 0, 1, 2, \ldots, \infty$,

$$c_i = \sum_{\substack{j+k=i \\ 0 \le j,k \le i}} a_j b_k = \sum_{j=0}^{i} a_j b_{i-j}.$$

From the closedness for multiplication on R, the closedness for $*$ on S_R is easily seen. Thus, $*$ can be dealt with as the multiplication on S_R.

For $\mathbf{a}, \mathbf{b}, \mathbf{c} \in S_R$, let $\mathbf{h} = \mathbf{a} * (\mathbf{b} * \mathbf{c})$ and $\mathbf{g} = (\mathbf{a} * \mathbf{b}) * \mathbf{c}$. Since, for $i = 0, 1, 2, \ldots, \infty$,

$$\begin{aligned} h_i &= \sum_{j=0}^{i} a_j \left(\sum_{k=0}^{i-j} b_k c_{i-j-k} \right), \quad \text{by the associative law on } R, \\ &= \sum_{j=0}^{i} \sum_{k=0}^{i-j} a_j b_k c_{i-j-k} \end{aligned}$$

and

$$\begin{aligned} g_i &= \sum_{j=0}^{i} \left(\sum_{k=0}^{j} a_k b_{j-k} \right) c_{i-j}, \quad \text{by exchanging } \Sigma_j \text{ and } \Sigma_k, \\ &= \sum_{j=0}^{i} \left(\sum_{k=0}^{i-j} a_j b_k \right) c_{i-j-k}, \quad \text{by the associative law on } R, \\ &= \sum_{k=0}^{i} \left(\sum_{j=k}^{i} a_k b_{j-k} \right) c_{i-j}, \quad \text{by substituting } l = j - k \text{ for } j, \\ &= \sum_{k=0}^{i} a_k \left(\sum_{j=k}^{i} b_{j-k} c_{i-j} \right), \quad \text{by the associative law on } R, \\ &= \sum_{k=0}^{i} a_k \left(\sum_{l=0}^{i-k} b_l c_{i-l-k} \right), \end{aligned}$$

we have $\mathbf{h} = \mathbf{g}$. This shows that $*$ obeys the associative law on S_R.

Let $\mathbf{1}_1 = (1, 0, 0, \ldots, 0)$. Since for any $\mathbf{a} \in S_R$ $(\mathbf{a} \neq \mathbf{0})$, $\mathbf{a} * \mathbf{1} = \mathbf{1} * \mathbf{a} = \mathbf{1}$, we find that $\mathbf{1}_1$ is the identity on S_R.

Theorem 1.3.1. *Let $\mathcal{R}\{x\} = (S_R; +, *, \mathbf{0}, \mathbf{1})$, then $\mathcal{R}\{x\}$ is a ring if, and only if, R is a ring.*

Proof. Sufficiency. On the basis of what discussed above, only necessary to observe if the two operations satisfy the distributive law.

For any $\mathbf{a}, \mathbf{b}, \mathbf{c} \in S_R$, let $\mathbf{h} = \mathbf{a} * (\mathbf{b} + \mathbf{c})$ and $\mathbf{g} = \mathbf{a} * \mathbf{b} + \mathbf{a} * \mathbf{c}$. Because of

$$h_i = \sum_{j=0}^{i} a_j(b_{i-j} + c_{i-j}) = \sum_{j=0}^{i}(a_j b_{i-j} + a_j c_{i-j}) = (\mathbf{a} * \mathbf{b})_i + (\mathbf{a} * \mathbf{c})_i = g_i,$$

where the second equality is from the distributive law on R, we find that the distributive law holds on S_R as well.

Necessity. It is easily seen that we have a relationship between $(S_R; +, \mathbf{0})$ and R as $S_R \Leftrightarrow R$, $+$ on $S_R \Leftrightarrow +$ on R and hence the closed law, the associate law, and the commutative law. By considering $\mathbf{0} \in S_R \Rightarrow 0 \in R$ as the zero and $-\mathbf{a} \in S_R \Rightarrow -a \in R$ as the inverse, from the four axioms we use Group 2–Group 4, and R induced from S_R is an Abelian group. On the basis of the relationship between $*$ on S_R and $\cdot$ on R, for multiplication, the closed law, associative law and the identity law on R are established from those on S_R. Further, for $+$ and $\cdot$ on R, the distributive law is from that on S_R. Thus, the ring R is derived from the ring S_R. This is the necessity. □

On the basis of Theorem 1.3.1, from $R = R\{0\} \subset R\{x\}$, $R\{x\}$ is the *extension* of the ring R via a series with only one undeterminate, which is still a ring. Similarly, the extensions $\mathcal{R}\{\mathbf{x}\}$, $\mathbf{x} = (x_1, x_2, x_3, \ldots)$ of ring R can be established from $\{x\}$ as a starting point. For example, $R\{x_1, x_2\} = R\{x_1\}\{x_2\}$ and for any integer $n \geq 3$, $R\{\mathbf{x}_n\} = R\{\mathbf{x}_{n-1}\}\{x_n\}$.

From the procedure of proving Theorem 1.3.1, all the extensions obtained in this way are a ring if R is commutative.

More generally, Theorem 1.2.5 enables us to extract the extensions for the case that some undeterminates make it allowable to generate polynomials but not series. For example, $R[x_1, x_2\}$ represents the case that $S(x_1, x_2) = S[x_1]\{x_2\} \in \mathcal{R}[x_1, x_2\}$ is a polynomial of x_1 and a series of x_2. Thus, the meaning of $\mathcal{R}[x_1, \ldots, x_k]\{x_{k+1}, \ldots\}$ is known for $n \geq 3$.

Let $L(x) = a_{-l_0}x^{-l_0} + a_{-l_0+1}x^{-l_0+1} + a_{-l_0+2}x^{-l_0+2} + \cdots + a_\infty x^\infty$ be a Laurent series of x with the least power $-l_0$, integer $l_0 \geq 0$, where $a_i \in R$, $-l_0 \leq i \leq \infty$, and R a ring. Denote by $\mathcal{L}$ the set of all Laurent series. Similarly to S, $L(x)$ is represented by $\mathbf{a} = (a_{-l_0}, a_{-l_0+1}, a_{-l_0+2}, \ldots, a_\infty)$.

If $\mathbf{a} + \mathbf{b} = \mathbf{c}$ is defined by

$$c_i = a_i + b_i, \quad -l_0 \leq i \leq \infty,$$

then the addition is established on $\mathcal{L}$.

Let $\mathbf{a} * \mathbf{b} = \mathbf{c}$ be such that

$$c_i = \sum_{j=-l_0}^{i} a_j b_{i-j}, \quad -l_0 \leq i \leq \infty.$$

The multiplication is established on $\mathcal{L}$.

Theorem 1.3.2. *Let* $\mathcal{L}\{R;x\} = (\mathcal{L};+,*,\mathbf{0},\mathbf{1})$*, then* $\mathcal{L}\{R;x\}$ *is a ring if, and only if,* R *is a ring.*

Proof. It is only necessary to pay attention to the fact that any ordinary series $S(y)$ can be deduced from a Laurent series $L(x)$ by the substitution $x^i = y^{i+l_0}$, $-l_0 \leq i \leq \infty$. Therefore, the conclusion is derived from Theorem 1.3.1. □

A ring which obeys the cancelation law is called an *integral domain.*

Theorem 1.3.3. *Both* $\mathcal{L}\{R;x\}$ *and* $\mathcal{R}\{x\}$ *are integral domains.*

Proof. According to what has been mentioned, it is only necessary to prove the theorem for $\mathcal{R}\{x\}$.

First, we prove that the ring R is actually an integral domain. Since R is an integer ring, it is easily shown from arithmetics that, for any integer $a,b,c \in R$, $c \neq 0$, $ac = bc \Leftrightarrow a = b$, hence the cancelation law. Therefore, R is an integral domain.

Then, we derive that $\mathcal{R}\{x\}$ is also an integral domain from the integral domain R. From Theorem 1.3.1, $\mathcal{R}\{x\} = (S_R;+,*,\mathbf{0},\mathbf{1})$ is a ring.

For the cancelation law, a polynomial without $\mathbf{0}$ has to be investigated. Two cases should be considered. Let $\mathbf{c} \neq \mathbf{0}$.

Case 1 $c_0 \neq 0$. Because of $\mathcal{R}\{x\}$ being a ring, there exists $\mathbf{c}' \in S_R$ such that $S(x)S'(x) = 1$, *i. e.*, $\mathbf{c}'$ is the inverse of $\mathbf{c}$. This implies the cancelation law.

Case 2 $c_0 = 0$. Let an integer, $\alpha \geq 1$, be the minimal power of all terms in $S_c(x)$ determined by $\mathbf{c}$. From $S_c(x) = c_\alpha x^\alpha T(x)$ with $T(x)$ in Case 1, the minimality of α leads to the existence of the inverse $T^{-1}(x)$ for $T(x)$. For any $S_a(x), S_b(x) \in S_R$, we have

$$\begin{aligned} S_a(x)S_c(x) = S_b(x)S_c(x) &\Rightarrow S_a(x)c_\alpha x^\alpha = S_b(x)c_\alpha x^\alpha, \\ &\quad \text{by } c_\alpha \neq 0 \text{ in } R, \\ &\Rightarrow S_a(x)x^\alpha = S_b(x)x^\alpha, \\ &\quad \text{by } x \text{ not being zero}, \\ &\Rightarrow S_a(x) = S_b(x). \end{aligned}$$

This is the cancelation law. Hence, $\mathcal{R}(x)$ is an integral domain. □

For the functions considered in this book, their coefficients are all in the integral domain $\mathcal{R}$. Particularly, they are in the extensions $R\{\mathbf{x}\}$ unless specifically indicated otherwise. Occasionally, $\mathcal{L}\{x\} = \mathcal{L}\{R;\mathbf{x}\}$ if necessary.

Let $a(x) \in R\{x\}$. For convenience, denote by $\mathbf{a}_{\{x\}} = (a_0, a_1, a_2, \ldots, \infty)_{\{x\}}$ the function $a(x)$, as a series of the determinate x.

The *differential* of the series $a(x)$ with respect to x is defined to be a transformation such that

$$\frac{\mathrm{d}a}{\mathrm{d}x} = \mathbf{a}_{\{x\}} A \tag{1.3.3}$$

where the matrix $A = (a_{i,j})_{0 \le i,j \le \infty}$, $\mathbf{a}_i = (a_{i,0}, a_{i,1}, a_{i,2}, \ldots, a_{i,\infty}) = \mathbf{0}_{[i]}$, *i. e.*, in $\mathbf{0}$, only the ith component 1 changed from 0. Attention: $\mathbf{0}_{[0]} = \mathbf{0}$.

Let $\mathbf{n}_{0 \to i} = (0, 1, 2, 3, \ldots, i)$ and $I_{k \to \infty} = (a_{i,j})_{k \le i,j \le \infty}$ where

$$a_{i,j} = \begin{cases} 1, & \text{when } k \le i = j \le \infty; \\ 0, & \text{otherwise.} \end{cases}$$

Simply, $\mathrm{Diag}\, \mathbf{1}_{k \to \infty} = \mathrm{Diag}[1_k, 1_{k+1}, 1_{k+2}, \ldots) = I_{k \to \infty}$ for $k \ge 0$. Thus, we have

$$A = \mathbf{n}_{0 \to \infty}\, \mathrm{Diag}[0, \mathrm{Diag}\, \mathbf{1}_{1 \to \infty}).$$

Therefore, we have

$$\frac{\mathrm{d}a}{\mathrm{d}x} = \mathbf{a}_{\{x\}} (\mathbf{n}_{0 \to \infty}\, \mathrm{Diag}[0, \mathrm{Diag}\, \mathbf{1}_{1 \to \infty})). \tag{1.3.4}$$

Theorem 1.3.4. $\frac{\mathrm{d}a}{\mathrm{d}x} \in \mathcal{R}\{x\}$ *if, and only if,* $a(x) \in \mathcal{R}\{x\}$.

Proof. On the basis of (1.3.4), the conclusion is straightforwardly obtained. □

The *integral* of a series $a(x) \in S_R$ with respect to x is defined as

$$\int a(x)\mathrm{d}x = \mathbf{a}_{\{x\}} B \tag{1.3.5}$$

where $B = (b_{i,j})_{0 \le i,j \le \infty}$, $\mathbf{b}_i = (b_{i,0}, b_{i,1}, b_{i,2}, \ldots, b_{i,\infty}) = \mathbf{0}_{[1/(i+1)]}$, *i. e.*, the ith component 0 of $\mathbf{0}$ replaced by $1/(i+1)$, $0 \le i \le \infty$.

Let us introduce another vector,

$$\mathbf{1/n}_{1 \to n-1} = \left(1, \frac{1}{2}, \frac{1}{3}, \ldots, \frac{1}{n}\right)$$

for $n \ge 1$. By employing the symbols used in (1.3.4), because of $B = \mathbf{1}/\infty\, \mathrm{Diag}\, \mathbf{1}_{0 \to \infty}$, (1.3.5) becomes

$$\int a(x)\mathrm{d}x = \mathbf{a}_{\{x\}} (\mathbf{1}/(\mathbf{n}+\mathbf{1})_{0 \to \infty}\, \mathrm{Diag}\, \mathbf{1}_{0 \to \infty}). \tag{1.3.6}$$

For convenience, the integral of any constant is always pre-assumed to be the constant itself.

Theorem 1.3.5. $\int a(x)\mathrm{d}x \in \mathcal{R}\{x\}$ *if, and only if,* $a(x) \in \mathcal{R}\{x\}$.

Proof. On the basis of (1.3.6), the conclusion is easily drawn. □

1.4 Functionals and functional equations

For a set S and P, a partition of S, let $P = \{P_0, P_1, P_2, P_3, \dots\}$, then we find that $P_i \in S$ for any integer $i \geq 0$; $P_i \cap P_j = \emptyset$ for any integers $i, j \geq 0$; and $\bigcup_{i\geq 1} P_i = S$. The last two statements are always written as

$$\sum_{i\geq 0} P_i = S,$$

the *sum*(or, as we say, disjoint union) of all $P_i, i \geq 0$.

Let $s_i, i \geq 1$, be the cardinality $|\{P_j \mid \forall j \geq 1, |P_j| = i, P_j \in P\}|$, then $\mathbf{s} = (s_1, s_2, s_3, \dots)$ is called the *partition vector*, while P_0 is taken for certain technical usage as its cardinality $|P_0| = m$.

For a set S, let $\mathcal{P}$ be all the partitions of S; the function

$$f_S = \sum_{P\in\mathcal{P}} x^{m(P)} \mathbf{y}^{\mathbf{s}(P)}, \tag{1.4.1}$$

where

$$\mathbf{y}^{\mathbf{s}(P)} = \prod_{i\geq 1} y_i^{s_i(P)},$$

is called the *partition function* of S.

Let $n = 1 + \sum_{i\geq 1} s_i$ and $s = m + \sum_{i\geq 1} is_i$. Because of the nonnegativity of m and s_i for all $i \geq 1$, if m and s are given, the function f_s in (1.4.1) becomes a polynomial. This enables us to determine f_S from $m + s$ smaller to greater.

A *functional* is considered in this book as a transformation from the function space with the basis, *e.g.*, $\{1, x, x^2, x^3, \dots\}$ to itself or another space, *i.e.*, an abstract linear space.

For a function $f \equiv f(x) \equiv f(x, \mathbf{y})$, the operation

$$\delta_{x,z} f|_{x=u} = \frac{f(x) - f(z)}{x - z} \tag{1.4.2}$$

is called a *straight difference* of f. Denote by $[f]_x^i = \partial_x^i f$ the coefficient of x^i in $f \in \mathcal{R}\{x, \mathbf{y}\}$ for $i \geq 0$.

Theorem 1.4.1. *The operator $\delta_{x,z}$ on f defined by* (1.4.2) *is a linear functional.*

Proof. Let $F_i = [f]_x^i$, $i \geq 0$ and $F_{\delta i} = [\delta_{x,z} f|_{x=u}]_x^i$, $i \geq 0$. Because of

$$\begin{aligned} f(x) - f(z) &= (x - z) \sum_{\mathbf{i}\geq 0} F_i \left(\sum_{j=0}^{i-1} x^j z^{i-1-j} \right) \\ &= (x - z) \sum_{j\geq 0} x^j \left(\sum_{i\geq j+1} F_i z^{i-1-j} \right), \end{aligned}$$

we have

$$\delta_{x,z} f|_{x=u} = \sum_{j\geq 0} x^j \Big(\sum_{i\geq j+1} F_i z^{i-1-j} \Big)$$

and hence

$$F_{\delta j} = \sum_{i\geq 0} F_{i+j+1} z^i, \quad j \geq 0.$$

This implies that the straight difference $\delta_{x,z}$ determines a transformation given by $Y_\delta = (y_{i,j}^{[\delta]})_{i,j\geq 0}$ where

$$y_{i,j}^{[\delta]} = \begin{cases} 0, & \text{when } i > j; \\ 1, & \text{when } i = j; \\ z^{j-i}, & \text{otherwise}, \end{cases}$$

in the function space with basis $\{1, x, x^2, x^3, \ldots\}$. On the basis of Theorem 1.2.1, the conclusion is drawn. □

For a function $f \equiv f(x) \equiv f(x, \mathbf{y})$, the operation

$$\partial_{x,z} f|_{x=u} = \frac{zf(x) - xf(z)}{x - z} \tag{1.4.3}$$

is called the *slope difference* of f.

Theorem 1.4.2. *The operator $\partial_{x,z}$ on f $(f(0) = 0)$ defined by* (1.4.3) *is a linear functional.*

Proof. Let $F_i = [f]_x^i$, $i \geq 0$ and $F_{\partial i} = [\partial_{x,z} f|_{x=u}]_x^i$, $i \geq 0$. Because of

$$\partial_{x,z} f|_{x=u} = \sum_{\mathbf{i}\geq 0} F_i \Big(\frac{zx^i - xz^i}{x - z} \Big) \tag{1.4.4}$$

we have $F_{\partial 0} = -F_0 = f(0) = 0$, $F_{\partial 1} = 0$. For $i \geq 2$,

$$F_{\partial i} = \Big[\sum_{i\geq 2} F_i \frac{zx^i - xz^i}{x - z} \Big]_x^i.$$

Because of $i \geq 2$,

$$\begin{aligned} zx^i - xz^i &= xz(x^{i-1} - z^{i-1}) \\ &= (x - z) \sum_{j=0}^{i-2} x^{j+1} z^{i-1-j} \\ &= (x - z) \sum_{j=1}^{i-1} x^j z^{i-j} \end{aligned}$$

and hence

$$\partial_{x,z} u^i = \frac{zx^i - xz^i}{x - z} = \sum_{j=1}^{i-1} x^j z^{i-j}.$$

From (1.4.4),

$$\begin{aligned} F_{\partial i} &= \left[\sum_{i \geq 2} F_i \sum_{j=1}^{i-1} x^j z^{i-j} \right]_x^i \\ &= \sum_{j \geq i+1} F_j z^{j-i} \end{aligned} \tag{1.4.5}$$

for $i \geq 1$.

On the basis of (1.4.5), it is seen that the slope difference $\partial_{x,z}$ is a transformation given by $Y_\partial = (y_{i,j}^{[\partial]})_{i,j \geq 1}$ where

$$y_{i,j}^{[\partial]} = \begin{cases} 0, & \text{when } i > j; \\ z^{j-i+1}, & \text{otherwise,} \end{cases}$$

in the function space with basis $\{1, x, x^2, x^3, \ldots\}$. On the basis of Theorem 1.2.1, the conclusion is drawn. □

For a function $f(x)$ in the functional space with basis $\{1, x, x^2, x^3, \ldots\}$, the *differential* of f with respective to x, denoted by $\frac{df}{dx}$, is defined as a linear combination of

$$\frac{dx^i}{dx} = \begin{cases} 0, & \text{when } i = 0; \\ ix^{i-1}, & \text{when } i \geq 1, \end{cases} \tag{1.4.6}$$

for $i \geq 0$.

Theorem 1.4.3. *The operator* $\frac{d}{dx}$ *defined by* (1.4.6) *is a linear functional.*

Proof. Let $F_{\text{dff}\,i} = [\frac{df}{dx}]_i$, $i \geq 0$. From (1.4.6), $F_{\text{dff}\,i} = (i+1)F_i$, $i \geq 0$, we see that the differential is the transformation determined by $Y_{\text{dff}} = (y_{i,j}^{[\text{dff}]})_{i,j \geq 0}$, where

$$y_{i,j}^{[\text{dff}]} = \begin{cases} i, & \text{when } j = i - 1; \\ 0, & \text{otherwise,} \end{cases} \tag{1.4.7}$$

for $i, j \geq 0$. On the basis of Theorem 1.2.1, the theorem is done. □

For a function $f(x)$ in the functional space with basis $\{1, x, x^2, x^3, \ldots\}$, the *integral* of f with respective to x, denoted by $\int f\,dx$, is defined as a linear combination of

$$\int x^i dx = \begin{cases} x, & \text{when } i = 0; \\ \dfrac{1}{i+1} x^{i+1}, & \text{when } i \geq 1, \end{cases} \tag{1.4.8}$$

for $i \geq 0$.

Theorem 1.4.4. *The operator* $\int \mathrm{d}x$ *defined by* (1.4.8) *is a linear functional.*

Proof. Let $F_{\text{int}\,i} = [\int f\mathrm{d}x]_i$, $i \geq 0$. From (1.4.8), $F_{\text{int}\,i} = \frac{1}{i+1}F_{i+1}$, $i \geq 0$, we see that the integral is the transformation determined by $Y_{\text{int}} = (y_{i,j}^{[\text{int}]})_{i,j\geq 0}$, where

$$y_{i,j}^{[\text{int}]} = \begin{cases} \dfrac{1}{i+1}, & \text{when } j = i+1; \\ 0, & \text{otherwise}, \end{cases} \tag{1.4.9}$$

for $i, j \geq 0$. On the basis of Theorem 1.2.1, the theorem is done. □

Let us write $(x)_n = x(x-1)\cdots(x-n+1)$, for integer $n \geq 1$, called the *decrease factorial function*. As shown in Rota GC [75] (1964), the *umbrella operation* (or *shadow operator*) denoted by L is the linear extension by transforming x^i into a polynomial $p_i(x)$ of degree i for $i \geq 1$. Because of the constant term, which is allowed to be missing, only homogeneous polynomials are considered with $p_i(0) = 0$. For example, $p_i(x) = (x)_i$.

Theorem 1.4.5. *The shadow operator is a linear functional.*

Proof. Let $F_{\mathrm{L}i} = [\mathrm{L}f]_i$, the coefficient of $(x)_i$, $i \geq 0$, in Lf. L is determined by the matrix $Y_{\mathrm{L}} = (y_{i,j}^{[\mathrm{L}]})_{i,j\geq 1}$ of infinite dimension by

$$y_{1,j}^{[\mathrm{L}]} = \begin{cases} 1, & \text{when } j = 1; \\ 0, & \text{otherwise}, \end{cases} \tag{1.4.10}$$

and, for $i \geq 2$,

$$y_{i,j}^{[\mathrm{L}]} = \begin{cases} -(i-1)y_{i-1,1}^{[\mathrm{L}]}, & \text{when } j = 1; \\ -(i-1)y_{i-1,j}^{[\mathrm{L}]} + y_{i-1,j-1}^{[\mathrm{L}]}, & \text{when } 2 \leq j \leq i-1; \\ 1, & \text{when } j = i; \\ 0, & \text{otherwise}, \end{cases} \tag{1.4.11}$$

the transformation is a functional. On the basis of Theorem 1.2.1, the conclusion is drawn. □

All functionals mentioned above are transformations from the space of functions into itself.

However, in Chapter 2, a new functional, called the *meson functional*, is going to be investigated from the space of functions into a general abstract linear space instead of the function space itself.

An equation of functions which involves a certain functional is concisely called a *functional equation*. Of the two volumes of the book, Volume II is concentrated on meson functional equations.

1.5 Notes

1.5.1. The basic knowledge needed in the whole book is explained in this chapter to make the book, in principle, self-contained. All material mentioned is particularly designed for use in the relevant context.

1.5.2. The system of sets using $(\Omega; \cup, \cap, ^{-})$ forms a lattice, or from another point of view, a Boolean algebra. As regards the algebraic structure, one might have a look in Birkhoff G [2] for details.

1.5.3. In §1.2, theorems 1.2.1 and 1.2.2 are not necessarily restricted for real functions. They both have a more general form for mappings on an abstract symbol set.

1.5.4. For more about the algebraic notions in §1.3, one might wish to read [79] (Shafarevich IR).

1.5.5. On umbrella calculus under a shadow functional, a number of topics involved with polynomials are particularly investigated in [75] (Rota GC), [74] (Roman SM, Rota GC), [78] (Rota GC, Shen J, Taylor BD), [76] (Rota GC, Kahaner D, Odlyzko A), [77] (Rota GC, Taylor BD), [63] (Mullin RC, Rota GC), *et al.*

1.5.6. In [19] (Liu YP, 1986), or [60] (Liu YP, Book 3: 1163–1167), the meson functional is used as the shadow operator. However, in [24] (Liu YP, 1986), or [60] (Liu YP, Book 3: 1175–1179), because of the shadow operator going from a basis to another on the function space itself, the linear operator which transforms a function space to a vector space (or abstract linear space) should be employed under a name different from that of the shadow. By [36] (Liu YP, 1990), or [60] (Liu YP, Book 3: 1326–1331), this operator is denoted $\int_y$.

Because the above operators mentioned in Section 1.4 are all seen as different types of functionals, by [57] (Liu YP, 2012), or [60] (Liu YP, Book 23: 11223–11230) and [58] (Liu YP, 2012), or [60] (Liu YP, Book 23: 11276–11283), this operator is named a meson functional.

2 Meson functional

2.1 Basic concepts

Let $\mathcal{V}$ be an abstract linear space with its basis $\{y_1, y_2, y_3, \ldots\}$ over a field F and $\mathcal{F}$ the space of functions with its basis $\{y, y^2, y^3, \ldots\}$ over the field F. The transformation from $\mathcal{F}$ to $\mathcal{V}$, denoted by $\int_y$, such that

$$\int_y y^i = y_i, \quad \text{for } i \geq 1, \tag{2.1.1}$$

is called a *meson functional.*

Attention: for an element $c \in F$, $\int_y c = c$ is always ensured and hence, $\int_y y^0 = 1$.

Because for $a_i \in F$

$$\int_y a_i y^i = a_i \int_y y^i$$

and for $a_i, a_j \in F$

$$\int_y (a_i y^i + a_j y^j) = a_i \int_y y^i + a_j \int_y y^j,$$

we see that, for any

$$f_i = \sum_{j\geq 0} a_{ij} y^j, \quad i = 1, 2,$$

on the basis of

$$\begin{aligned}
\int_y af_i &= a \int_y \sum_{j\geq 0} a_{ij} y^j \\
&= a \sum_{j\geq 0} a_{ij} \int_y y^j \\
&= a \int_y f_i, \quad a \in F; \\
\int_y (af_1 + bf_2) &= \sum_{j\geq 0} (aa_{1j} + ba_{2j}) \int_y y^j \\
&= a \sum_{j\geq 0} a_{1j} y_j + b \sum_{j\geq 0} a_{2j} y_j \\
&= a \int_y f_1 + b \int_y f_2, \quad a, b \in F,
\end{aligned} \tag{2.1.2}$$

the functional is linear on the space $\mathcal{F}$.

https://doi.org/10.1515/9783110625837-002

On the other hand, the transformation from $\mathcal{V}$ to $\mathcal{F}$, denoted by $\int^y$, such that

$$\int^y y_i = y^i, \quad \text{for } i \geq 1, \tag{2.1.3}$$

is called *anti-meson*.

Attention: for an element $c \in F$, $\int^y c = c$ is always ensured and hence $\int^y y_0 = 1$.

For the anti-meson, because, for $a_i \in F$,

$$\int^y a_i y_i = a_i \int^y y_i$$

and, for $a_i, a_j \in F$,

$$\int^y (a_i y_i + a_j y_j) = a_i \int^y y_i + a_j \int^y y_j,$$

we see that, for any

$$v_i = \sum_{j\geq 0} a_{ij} y^j, \quad i = 1, 2,$$

on the basis of

$$\begin{aligned}
\int^y a v_i &= a \int^y \sum_{j\geq 0} a_{ij} y_j \\
&= a \sum_{j\geq 0} a_{ij} \int^y y_j \\
&= a \int^y v_i, \quad a \in F; \\
\int^y (a v_1 + b v_2) &= \sum_{j\geq 0} (a a_{1j} + b a_{2j}) \int^y y_j \\
&= a \sum_{j\geq 0} a_{1j} y_j + b \sum_{j\geq 0} a_{2j} y_j \\
&= a \int^y v_1 + b \int^y v_2, \quad a, b \in F,
\end{aligned} \tag{2.1.4}$$

this functional is also linear on the space $\mathcal{V}$.

Theorem 2.1.1. *The meson $\int_y$ and anti-meson $\int^y$ functionals are mutually inverse.*

Proof. For any $s \in \mathcal{F}$, $s = s_0 + s_1 y + s_2 y^2 + \cdots$,

$$v = \int_y s = \sum_{i\geq 0} s_i y_i \in \mathcal{V},$$

we have

$$
\begin{aligned}
\int^{y} v &= \int^{y}\Big(\Big(\int_{y} \sum_{i\ge 0} s_i y^i\Big)\Big), \quad \text{by (2.1.2)},\\
&= \int^{y}\Big(\sum_{i\ge 0} s_i \int_{y} y^i\Big), \quad \text{by (2.1.1)},\\
&= \int^{y}\Big(\sum_{i\ge 0} s_i y_i\Big), \quad \text{by (2.1.4)},\\
&= \sum_{i\ge 0} s_i \int^{y} y^i, \quad \text{by (2.1.3)},\\
&= s \in \mathcal{F}.
\end{aligned}
$$

Hence, $\int^{y}$ is the inverse of $\int_{y}$.

Conversely, for any $\mathbf{v} \in \mathcal{V}$, $\mathbf{v} = v_0 + v_1 y_1 + v_2 y_2 + \cdots$,

$$
s = \int^{y} \mathbf{v} = \sum_{i\ge 0} v_i y^i \in \mathcal{F},
$$

we have

$$
\begin{aligned}
\int_{y} s &= \int_{y}\Big(\Big(\int^{y} \sum_{i\ge 0} v_i y_i\Big)\Big), \quad \text{by (2.1.4)},\\
&= \int_{y}\Big(\sum_{i\ge 0} v_i \int^{y} y_i\Big), \quad \text{by (2.1.3)},\\
&= \int_{y}\Big(\sum_{i\ge 0} v_i y^i\Big), \quad \text{by (2.1.2)},\\
&= \sum_{i\ge 0} v_i \int_{y} y^i, \quad \text{by (2.1.1)},\\
&= \mathbf{v} \in \mathcal{V}.
\end{aligned}
$$

Therefore, $\int_{y}$ is the inverse of $\int^{y}$. □

2.2 Shift

For $\mathbf{a}, \mathbf{b} \in \mathcal{V}$, assume $\mathbf{a} = (a_0, a_1, a_2, \dots)$ and $\mathbf{b} = (b_0, b_1, b_2, \dots)$. The transformation from $\mathbf{a}$ to $\mathbf{b}$ such that

$$
b_i = a_{i+1}, \quad i \ge 0, \tag{2.2.1}
$$

is called a *left shift*, denoted by $\mathbf{L}$.

In fact, it is seen that

$$\mathbf{La}^{\mathrm{T}} = L\mathbf{a}^{\mathrm{T}}, \quad {}^{\mathrm{T}} \text{ is the transpose,}$$

where $L = (l_{i,j})_{0\le i,j\le\infty}$ and

$$l_{i,j} = \begin{cases} 1, & \text{when } j = i + 1; \\ 0, & \text{otherwise,} \end{cases}$$

$0 \le i, j \le \infty$.

Theorem 2.2.1. *The left shift* **L** *obeys the following rules:*
(i) $L^n = (l^n_{i,j})_{0\le i,j\le\infty}$ *where*

$$l^n_{i,j} = \begin{cases} 1, & \textit{when } j = i + n; \\ 0, & \textit{otherwise.} \end{cases}$$

(ii) $LL^{\mathrm{T}} = I \neq L^{\mathrm{T}}L$ *where* $I = (e_{i,j})_{0\le i,j\le\infty}$, *i. e., the identity,*

$$e_{i,j} = \begin{cases} 1, & \textit{when } j = i; \\ 0, & \textit{otherwise.} \end{cases}$$

(iii) *For any* $\mathbf{a} \in \mathcal{V}$,

$$\int^{y} (\mathbf{La}) = y \int^{y} \mathbf{a}.$$

(iv) *For any* $s(y) \in \mathcal{F}$,

$$\int_{y} ys(y) = L \int_{y} s(y).$$

Proof. Different from (i) proved by induction, (ii)–(iv) are all from the definition. □

For **a**, **b** $\in \mathcal{V}$, assume $\mathbf{a} = (a_0, a_1, a_2, \ldots)$ and $\mathbf{b} = (b_0, b_1, b_2, \ldots)$, the transformation from **a** to **b** such that

$$b_i = \begin{cases} a_{i-1}, & \text{when } i \ge 1; \\ 0, & \text{otherwise } i = 0, \end{cases} \tag{2.2.2}$$

is called the *right shift*, denoted by **R**.

It is easily seen that

$$\mathbf{Ra}^{\mathrm{T}} = R\mathbf{a}^{\mathrm{T}}, \quad {}^{\mathrm{T}} \text{ is the transpose,}$$

where $R = (r_{i,j})_{0\le i,j\le\infty}$ and

$$r_{i,j} = \begin{cases} 1, & \text{when } j = i-1; \\ 0, & \text{otherwise}, \end{cases}$$

$0 \le i, j \le \infty$.

Theorem 2.2.2. *The right shift* **R** *obeys the following rules:*
(i) $R^n = (l^n_{i,j})_{0\le i,j\le\infty}$ *where*

$$l^n_{i,j} = \begin{cases} 1, & \textit{when } j = i-n; \\ 0, & \textit{otherwise}. \end{cases}$$

(ii) $R = L^{\mathrm{T}}$.
(iii) $RR^{\mathrm{T}} = I_{\langle 1_0\rangle} = R^{\mathrm{T}}R = I$, *where* $I_{\langle 1_0\rangle}$ *is the identity (the matrix with all entries on the diagonal* 1*; all other entries* 0*).*
(iv) *For any* $\mathbf{a} \in \mathcal{V}$,

$$\int^{y}(R\mathbf{a}) = y^{-1}\int^{y}(\mathbf{a} - a_0 y_0).$$

(v) *For any* $s(y) \in \mathcal{F}$,

$$\int_{y} y^{-1}(s(y) - s_0) = R\int_{y} s(y).$$

Proof. Different from (i) proved by induction, (ii) and (iv)–(v) are all from the definition. Then (iii) is from (ii) and Theorem 2.2.1(ii). □

2.3 Truncation

The operation of the first i components on a vector put all 0 is called a *truncation*, denoted by **J**, or precisely an *i-truncation*.

For $\mathbf{a} \in \mathcal{V}$, let $\mathbf{Ja} = J\mathbf{a}$, $J = (c_{i,j})_{0\le i,j\le\infty}$, then

$$c_{i,j} = \begin{cases} 1, & \text{when } j = i,\ i \ge 1; \\ 0, & \text{otherwise}. \end{cases} \tag{2.3.1}$$

Let $I_{i,j}$ be the matrix in which only the entry at (i,j) is 1 with all others 0, then the left shift matrix

$$L = \sum_{i\ge 0} I_{i,i+1};$$

$$L^2 = \Big(\sum_{i\geq 0} I_{i,i+1}\Big)\Big(\sum_{i\geq 0} I_{i,i+1}\Big) = \sum_{i\geq 0} I_{i,i+2}.$$

Further, for any integer $n \geq 1$,

$$\begin{aligned} L^n &= \Big(\sum_{i\geq 0} I_{i,i+1}\Big)^{n-1}\Big(\sum_{i\geq 0} I_{i,i+1}\Big) \\ &= \Big(\sum_{i\geq 0} I_{i,i+n-1}\Big)\Big(\sum_{i\geq 0} I_{i,i+1}\Big) \\ &= \sum_{i\geq 0} I_{i,i+n}. \end{aligned} \tag{2.3.2}$$

Similarly, for the right shift matrix R, we have

$$\begin{aligned} R^n &= \Big(\sum_{i\geq 0} I_{i,i-1}\Big)^{n-1}\Big(\sum_{i\geq 0} I_{i,i-1}\Big) \\ &= \Big(\sum_{i\geq 0} I_{i,i-n+1}\Big)\Big(\sum_{i\geq 0} I_{i,i-1}\Big) \\ &= \sum_{i\geq 0} I_{i,i-n}. \end{aligned} \tag{2.3.3}$$

Theorem 2.3.1. *For any integer $n \geq 1$, the n-truncation matrix*

$$J^{(n)} = R^n L^n, \tag{2.3.4}$$

where $J^{(1)} = J$.

Proof. From (2.3.2) and (2.3.3),

$$R^n L^n = \Big(\sum_{i\geq 0} I_{i,i-n}\Big)\Big(\sum_{i\geq 0} I_{i,i+n}\Big) = \sum_{i\geq n} I_{i,i}.$$

This is the conclusion. □

2.4 Projection

On $\mathcal{V}$, let $\mathbf{a} = \sum_{j\geq 0} a_j y_j \in \mathcal{V}$, then the operation $\mathbf{P}$ and $\mathbf{Q}$ deduced from

$$\begin{cases} \mathbf{Pa} = \sum_{j\geq 0} (j+1) a_{j+1} y_j; \\ \mathbf{Qa} = \sum_{j\geq 1} \frac{1}{j} a_{j-1} y_j, \end{cases} \tag{2.4.1}$$

is, respectively, called *left projection* and *right projection*.

It is easily seen that if $\mathbf{Pa} = P\mathbf{a}^{\mathrm{T}}$ and $P = (p_{i,j})_{0\le i,j\le\infty}$, then we have

$$p_{i,j} = \begin{cases} i+1, & \text{when } j = i+1; \\ 0, & \text{otherwise.} \end{cases} \tag{2.4.2}$$

Similarly, if $\mathbf{Qa} = Q\mathbf{a}^{\mathrm{T}}$ and $Q = (q_{i,j})_{0\le i,j\le\infty}$, then we have

$$q_{i,j} = \begin{cases} \frac{1}{i}, & \text{when } j = i-1,\ i \ge 1; \\ 0, & \text{otherwise.} \end{cases} \tag{2.4.3}$$

Theorem 2.4.1. *If* $\mathbf{n} = (1, 2, 3, \ldots, \infty)$ *and* $\mathbf{u} = (1, 1/2, 1/3, \ldots, \infty)$, *then the left projection matrix and the right projection matrix are, respectively,*

$$P = \mathbf{n}L \quad \text{and} \quad Q = R\mathbf{u}^{\mathrm{T}} \tag{2.4.4}$$

where L and R are, respectively, the left shift matrix and the right shift matrix.

Proof. From (2.3.2),

$$\begin{aligned} \mathbf{n}L &= \mathbf{n}\Big(\sum_{i\ge 0} I_{i,i+1}\Big), && \text{by the distributive law,} \\ &= \sum_{i\ge 0} \mathbf{n}I_{i,i+1}, && \text{by the definition of } I_{i,j}, \\ &= \sum_{i\ge 0} (i+1)I_{i,i+1}, && \text{from (2.4.2),} \\ &= P. \end{aligned}$$

The first conclusion is done.

We proceed similarly for the last conclusion. □

Attention: P and Q are not commutative under multiplication, because of

$$PQ = \begin{pmatrix} I & \mathbf{0}^{\mathrm{T}} \\ \mathbf{0} & 0 \end{pmatrix} \quad \text{and} \quad QP = \begin{pmatrix} 0 & \mathbf{0} \\ \mathbf{0}^{\mathrm{T}} & I \end{pmatrix}. \tag{2.4.5}$$

Theorem 2.4.2. *For integer* $n \ge 2$, *let* $P^n = PP^{n-1} = (p_{i,j}^{(n)})_{0\le i,j\le\infty}$ *and* $Q^n = QQ^{n-1} = (q_{i,j}^{(n)})_{0\le i,j\le\infty}$, *then we have*

$$p_{i,j}^{(n)} = \begin{cases} \prod_{l=0}^{n-1}(i+n-l), & \text{when } j = i+n; \\ 0, & \text{otherwise,} \end{cases} \tag{2.4.6}$$

and

$$q_{i,j}^{(n)} = \begin{cases} \prod_{l=0}^{n-1} \frac{1}{i+n-l}, & \text{when } j = i-n,\ i \ge n; \\ 0, & \text{otherwise.} \end{cases} \tag{2.4.7}$$

Proof. When $n = 2$, because of

$$
\begin{aligned}
PP &= \Big(\sum_{i\geq 0}(i+1)I_{i,i+1}\Big)\Big(\sum_{i\geq 0}(i+1)I_{i,i+1}\Big), \\
&\quad \text{the first conclusion of Theorem 2.4.1 is employed,} \\
&= \Big(\sum_{i\geq 0}(i+1)(i+2)I_{i,i+2}\Big), \\
&\quad \text{the definition of matrix } I_{i,j} \text{ is employed,} \\
&= P^2,
\end{aligned}
$$

this is the case of the first conclusion when $n = 2$.

For $n \geq 3$, we have

$$
\begin{aligned}
PP^{n-1} &= \Big(\sum_{i\geq 0} I_{i,i+1}\Big)\Big(\sum_{i\geq 0}\prod_{l=0}^{n-2}(i+n-1-l)I_{i,i+n-1}\Big), \\
&\quad \text{induction is employed,} \\
&= \sum_{i\geq 0}(i+n)\prod_{l=0}^{n-2}(i+n-1-l)I_{i,i+n} \\
&= \sum_{i\geq 0}\prod_{l=0}^{n-1}(i+n)I_{i,i+n} \\
&= P^n.
\end{aligned}
$$

Therefore, the first conclusion is drawn.

Similarly, by considering Theorem 2.2.2(ii) and Theorem 2.4.2, the second conclusion is drawn from the first one. □

2.5 Convolution

The *convolution* of two vectors $\mathbf{a}, \mathbf{b} \in \mathcal{V}$, denoted by $\mathbf{a} \otimes \mathbf{b}$, is

$$\mathbf{a} \otimes \mathbf{b} = \mathbf{c} = (c_0, c_1, c_2, c_3, \ldots) \tag{2.5.1}$$

where, for integer $j \geq 0$,

$$c_j = \sum_{i=0}^{j} a_i b_{j-i}. \tag{2.5.2}$$

Let $s, h \in \mathcal{F}$. For integer $i \geq 0$, let $S_i = \partial_y^i s$ and $H_i = \partial_y^i h$, then we have

$$
\begin{cases}
\mathbf{S} = \int_y s = (S_0, S_1, S_2, \ldots) \in \mathcal{V}; \\
\mathbf{H} = \int_y h = (H_0, H_1, H_2, \ldots) \in \mathcal{V}.
\end{cases} \tag{2.5.3}
$$

Theorem 2.5.1. *Let* $\Psi = \mathbf{S}$ *and* $\Phi = \mathbf{H}$, *then*

$$\Psi \otimes \Phi = \int_y (sh). \tag{2.5.4}$$

Proof. Since

$$\begin{aligned} \partial_y^i(sh) &= \sum_{j=0}^{i} S_j H_{i-j}, \quad \text{by (2.3.2)}, \\ &= \mathbf{S} \otimes \mathbf{H}, \end{aligned}$$

from (2.3.3), the theorem is done. □

From the commutativity of $\mathcal{F}$ under multiplication, it is seen that the convolution obeys the commutative law on $\mathcal{V}$; this implies that, for any $s, h \in \mathcal{F}$,

$$\int_t s \otimes \int_y h = \int_t h \otimes \int_y s. \tag{2.5.5}$$

For any integer $k \geq 1$ and $h \in \mathcal{F}$, let $H_i^{[k]} = \partial_y^i h^k$, $i \geq 0$; then we have

$$H_i^{[k]} = \begin{cases} H_i, & k = 1; \\ \sum_{j=0}^{i} H_{i-j}^{[k-1]} H_j, & \text{when } k \geq 1. \end{cases} \tag{2.5.6}$$

For any integer $k \geq 1$, let

$$\Phi^{[k]} = \int_y h^k, \tag{2.5.7}$$

then from (2.3.3), we have

$$\Phi^{[k]} = (H_0^{[k]}, H_1^{[k]}, H_2^{[k]}, \ldots). \tag{2.5.8}$$

Theorem 2.5.2. *For any integers* $k \geq 1$ *and* $i \geq 0$, $\Phi_i^{[k]} = H_i^{[k]}$, *i. e.,*

$$\Phi_i^{[k]} = \begin{cases} \Phi_i, & \text{when } k = 1; \\ \sum_{j=0}^{i} \Phi_{i-j}^{[k-1]} \Phi_j, & \text{when } k \geq 2. \end{cases} \tag{2.5.9}$$

Proof. From (2.5.6) and (2.5.8), (2.5.9) is derived. □

2.6 Differential and integral

For the function y^n, integer $n \geq 0$, the following operation:

$$\frac{\mathrm{d}}{\mathrm{d}y} y^n = \begin{cases} n y^{n-1} & \text{when } n \geq 1; \\ 0, & \text{when } n = 0, \end{cases} \tag{2.6.1}$$

is called the *differential*. The differential on the space $\mathcal{F}\{y;\mathbf{y}\}$ over the field $F(\mathbf{y})$, is determined by the linear extension, *i. e.*, for any $a \in F$,

$$\frac{\mathrm{d}}{\mathrm{d}y}(ay^n) = a\frac{\mathrm{d}}{\mathrm{d}y}y^n$$

and

$$\frac{\mathrm{d}}{\mathrm{d}y}(y^m + y^n) = \frac{\mathrm{d}}{\mathrm{d}y}y^m + \frac{\mathrm{d}}{\mathrm{d}y}y^n.$$

Theorem 2.6.1. *For any* $s = s(y), t = t(y) \in \mathcal{F}\{y;\mathbf{y}\}$,

$$\frac{\mathrm{d}(st)}{\mathrm{d}y} = \frac{\mathrm{d}s}{\mathrm{d}y}t + s\frac{\mathrm{d}t}{\mathrm{d}y}. \tag{2.6.2}$$

Proof. Let $s_i = \partial_y^i s(y)$ and $t_i = \partial_y^i t(y)$, $i \geq 0$. Because of

$$\partial_y^n\left(\frac{\mathrm{d}s}{\mathrm{d}y}t\right) = \sum_{k=0}^{n}(k+1)s_{k+1}t_{n-k} \quad \text{and}$$
$$\partial_y^n\left(s\frac{\mathrm{d}t}{\mathrm{d}y}\right) = \sum_{k=0}^{n}(k+1)t_{k+1}s_{n-k},$$

we have

$$\begin{aligned}
\partial_y^n\left(\frac{\mathrm{d}s}{\mathrm{d}y}t\right) + \partial_y^n\left(s\frac{\mathrm{d}t}{\mathrm{d}y}\right) &= \sum_{k=0}^{n}(k+1)s_{k+1}t_{n-k} + \sum_{k=0}^{n}(k+1)t_{k+1}s_{n-k} \\
&= \sum_{k=0}^{n}(k+1)s_{k+1}t_{n-k} + \sum_{k=0}^{n}(n-k+1)t_{n-k+1}s_k \\
&= (n+1)s_{n+1}t_0 + \sum_{k=0}^{n-1}(k+1)s_{k+1}t_{n-k} \\
&\quad + \sum_{k=1}^{n}(n-k+1)t_{n-k+1}s_k + (n+1)t_{n+1}s_0 \\
&= (n+1)s_{n+1}t_0 + \sum_{k=0}^{n-1}((k+1)s_{k+1}t_{n-k} + (n-k)t_{n-k}s_{k+1}) \\
&\quad + (n+1)t_{n+1}s_0 \\
&= (n+1)s_{n+1}t_0 + (n+1)\sum_{k=1}^{n}s_k t_{n-k+1} + (n+1)t_{n+1}s_0 \\
&= (n+1)\sum_{k=0}^{n+1}s_k t_{n-k+1} \\
&= \partial_y^n\frac{\mathrm{d}(st)}{\mathrm{d}y}.
\end{aligned}$$

Therefore, (2.6.2) is proven. □

Theorem 2.6.2. *For $s = s(y) \in \mathcal{F}\{y; \mathbf{y}\}$, and integer $n \geq 1$,*

$$\frac{\mathrm{d}s^n}{\mathrm{d}y} = ns^{n-1}\frac{\mathrm{d}s}{\mathrm{d}y}. \tag{2.6.3}$$

Proof. We proceed by induction. When $n = 0$ and $n = 1$, it is easily seen that the conclusion is true. For $n \geq 2$, because of

$$\begin{aligned}
\frac{\mathrm{d}s^n}{\mathrm{d}y} &= \frac{\mathrm{d}s}{\mathrm{d}y}s^{n-1} + s\frac{\mathrm{d}s^{n-1}}{\mathrm{d}y}, \quad \text{by Theorem 2.3.1,}\\
&= \frac{\mathrm{d}s}{\mathrm{d}y}s^{n-1} + s(n-1)s^{n-2}\frac{\mathrm{d}s}{\mathrm{d}y},\\
&\qquad \text{by the induction hypothesis,}\\
&= s^{n-1}\frac{\mathrm{d}s}{\mathrm{d}y} + (n-1)s^{n-1}\frac{\mathrm{d}s}{\mathrm{d}y},\\
&\qquad \text{by the commutative law for multiplication,}\\
&= (1+n-1)s^{n-1}\frac{\mathrm{d}s}{\mathrm{d}y},\\
&\qquad \text{by the distributive law for multiplication and addition,}\\
&= ns^{n-1}\frac{\mathrm{d}s}{\mathrm{d}y},
\end{aligned}$$

then (2.6.3) is obtained. □

On the basis of $\mathcal{F}\{y; \mathbf{y}\}$, for any integer $n \geq 1$, the operation defined by

$$\int y^n \mathrm{d}y = \frac{1}{n+1}y^{n+1} \tag{2.6.4}$$

is called *integration*.

Then, by linear extension, we have, for any $a \in F$,

$$\int (ay^n)\mathrm{d}y = a\int y^n \mathrm{d}y$$

and

$$\int (y^m + y^n)\mathrm{d}y = \int y^m \mathrm{d}y + \int y^n \mathrm{d}y$$

on $\mathcal{F}\{y; \mathbf{y}\}$.

Theorem 2.6.3. *For any $s = s(y)$, $t = t(y) \in \mathcal{F}\{y; \mathbf{y}\}$,*

$$\int \frac{\mathrm{d}s}{\mathrm{d}y}t\mathrm{d}y = st - \int s\frac{\mathrm{d}t}{\mathrm{d}y}\mathrm{d}y. \tag{2.6.5}$$

Proof. On the basis of the cancellation law on the integral domain $\mathcal{F}\{y;\mathbf{y}\}$, it is only necessary to prove

$$\int \frac{\mathrm{d}s}{\mathrm{d}y} t\mathrm{d}y + \int s\frac{\mathrm{d}t}{\mathrm{d}y}\mathrm{d}y = \int \frac{\mathrm{d}(st)}{\mathrm{d}y}\mathrm{d}y.$$

For any integer $n \geq 1$, because of

$$\partial_y^n\left(\int \frac{\mathrm{d}s}{\mathrm{d}y} t\mathrm{d}y\right) = \frac{1}{n}\partial_y^{n-1}\left(\frac{\mathrm{d}s}{\mathrm{d}y}t\right) \quad \text{and} \quad \partial_y^n\left(\int s\frac{\mathrm{d}t}{\mathrm{d}y}\mathrm{d}y\right) = \frac{1}{n}\partial_y^{n-1}\left(s\frac{\mathrm{d}t}{\mathrm{d}y}\right),$$

we have

$$\begin{aligned}\partial_y^n\left(\int \frac{\mathrm{d}s}{\mathrm{d}y} t\mathrm{d}y\right) + \partial_y^n\left(\int s\frac{\mathrm{d}t}{\mathrm{d}y}\mathrm{d}y\right) &= \frac{1}{n}\left(\partial_y^{n-1}\left(\frac{\mathrm{d}s}{\mathrm{d}y}t\right) + \partial_y^{n-1}\left(s\frac{\mathrm{d}t}{\mathrm{d}y}\right)\right)\\ &= \frac{1}{n}\partial_y^{n-1}\left(\frac{\mathrm{d}(st)}{\mathrm{d}y}\right), \quad \text{Theorem 2.6.1 is employed,}\\ &= \partial_y^n\left(\int \frac{\mathrm{d}(st)}{\mathrm{d}y}\mathrm{d}y\right).\end{aligned}$$

Therefore, (2.6.5) is obtained. □

Theorem 2.6.4. *For any integer $n \geq 0$ and $s \in \mathcal{F}\{y;\mathbf{y}\}$,*

$$\int\left(s^n\frac{\mathrm{d}s}{\mathrm{d}y}\right)\mathrm{d}y = \frac{s^{n+1}}{n+1}. \tag{2.6.6}$$

Proof. On the basis $\{y, y^2, y^3, \ldots\}$, from (2.3.4), for any integer $n \geq 1$, we have

$$\begin{aligned}\frac{\mathrm{d}}{\mathrm{d}y}\int y^n\mathrm{d}y &= \frac{\mathrm{d}}{\mathrm{d}y}\left(\frac{y^{n+1}}{n+1}\right), \quad \text{by (2.6.1),}\\ &= y^n.\end{aligned}$$

This implies that on the basis of $\mathcal{F}\{y;\mathbf{y}\}$, integration is the inverse of differentiation. From the linearity of the two operations on $\mathcal{F}\{y;\mathbf{y}\}$, for any $s \in \mathcal{F}\{y;\mathbf{y}\}$, by (2.6.3), we have

$$s^{n+1} = (n+1)\int\left(s^n\frac{\mathrm{d}s}{\mathrm{d}y}\right)\mathrm{d}y.$$

By dividing $n+1$ on the two sides, (2.6.6) is obtained. □

Theorem 2.6.5. *For any $s = s(z) \in \mathcal{F}\{z;\mathbf{y}\}$,*

$$\left(\int_z \frac{\mathrm{d}s}{\mathrm{d}z}\right)^{\mathrm{T}} = L\left(P\left(\int_z s\right)^{\mathrm{T}}\right) \tag{2.6.7}$$

where P is the left projection matrix and

$$\int^{z} LP\mathbf{s}^{\mathrm{T}} = \frac{\mathrm{d}}{\mathrm{d}z}\int^{z} \mathbf{s} \tag{2.6.8}$$

where $\mathbf{s} = \int_z s \in \mathcal{V}$.

Proof. First, we prove (2.6.7). Let $s_i = \partial_z^i$, $i \geq 0$, then

$$\int_z s = (s_0, s_1, s_3, \ldots) = \mathbf{s}.$$

Since the right hand side of (2.6.7) is

$$P\mathbf{s}^{\mathrm{T}} = (0, s_1, 2s_2, 3s_3, \ldots) \Rightarrow L\left(P\left(\int_z s\right)^{\mathrm{T}}\right) = (s_1, 2s_2, 3s_3, \ldots)$$

and the left hand side of (2.6.7) is

$$\begin{aligned}\int_z \frac{\mathrm{d}s}{\mathrm{d}z} &= \int_z\left(\sum_{n\geq 0}(n+1)s_{n+1}z^n\right)\\ &= \sum_{n\geq 0}(n+1)s_{n+1}\int_z z^n\\ &= (s_1, 2s_2, 3s_3, \ldots),\end{aligned}$$

by comparing the two sides, (2.6.7) is found.

Then, we prove (2.6.8). Because of $LP\mathbf{s}^{\mathrm{T}} = (s_1, 2s_2, 3s_3, \ldots)$,

$$\begin{aligned}\int^{z} LP\mathbf{s}^{\mathrm{T}} &= \int^{z} \sum_{n\geq 0}(n+1)s_{n+1}z_n\\ &= \sum_{n\geq 0}(n+1)s_{n+1}\int^{z} z_n\\ &= \sum_{n\geq 0}(n+1)s_{n+1}z^n.\end{aligned}$$

The left hand side is the differential of s. On the other hand, because of $\mathbf{s} = \int_z s$,

$$\int^{z}\mathbf{s} = \int^{z}\left(\int_z s\right) = \int^{z}\int_z s = s.$$

The left hand side is the differential of s as well. Therefore, (2.6.8) is obtained. □

Theorem 2.6.6. *For any* $\mathbf{v} = v(v_0, v_1, v_2, \ldots) \in \mathcal{V}$, *let* $v = v(z) = \int^z \mathbf{v} \in \mathcal{F}(z; \mathbf{y})$, *then*

$$\frac{\mathrm{d}v}{\mathrm{d}z} = \int^z \mathbf{v} P^{\mathrm{T}} L^{\mathrm{T}} \tag{2.6.9}$$

and

$$\int v \mathrm{d}z = \int^z \mathbf{v} Q^{\mathrm{T}}. \tag{2.6.10}$$

Proof. First, we prove (2.6.9). Because of $\mathbf{v} = \int_z s$, by transposing the two sides of (2.6.7), and then invoking $\int^z$, (2.6.9) is found.

Then, we prove (2.6.10). Because of

$$\begin{aligned} \mathbf{v} Q^{\mathrm{T}} &= \left(0, v_0, \frac{1}{2} v_1, \frac{1}{3} v_2, \frac{1}{4} v_3, \ldots\right), \quad \text{by } \mathbf{v} \in \mathcal{V}, \\ &= \sum_{n \geq 0} \frac{1}{n+1} v_n z_n, \end{aligned}$$

we have

$$\begin{aligned} \int^z \mathbf{v} Q^{\mathrm{T}} &= \int^z \sum_{n \geq 0} \frac{1}{n+1} v_n z_n, \quad \text{by the linearity of the anti-meson functional,} \\ &= \sum_{n \geq 0} \frac{1}{n+1} v_n \int^z z_n, \quad \text{by } \partial_z^n s = v_n (n \geq 0), \\ &= \int s \mathrm{d}z. \end{aligned}$$

This is (2.6.10). □

2.7 Differences

We proceed on the basis of $\mathcal{R}\{z\} \subseteq \mathcal{F}\{z, \mathbf{y}\}$. For integer $n \geq 1$, two operations are established as

$$\delta_{x,y} z^n = \frac{x^n - y^n}{x - y} \tag{2.7.1}$$

and

$$\partial_{x,y} z^n = \frac{y x^n - x y^n}{x - y}, \tag{2.7.2}$$

called *straight difference* and *slope difference*, respectively.

For any function $f(z) \in \mathcal{R}\{z\}$, by the linear extension from (2.7.1) and (2.7.2), the straight difference and the slope difference of $f = f(z)$ are, respectively, obtained:

$$\delta_{x,y}f = \frac{f(x)-f(y)}{x-y}, \tag{2.7.3}$$

$$\partial_{x,y}f = \frac{yf(x)-xf(y)}{x-y}. \tag{2.7.4}$$

Theorem 2.7.1. *For any $f \in \mathcal{R}\{z\}$, let $f = f(z)$, then*

$$\partial_{x,y}(zf) = xy\delta_{x,y}f. \tag{2.7.5}$$

Proof. From the linearity of $\partial_{x,y}$ and $\delta_{x,y}$, it is only necessary to discuss $f(z) = z^n$, $n > 0$. Because of

$$\begin{aligned}\partial_{x,y}zf &= \partial_{x,y}z^{n+1} \\ &= \frac{yx^{n+1}-xy^{n+1}}{x-y} \\ &= xy\frac{x^n-y^n}{x-y} \\ &= xy\delta_{x,y}z^n \\ &= xy\delta_{x,y}f,\end{aligned}$$

the conclusion is drawn. □

Theorem 2.7.2. *For any $f \in \mathcal{R}\{z\}$,*

$$x^2y^2\delta^2_{x^2,y^2}(zf) - \partial^2_{x^2,y^2}(zf) = x^2y^2\delta_{x^2,y^2}(zf^2). \tag{2.7.6}$$

Proof. From (2.7.3) and (2.7.4), the left hand side of (2.7.6) is

$$\begin{aligned}&\frac{x^2y^2((x^2f(x^2)-y^2f(y^2))^2 - x^2y^2(f(x^2)-f(y^2))^2)}{x^2-y^2} \\ &= \frac{x^2y^2(x^2f^2(x^2)-y^2f^2(y^2))}{x^2-y^2}.\end{aligned}$$

From (2.7.3), this is the right hand side of (2.7.6). □

For a set $\mathcal{A}$ of configurations, let

$$f_{\mathcal{A}}(x,\mathbf{y}) = \sum_{A\in\mathcal{A}} x^{m(A)}\mathbf{y}^{\mathbf{n}(A)} \tag{2.7.7}$$

where $m(A) \geq 0$ and $\mathbf{n}(A) \geq \mathbf{0}$ are, respectively, an invariant number and an invariant vector of A. Let $F_{\mathcal{A}}(x,y)$ be a function of two variables such that

$$F_{\mathcal{A}}(x,y) = \int^{y} f_{\mathcal{A}}(x,\mathbf{y}). \tag{2.7.8}$$

The powers of x and y on $F_{\mathcal{A}}(x,y)$ are, respectively, called the *first parameter* and the *second parameter*.

Theorem 2.7.3. *Let $\mathcal{S}$ and $\mathcal{T}$ be two sets of configurations. If for $T \in \mathcal{T}$, there exists a mapping from $\mathcal{T}$ to $\mathcal{S}$ as $\lambda(T) = \{S_1, S_2, \dots, S_{m(T)+1}\}$ such that S_i corresponds 1–1 to $\{i, m(T)+2-i\}$, where i and $m(T)+2-i$ are the contributions to, respectively, the first parameter and the second parameter, $i = 1, 2, \dots, m(T)+1$, satisfying the condition*

$$\mathcal{S} = \sum_{T \in \mathcal{T}} \lambda(T),$$

then

$$F_{\mathcal{S}}(x, y) = xy\delta_{x,y}(zf_{\mathcal{T}}) \tag{2.7.9}$$

where $f_{\mathcal{T}} = f_{\mathcal{T}}(z) = f_{\mathcal{T}}(z, \mathbf{y})$.

Proof. On the basis of λ,

$$\begin{aligned} F_{\mathcal{S}}(x, y) &= \sum_{T \in \mathcal{T}} \sum_{i=1}^{m(T)+1} x^i y^{m(T)-i+2} \mathbf{y}^{\mathbf{n}(T)} \\ &= xy \sum_{T \in \mathcal{T}} \frac{x^{m(T)+1} - y^{m(T)+1}}{x - y} \mathbf{y}^{\mathbf{n}(T)} \\ &= xy\delta_{x,y}(zf_{\mathcal{T}}). \end{aligned}$$

This is (2.7.9). □

Corollary 2.7.4. *Let $f_{\mathcal{T}}$ and $F_{\mathcal{S}}$ be, respectively, given by (2.7.7) and (2.7.8), then*

$$F_{\mathcal{S}}(x, \mathbf{y}) = \int_y xy\delta_{x,y}(zf_{\mathcal{T}}).$$

Proof. From (2.7.8) and Theorem 2.7.3, the conclusion is derived. □

Theorem 2.7.5. *Let $\mathcal{S}$ and $\mathcal{T}$ be two sets of configurations. If for $T \in \mathcal{T}$, there exists a mapping from $\mathcal{T}$ to $\mathcal{S}$ as $\lambda(T) = \{S_1, S_2, \dots, S_{m(T)-1}\}$ such that S_i corresponds 1–1 to $\{i, m(T)-i\}$ where i and $m(T)+2-i$ are the contributions to, respectively, the first parameter and the second parameter, $i = 1, 2, \dots, m(T)-1$, satisfying the condition*

$$\mathcal{S} = \sum_{T \in \mathcal{T}} \lambda(T),$$

then

$$F_{\mathcal{S}}(x, y) = \partial_{x,y}(f_{\mathcal{T}}) \tag{2.7.10}$$

where $f_{\mathcal{T}} = f_{\mathcal{T}}(z) = f_{\mathcal{T}}(z, \mathbf{y})$.

Proof. On the basis of the determination of λ,

$$F_{\mathcal{S}}(x, y) = \sum_{T \in \mathcal{T}} \sum_{i=1}^{m(T)-1} x^i y^{m(T)-i} \mathbf{y}^{\mathbf{n}(T)}$$

$$\begin{aligned}&= xy \sum_{T \in \mathcal{T}} \frac{yx^{m(T)} - xy^{m(T)}}{x - y} \mathbf{y}^{\mathbf{n}(T)} \\ &= \partial_{x,y}(f_{\mathcal{T}}).\end{aligned}$$

This is (2.7.10). □

Corollary 2.7.6. *Let $f_{\mathcal{T}}$ and $F_{\mathcal{S}}$ be, respectively, determined by* (2.7.7) *and* (2.7.8), *then*

$$F_{\mathcal{S}}(x, \mathbf{y}) = \int_y \partial_{x,y}(f_{\mathcal{T}}).$$

Proof. From (2.7.8) and Theorem 2.7.5, the conclusion is drawn. □

2.8 Meson equations

If an equation involves the meson functional, probably companied by some of differences, differentiations, and/or integrations, it is in short called a *meson equation*.

In this book, all meson equations are considered to be of the form

$$\begin{cases} f = a + bx^{\gamma} A(x; f) + cx^{\alpha} \int_y (y^{\beta} B(y : f, f|_{x=y}, O)); \\ f|_{\mathbf{y}=0 \Rightarrow x=0} = d, \end{cases} \tag{2.8.1}$$

where $\alpha \geq 0$ and $\beta, \gamma \geq 1$ are integers, $a, b, c, d \in \mathcal{R}$, $f, A, B \in \mathcal{R}\{x, \mathbf{y}\}$ and O is a set of operations as regards functionals including probably a meson functional itself.

Observation 2.8.1. *If $a \neq d$, then equation* (2.8.1) *is not consistent.*

Proof. Because of $\beta, \gamma \geq 1$, no constant term is in $f - a$ and hence the conclusion of the observation. □

This observation enables us to restrict ourselves to considering equation (2.8.1) always with the condition: $a = d$.

Because of f, $A = A(x; f)$ and $B = B(y : f, f|_{x=y}, O)$ all in $\mathcal{R}\{x; \mathbf{y}\}$, we are allowed to write it as a sum of homogeneous functions of a parameter chosen beforehand in $\mathcal{R}\{x; \mathbf{y}\}$.

Let $n = |\mathbf{n}| = |(n_1, n_2, n_3, \ldots)|$ and $s = \mathbf{i}\mathbf{n}^{\mathrm{T}}$ be, respectively, called the *pan-order* and the *pan-size* where $\mathbf{i} = (1, 2, 3, \ldots)$ and ${}^{\mathrm{T}}$ is the transpose of a vector, or generally a matrix.

Observation 2.8.2. *For two sizes s_1 and s_2, $s_1 + s_2$ is a size as well.*

Proof. Let $s_1 = \mathbf{i}\mathbf{n}_1$ and $s_2 = \mathbf{i}\mathbf{n}_2^{\mathrm{T}}$. Because of $s_1 + s_2 = \mathbf{i}(\mathbf{n}_1^{\mathrm{T}} + \mathbf{n}_2)^{\mathrm{T}}$, $s = \mathbf{i}\mathbf{n}^{\mathrm{T}}$, $\mathbf{n} = \mathbf{n}_1^{\mathrm{T}} + \mathbf{n}_2^{\mathrm{T}}$, is a size as well. □

This observation enables us to introduce two types of partitioning of the set

$$\mathcal{A}_f = \{\mathbf{n} \mid \text{a power index of } f \in \mathcal{R}\{\mathbf{y}\}\}$$

for pan-order and for pan-size. Thus, we have

$$f(x;\mathbf{y}) = \begin{cases} \sum_{n\geq 0} F_{*,n} & \text{for } n \text{ as the pan-order;} \\ \sum_{n\geq 0} F_{s,*} & \text{for } s \text{ as the pan-size,} \end{cases} \tag{2.8.2}$$

where $F_{*,n}$ and $F_{s,*}$ are, respectively, homogeneous in n and s for $\mathbf{y}$.

Observation 2.8.3. *For two functions $f, g \in \mathcal{R}\{x;\mathbf{y}\}$ determined by, respectively, $F_i(i \geq 0)$ and $G_j(j \geq 0)$ as in (2.8.2), their product is determined by*

$$C_l = \sum_{k=0}^{l} F_k G_{l-k} \tag{2.8.3}$$

for $l \geq 0$.

Proof. For pan-order, it is natural. For pan-size, it is from Observation 2.8.2. □

This observation shows that C_l is determined by $F_i(i \leq l)$ and $G_j(j \leq l)$. If $f = g$, then $fg = f^2$ and $C_l = F_l^{[2]}$ for $l \geq 0$. Furthermore, for any integer $n \geq 3$, f^n is the sum of

$$F_l^{[n]} = \sum_{k=0}^{l} F_k F^{[n-1]_{l-k}}$$

over $l \geq 0$.

On the basis of equations (2.8.1) and (2.8.2), two infinite equation systems can be extracted each of which is equivalent to equation (2.8.1) in $\mathcal{R}\{x;\mathbf{y}\}$.

Because of the additivity for the pan-order and pan-size, from $A(x,f) \in \mathcal{R}\{x,\mathbf{y}\}$, we are allowed to write

$$A(x,f) = \begin{cases} \sum_{s\geq 0} A_{s,*}, & \text{when } s \text{ is the pan-size;} \\ \sum_{m\geq 0} A_{*,n}, & \text{when } n \text{ is the pan-order.} \end{cases} \tag{2.8.4}$$

Because of $B(y : f, f|_{x=y}, O) \in \mathcal{R}\{x,\mathbf{y}\}$, let $B(y : f, f|_{x=y}, O) = B_0 + yB_1 + y^2B_2 + \cdots$, then we have

$$\int_f (y^\beta B(y : f, f|_{x=y}, O)) = \sum_{i\geq 0} y_{\beta+i} B_i$$

where

$$B_i = \begin{cases} \sum_{s\geq 0} B_{i[s,*]}, & \text{when } s \text{ is the pan-size;} \\ \sum_{n\geq 0} B_{i[*,n]}, & \text{when } n \text{ is the pa-order.} \end{cases} \tag{2.8.5}$$

Lemma 2.8.4. *Equation* (2.8.1) *for $a = d$ is equivalent to the equation system of infinite number of undeterminates $F_{s,*}$, $s \geq 0$,*

$$\begin{cases} F_{s,*} = a + bx^{\gamma}A_{s,*} + cx^{\alpha}\sum_{i\geq 0} y_{\beta+i}B_{i[s-1,*]}, & \text{for } s \geq 1; \\ F_{s,*} = a, & \text{for } s = 0. \end{cases} \tag{2.8.6}$$

Proof. Because of the equality for two subfunctions with the same pan-size on the two sides of equation (2.8.1), a solution of equation (2.8.6) determines a solution of equation (2.8.1) □

Similarly, for order, we have the following.

Lemma 2.8.5. *Equation* (2.8.1) *for $a = d$ is equivalent to the equation system of an infinite number of undeterminates $F_{*,n}$, $n \geq 0$,*

$$\begin{cases} F_{*,n} = a + bx^{\gamma}A_{*,n} + cx^{\alpha}\sum_{i\geq 0} y_{\beta+i}B_{i[*,n-1]}, & \text{for } n \geq 1; \\ F_{*,n} = a, & \text{for } n = 0. \end{cases} \tag{2.8.7}$$

Proof. Because of the equality for two subfunctions with the same pan-order on the two sides of equation (2.8.1), a solution of equation (2.8.6) determines a solution of equation (2.8.1). □

A function $f \in \mathcal{R}\{\mathbf{y}\}$ with all terms $x^m\mathbf{y}^{\mathbf{i}}$ for $m \geq 0$ and $\mathbf{i} \geq \mathbf{0}$ having $m \in \{i_j \mid j \geq 1\}$ is said to be a *partition function*. Let $\mathcal{P}$ be the set of all partition functions.

Observation 2.8.6. *Any partition function is in $\mathcal{R}\{x; \mathbf{y}\}$.*

Proof. The reason is that polynomials are seen as special cases of series. □

Observation 2.8.7. *For any partition function $f \in \mathcal{R}\{x; \mathbf{y}\}$, $f|_{\text{pan-size}=s} = \langle f\rangle_s$ is a homogeneous polynomial of pan-size s.*

Proof. Because of the limitation $\mathbf{y}_s$ allowable from s given, $\langle f\rangle_s$ in the partition function has a finite number of terms and hence is a homogeneous polynomial of pan-size s. □

Observation 2.8.8. *If f is a solution of equation* (2.8.1), *then, for any integer $s \geq 1$, the homogeneous polynomial $\langle f\rangle_s$ for pan-size can be determined by $\langle f\rangle_t$, $t \leq s - 1$.*

Proof. On the basis of Observation 2.8.7, by considering $\alpha, \beta \geq 1$, the conclusion is drawn. □

Theorem 2.8.9. *Equation* (2.8.1) *is well-defined on $\mathcal{R}\{x; \mathbf{y}\}$ for $f \in \mathcal{P}$ if, and only if, $a = d$.*

Proof. Observation 2.8.1 provides the necessity. Only the sufficiency is considered. Because of the initial condition of equation (2.8.1) and the three observations just mentioned above, we are allowed to establish a procedure for reaching the solution of equation (2.8.1). □

2.9 Notes

2.9.1. Since the series of articles by Blissard (see [3, 4], 1861–1862), the Blissard operation dealing with a symbol operator has been established, it looks no further attention has been paid to it until Rota dealt with a functional, called shadow (or umbrella) functional, to extract a summation free explicit form of the Bell number for enumerating partitions of a set (or an integer), as shown in Rota GC [75] a century ago.

2.9.2. Umbrella calculus is formulated on the basis of shadow functionals by the Rota group with relationships to Hopf algebra and the Möbius algebra. The reader interested in symbol algebra is referred to [63] (Mullin RC, Rota GC, 1970), [74] (Roman SM, Rota GC, 1978), [75] (Rota GC, 1964), [76] (Rota GC, Kahaner D, Odlyzko A, 1973), [77] (Rota GC, Tayler BD, 1994), [78] (Rota GC, Shen J, Tayler BD, 1997) *etc.*

2.9.3. In the beginning of our research on the enumeration of planar maps with a vertex partition vector as parameter, the shadow operator used to be misemployed as a particular type of Blissard operator denoted by φ as in [17–31, 34, 35] (Liu YP, 1985–1989).

2.9.4. Then, from 1990 on, as shown in [36–39, 41, 43] (Liu YP, 1990–1993), *etc.*, the operator has been ignored as a more general functional of transforming the space of functions to an abstract linear space denoted by $\int$ instead of the shadow functional.

2.9.5. Because of the distinction of the functional from the shadow, we have had to adopt in our work the meson functional as shown in [57, 58] (Liu YP, 2012) since 2012, particularly in [59] (Liu YP, 2015).

2.9.6. This book only concentrates on the property of being well-defined of the meson functional equations considered and their solutions extracted via constructions for realization on computers without much investigation on the inner structures of the functional itself. These structures hopefully are useful for evaluating the explicit forms of their solutions in the extended integral domain directly.

3 Function equations of one variable

3.1 First coefficient variable

Let $\mathcal{R}\{z\}$ be the extension of integral domain $\mathcal{F}\{z\}$ over the integer ring R with z as an undeterminate. Of course, $\mathcal{R}\{z\}$ is a ring as well. The equation considered in this section is about finding a function f of z, $f \in \mathcal{R}_+\{z\}$, such that

$$\begin{cases} azf^2 - bf + c = 0; \\ f|_{z=0} = d, \end{cases} \tag{3.1.1}$$

where $a, b, c, d \in R_+$ is satisfied. This is equation (1) in Introduction.

At first glance, it looks like an ordinary quadratic equation. However, its main differences from the quadratic equation in elementary algebra are: (1) at least one of the coefficients in the terms involving the undeterminate contains a variable; (2) the solution has to be in $\mathcal{R}_+\{x\}$.

Whenever noticing that if $a = b = c = d = 1$, the equation becomes what is obtained in enumerating the classes of non-isomorphic planted plane trees. We face the suggestion to call equation (3.1.1) the *model of planted trees*.

In what follows, some conditions have to be clarified.

Condition 1. Because of $f \in \mathcal{R}\{z\}$,

$$f = \sum_{n \geq 0} F_n z^n \tag{3.1.2}$$

where $F_n \in \mathcal{R}$, $n \geq 0$.

From the initial condition of (3.1.1), whenever

$$c = bd \tag{3.1.3}$$

is satisfied, equation (3.1.1) has the possibility of consistency, and hence only

$$\begin{cases} azf^2 - bf + bd = 0; \\ f|_{z=0} = d, \end{cases} \tag{3.1.4}$$

has to be considered.

Condition 2. If $d = 0$, equation (3.1.4) becomes

$$(azf - b)f = 0.$$

Because of the triviality that $f = 0$ is a degenerate case of a linear equation, it is only necessary to consider $azf - b = 0$. However, because of the non-existence of an inverse of az in $\mathcal{R}\{z\}$, this leads to the conclusion that equation (3.1.4) has no solution. Therefore, $d \neq 0$.

https://doi.org/10.1515/9783110625837-003

Condition 3. Because of that, $b = 0$ or $a = 0$ leads to the triviality of equation (3.1.4), and both $a \neq 0$ and $b \neq 0$ have to be pre-assumed.

Theorem 3.1.1. *Equation* (3.1.4) *has, and is the only one to have, a solution in* $\mathcal{R}_+\{z\}$ *if, and only if,* $abd \neq 0$.

Proof. Because of the infiniteness in the ring $\mathcal{R}$, no zero factor exists in $\mathcal{R}$. Thus, $b^{-1} \in \mathcal{R}$. By the cancelation law, equation (3.1.4) has the following equivalent form:

$$\begin{cases} f = ab^{-1}zf^2 + d; \\ f|_{z=0} = d, \end{cases} \tag{3.1.5}$$

in $\mathcal{R}\{z\}$.

From (3.1.2), by equating the coefficients of terms with the same degree on two sides, equation (3.1.5) becomes, for any integer $m \geq 0$,

$$\begin{cases} F_0 = d \quad \text{(the initial condition)}, \quad m = 0; \\ F_m = \frac{a}{b}\partial_z^{m-1} f^2 \quad \text{(two coefficients of } z^m \text{ equal)} \\ \quad = \frac{a}{b}\sum_{i=0}^{m-1} F_i F_{m-1-i}, \quad m \geq 1. \end{cases} \tag{3.1.6}$$

Since all F_l, $0 \leq l \leq m-1$, are known, F_m is determined by some of F_l, $0 \leq l \leq m-1$. On account of the principle shown in the proof of Theorem 2.8.9, the conclusion is drawn. □

In the proof of Theorem 3.1.1, it is seen that (3.1.6) has provided the solution of equation (3.1.5), hence equation (3.1.1), if any, has the form of a sum with all terms positive.

Moreover, when $a = b = d = 1$ ($c = bd = 1$), the F_m become the Catalan numbers

$$C_m = \frac{(2m)!}{m!(m+1)!} = \frac{1}{2m+1}\binom{2m+1}{m}, \tag{3.1.7}$$

or in the form of a recursion

$$C_m = \begin{cases} 1, & \text{when } m = 0; \\ \sum_{l=0}^{m-1} C_l C_{m-1-l}, & \text{when } m \geq 1. \end{cases} \tag{3.1.8}$$

Theorem 3.1.2. *The solution of equation* (3.1.4) *is determined by*

$$F_m = \frac{a^m d^{m+1}(2m)!}{b^m m!(m+1)!} \tag{3.1.9}$$

for $m \geq 0$.

Proof. We proceed by induction on m. When $m = 0$, from the initial condition of equation (3.1.4), (3.1.9) is checked to be true. When $m \geq 1$, on the basis of the assumption that, for any integer $0 \leq l \leq m-1$, (3.1.9) holds. From (3.1.6), we have

$$\begin{aligned} F_m &= \frac{a}{b}\sum_{l=0}^{m-1} \frac{a^l d^{l+1}(2l)!}{b^l l!(l+1)!} \frac{a^{m-1-l} d^{m-1-l+1}(2(m-1-l))!}{b^{m-1-l}(m-1-l)!(m-1-l+1)!} \\ &= \frac{a^m d^{m+1}}{b^m}\sum_{l=0}^{m-1} \frac{(2l)!}{l!(l+1)!} \frac{(2(m-1-l))!}{(m-1-l)!(m-1-l+1)!}, \\ &\quad \text{by (3.1.8) and (3.1.7),} \\ &= \frac{a^m d^{m+1}}{b^m} \frac{(2m)!}{m!(m+1)!}. \end{aligned}$$

Thus, (3.1.9) is true for $m \geq 1$. □

This is a summation-free explicision (*i. e.*, an explicit expression) of F_m.

Now, let us go back to the specific case of $a = b = c = d = 1$ in equation (3.1.1).

Theorem 3.1.3. *The equation*

$$\begin{cases} zf^2 - f + 1 = 0; \\ f|_{z=0} = 1, \end{cases} \tag{3.1.10}$$

for $f \in \mathcal{R}\{z\}$ is well-defined. The solution of equation (3.1.10) *is determined by*

$$\begin{cases} F_0 = 1 \quad (\textit{the initial condition}), & m = 0; \\ F_m = \sum_{i=0}^{m-1} F_i F_{m-1-i}, & m \geq 1, \end{cases} \tag{3.1.11}$$

for $F_m = \partial_z^m f$, $m \geq 0$. For $m \geq 0$,

$$F_m = \frac{(2m)!}{m!(m+1)!}. \tag{3.1.12}$$

Proof. The first conclusion is seen to be true from Theorem 3.1.1. The second conclusion is seen to be true from (3.1.6). The third conclusion is seen to be true from (3.1.8). □

From (3.1.7) and (3.1.7), F_m in (3.1.12) is just the Catalan number C_m. This explicision is, particularly, summation-free.

In what follows, some examples are given to show certain direct applications of equation (3.1.1) and other equations which are going to be discussed in this book.

Example 1. Topological classifications of binary trees. A *binary tree* is defined to be a tree such that it has exactly one vertex of valency 2 (*root-vertex*!) with all other vertices

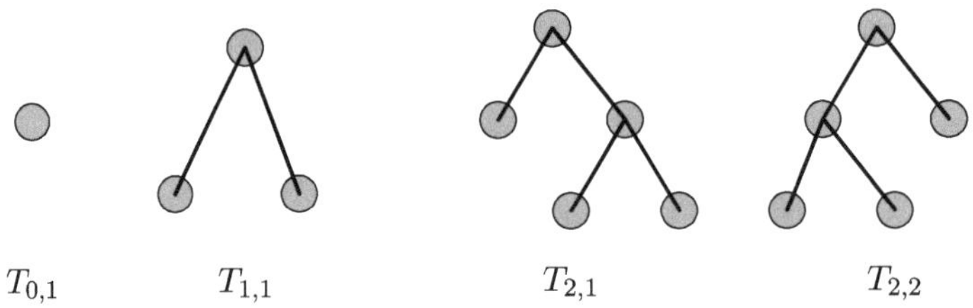

Figure 3.1.1: Binary trees of $n = 1$–2.

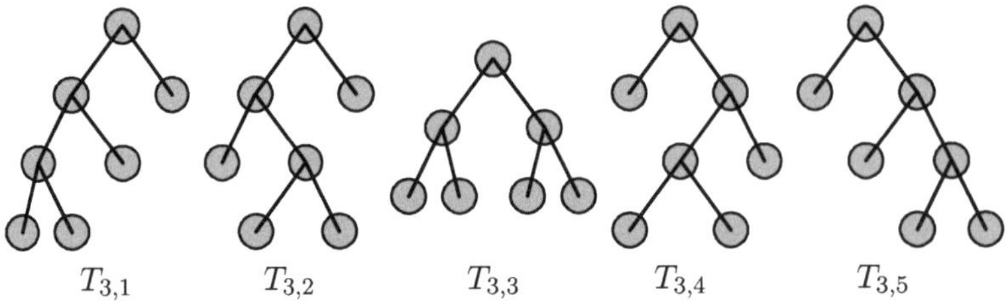

Figure 3.1.2: Binary trees of $n = 3$.

of valency either 1 (*end vertex*!) or 3 (*inner vertex*!). In some references, a root-vertex is called a *peak* and the end vertex is called an *articulate vertex*.

Two binary trees are said to be topologically equivalent if they are treated as planar embedding equivalents. In other words, edges at a vertex are certainly considered to have a rotation (a cyclic order).

In Figure 3.1.1 and Figure 3.1.2, non-equivalent binary trees with the number $n = 0$, 1 and 2 of non-end vertices are shown. For example, when $n = 0$, only one binary three $T_{0,1}$ as shown. When $n = 1$, only one as well, shown as $T_{1,1}$. When $n = 2$, two binary trees are shown as $T_{2,1}$ and $T_{2,2}$. When $n = 3$, five binary trees are as shown by $T_{3,1}$, $T_{3,2}, \ldots, T_{3,5}$.

Because two smaller binary trees are obtained by deleting the root-vertex (or the peak), the enumerating function $t_{\text{bint}} = t_{\text{bint}}(z)$ is checked to satisfy equation (3.1.10). From Theorem 3.1.3,

$$\partial_z^m t_{\text{bint}} = \frac{(2m)!}{m!(m+1)!} = T_m.$$

Therefore, $T_0 = 1$, $T_1 = 1$, $T_2 = 2$ and $T_3 = 5$.

Example 2. Classification of plane rooted trees by size $n \geq 0$. A plane tree is a planar embedding of a tree, because of two sides occurring at one end of an edge. If the pair at an end and a side for a chosen edge is marked on a plane tree, then the plane tree is called rooted. The symbol of the mark is said to be the *root*, denoted by a hollow. In Figure 3.1.3, $L_{n,i}$ denote the i ($i \geq 1$)th plane rooted tree of size n. Because of the

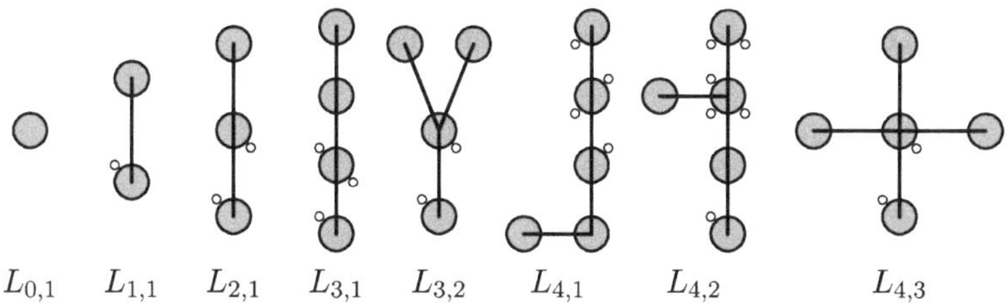

Figure 3.1.3: Plane rooted trees of $n = 0$–4.

asymmetry of a rooted plane tree, each hollow on a plane tree represents a rooted one. Rooted plane trees produced from a chosen plane tree are distinguished by i. For integer $n \geq 0$ given, the number of rooted isomorphic classes of plane trees is $\Lambda_n = l_{n,1} + l_{n,2} + l_{n,3} + \cdots$, where $l_{n,i}$, $i \geq 1$, is the number of hollows on $L_{n,i}$. For example, as shown in Figure 3.1.3, $\Lambda_0 = 1$, $\Lambda_1 = 1$, $\Lambda_2 = 2$, $\Lambda_3 = 5$ and $\Lambda_4 = 14$.

The recursion (3.1.8) is satisfied by Λ_n, $\Lambda_n = C_n$, $n \geq 0$.

Example 3. Classification of planar rooted petal bundles by size. In a planar embedding of a graph, each edge has two *sides* and two *ends*. A pair {end,side} marked is called a *root*. A graph with a singe vertex is said to be a *petal bundle*. The mark of a root is represented by a hollow on its figure as shown in Figure 3.1.4. $P_{n,i}$ is the ith ($i \geq 1$) isomorphic class of a planar petal bundle of size n. A hollow determines a rooted petal bundle. For given integer $n \geq 0$, the number of planar rooted petal bundles of size n is $P_n = l_{n,1} + l_{n,2} + l_{n,3} + \cdots$, where $l_{n,i}$, $i \geq 1$, is the number of hollows on $P_{n,i}$. For example, in Figure 3.1.4, $P_0 = 1$, $P_1 = 1$, $P_2 = 2$, $P_3 = 5$ and $P_4 = 14$. On $P_{n,i}$, each pair of two occurrences of a number represents an edge.

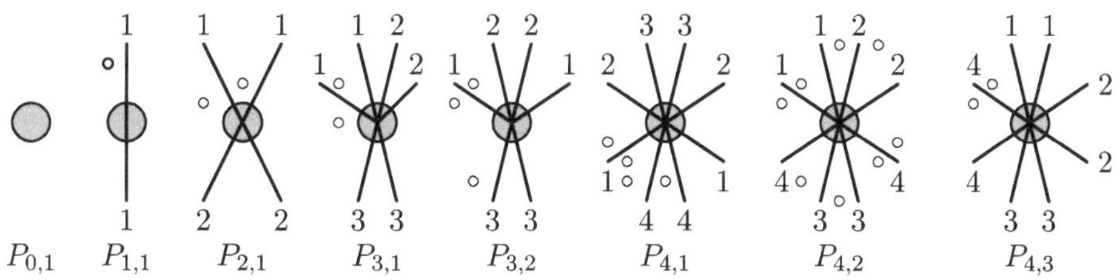

Figure 3.1.4: Planar petal bundles of size $n = 0$–4.

Similarly, it can be shown that the function t_{pet} determined by $\partial_z^n t_{\text{pet}} = P_n$ ($n \geq 0$) satisfies equation (3.1.1) for $a = b = c = d = 1$. From Theorem 3.1.3, $P_n = C_n$, $n \geq 0$.

Example 4. Sequence of integrals. Consider the function

$$\lambda(z) = \sum_{n\geq 0} z^n \frac{b^{2n+1}}{\pi} \int_{-1}^{1} \tau^{2n} \sqrt{1-\tau^2} \mathrm{d}\tau.$$

It is well known that, for any integer $n \geq 0$,

$$\frac{b^{2n+1}}{\pi} \int_{-1}^{1} \tau^{2n} \sqrt{1-\tau^2} \mathrm{d}\tau = C_n.$$

From Theorem 3.1.3,

$$\partial_z^m \lambda(z) = F_m$$

for $m \geq 0$. On account of equation (3.1.10) being well-defined, $\lambda(z)$ is the solution of equation (3.1.1) for $a = b = c = d = 1$ as well.

Example 5. Continued fraction. We address the continued fraction

$$\cfrac{1}{1 - \cfrac{\tau^2}{1 - \cfrac{\tau^2}{1 - \cfrac{\tau^2}{1 - \cfrac{\tau^2}{1 - \cfrac{\tau^2}{1 - \cdots}}}}}} = \sum_{m\geq 0} C_m \tau^{2m} = \psi(\tau);$$

let $\tau = \sqrt{z}$, then, from Theorem 3.1.3, $\partial_z^m \xi = \partial_z^m \psi(\sqrt{z}) = F_m$. From Theorem 3.1.1 (whose specific case of $a = b = d = 1$ is Theorem 3.1.3), $\xi(z)$ is the solution of equation (3.1.1) in the case of $a = b = c = d = 1$.

3.2 Last coefficient variable

Let $\mathcal{R}\{z\}$ be the extension of the integral domain $\mathcal{R}$ consisting of all Laurent series over the integer ring R with z as an undeterminate.

The purpose of this section is to find a function $f \in \mathcal{R}\{z\}$ with all coefficients positive over the non-negative part of the real field $\mathbb{R}_+$ such that the equation

$$\begin{cases} af^2 - bf + cz = 0; \\ f|_{z=0} = d, \end{cases} \tag{3.2.1}$$

where $a, b, c, d \in \mathbb{Z}_+$ is satisfied. This is equation (2) in Introduction.

Because of the constant term involving a variable, the equation is said to be of the type of *last coefficient variable*.

Observation 3.2.1. *If $d \neq 0$, then equation* (3.2.1) *is not consistent.*

Proof. Assume integer $d > 0$. Let $F_n = \partial_Z^n f$, $n \geq 0$, then from equation (3.2.1),

$$aF_0^2 - bF_0 = ad^2 - bd = 0 \Longrightarrow ad = b \Longrightarrow F_0 = \frac{b}{a}.$$

Because of $F_0 \in \mathbb{R}_+$, $a >$ and $b > 0$. However, for determining F_1, from equation (3.2.1), we have

$$2aF_0F_1 - bF_1 + c = 0 \Longrightarrow 2bF_1 + c = 0 \Longrightarrow F_1 = -\frac{c}{2b} < 0,$$

where the last $\Longrightarrow$ is reasoned from the fact that if $c = 0$ then the equation becomes trivial. A contradiction to $F_1 \in \mathbb{R}_+$ occurs. □

If a new function h of z is introduced with $f = zh$, then equation (3.2.1) becomes

$$\begin{cases} az^2h^2 - bzh + cz = 0; \\ h|_{z=0} = d. \end{cases}$$

This is just the type discussed in the last section.

So, another equation is considered in this section as

$$\begin{cases} af^2 - bf + cz = 0; \\ \partial_z^1 f = d, \end{cases} \tag{3.2.2}$$

for $f \in \mathcal{R}\{z\}$ with all coefficients in $\mathbb{R}_+$.

Observation 3.2.2. *In equation* (3.2.2), $c = bd$.

Proof. Because of $\partial_z^0 f = 0$, from equation (3.2.2), we have

$$-b\partial_z^1 f + c = 0 \Longrightarrow d = \frac{c}{b} \Longrightarrow b > 0,$$

and hence $c = bd$. □

On account of $a = 0$ leading to equation (3.2.2) being trivial, in what follows, only

$$\begin{cases} af^2 - bf + bdz = 0; \\ \partial_z^1 f = d, \end{cases} \tag{3.2.3}$$

where $abcd \neq 0$, $a, b, c, d \in \mathbb{Z}_+$ and $f \in \mathcal{R}\{z\}$ is considered.

Theorem 3.2.3. *Equation* (3.2.3) *has, and is the only one to have, a solution in $f \in \mathcal{R}\{z\}$.*

Proof. Let $f \in \mathcal{R}\{z\}$ be determined by $F_m = \partial_z^m f \in \mathbb{R}_+$, $m \geq 0$.

It is well known that $F_0 = 0$ and $F_1 = d$. For integer $m \geq 2$, from equation (3.2.3),

$$\begin{aligned} bF_m &= a \sum_{i=0}^{m} F_i F_{m-i}, \text{ by } F_0 = 0, \\ &= a \sum_{i=1}^{m-1} F_i F_{m-i}. \end{aligned}$$

Because of $b > 0$,

$$F_m = \frac{a}{b} \sum_{i=1}^{m-1} F_i F_{m-i}. \tag{3.2.4}$$

By the induction principle on n, on the basis of all F_i $(0 \leq i \leq m-1)$ being known, from (3.2.5), F_m is then determined.

On account of m being arbitrarily chosen, f is determined as the solution. □

On the basis of the proof of Theorem 3.2.1, a procedure is easily found to extract the solution:

$$F_m = \begin{cases} 0, & \text{when } m = 0; \\ d, & \text{when } m = 1; \\ \frac{a}{b} \sum_{i=1}^{m-1} F_i F_{m-i}, & \text{when } m \geq 2. \end{cases} \tag{3.2.5}$$

By employing (3.2.5), one might like to see what happens for m smaller. Then we have

$$\begin{aligned} F_0 &= 0, \quad \text{known from what we mentioned above;} \\ F_1 &= d, \quad \text{from the initial condition.} \end{aligned}$$

When $m = 2$, from (3.2.5),

$$\begin{aligned} F_2 &= \frac{a}{b} \sum_{i=1}^{1} F_i F_{2-i} = \frac{a}{b} F_1^2 = \frac{a}{b} d^2, \\ F_3 &= \frac{a}{b} \sum_{i=1}^{2} F_i F_{3-i} = \frac{a}{b} (2F_1 F_2) = 2\left(\frac{a^2 d^3}{b^2}\right), \\ F_4 &= \frac{a}{b} \sum_{i=1}^{3} F_i F_{4-i} = \frac{a}{b} (2F_1 F_3 + F_2^2) \\ &= \frac{a}{b} \left(4d\left(\frac{a^2 d^3}{b^2}\right) + \left(\frac{a^2 d^4}{b^2}\right)\right) \\ &= 5\left(\frac{a^3 d^4}{b^3}\right), \end{aligned}$$

$$F_5 = \frac{a}{b}\sum_{i=1}^{4} F_i F_{5-i} = 2\frac{a}{b}(F_1F_4 + F_2F_3)$$
$$= 2\left(\frac{a}{b}\right)\left(5\left(\frac{a}{b}\right)^3 d^5 + 2\left(\frac{a}{b}\right)^3\right)$$
$$= 14\left(\frac{a}{b}\right)^4 d^5.$$

Theorem 3.2.4. *The solution of equation* (3.2.3) *determined by* F_m, $m \geq 0$, *has a summation-free explicision in the form of*

$$F_m = \begin{cases} 0, & \text{when } m = 0; \\ d, & \text{when } m = 1; \\ (\frac{a}{b})^{m-1} d^m C_{m-1}, & \end{cases} \tag{3.2.6}$$

where C_{m-1} *is the Catalan number shown in* (3.1.7).

Proof. Because of $F_0 = 0$ and $F_1 = d$ as is known from the initial condition of equation (3.2.3), F_1 is checked from $C_0 = 1$.

For all $1 \leq l \leq m-1$,

$$F_l = \left(\frac{a}{b}\right)^{l-1} d^l C_{l-1}$$

are assumed to be true. By the principle of induction, we determine F_m.

On the basis of (3.2.5) and the assumption, we see that

$$F_m = \frac{a}{b}\sum_{i=1}^{m-1}\left(\frac{a}{b}\right)^{i-1} d^i C_{i-1}\left(\frac{a}{b}\right)^{m-i-1} d^{m-i} C_{m-i-1}$$
$$= \left(\frac{a}{b}\right)^{m-1} d^m \sum_{i=1}^{m-1} C_{i-1}C_{m-i-1}, \quad \text{by (3.1.8)},$$
$$= \left(\frac{a}{b}\right)^{m-1} d^m C_{m-1}.$$

This is the conclusion. □

This theorem enables us to find the condition for the solution of equation (3.2.3) with all coefficients integers.

Corollary 3.2.5. *The solution* f *of equation* (3.2.3) *holds for all* $\partial_z^m = F_m \in \mathbb{Z}_+$ *if, and only if,* $b/a \in \mathbb{Z}_+$.

Proof. A direct result of Theorem 3.2.4. □

Now, let us to go back to the case of $a = b = c = d = 1$ for equation (3.2.3), *i. e.*,

$$\begin{cases} f^2 - f + z = 0; \\ \partial_z^1 f = 1. \end{cases} \tag{3.2.7}$$

Theorem 3.2.6. *Equation* (3.2.7) *is well-defined on* $\mathcal{R}\{z\}$.

Proof. We have the case of $a = b = c = d = 1$ of Theorem 3.2.3. □

From (3.2.5) of $a = b = c = d = 1$, a procedure is easily found to extract the solution of equation (3.2.7) as

$$F_m = \begin{cases} 0, & \text{when } m = 0; \\ 1, & \text{when } m = 1; \\ \sum_{i=1}^{m-1} F_i F_{m-i}, & \text{when } m \geq 2. \end{cases} \tag{3.2.8}$$

On the basis of (3.2.8), a summation-free explicision is easily found as well.

Theorem 3.2.7. *The solution f of equation* (3.2.7) *is in the summation-free form of*

$$\partial_x^m f = \frac{(2m-2)!}{(m-1)!m!} \tag{3.2.9}$$

for integer $m \geq 1$.

Proof. In the case of $a = b = c = d = 1$ of Theorem 3.2.4, by (3.1.8), the conclusion is then drawn. □

For Theorems 3.2.6 and 3.2.6, some applications are shown via examples.

Example 1. Classification of planted trees by size. A *planted tree* is a plane rooted tree with rooted vertex of valency 1. The enumerating function of planted trees by size as the parameter is

$$t_{\text{plant}} = zt_{\text{tree}} = \sum_{n \geq 1} C_{n-1} z^n \tag{3.2.10}$$

where t_{tree} is the enumerating function of plane rooted trees by size (see Liu YP [46], Tutte WT [85], 1964).

Example 2. Classification of lei petal bundles. *Lei petal bundles* are plane petal bundles with root face of valency 1. The *enufunction* (*i. e.*, enumerating function!) of lei petal bundles by size is denoted by

$$t_{\text{lei}} = zt_{\text{pet}}$$

where t_{pet} is the enufunction of plane rooted petal bundles by size.

From Example 2 in 3.1 and Example 1 in this section,

$$t_{\text{lei}} = t_{\text{plant}} = \sum_{n \geq 1} C_{n-1} z^n.$$

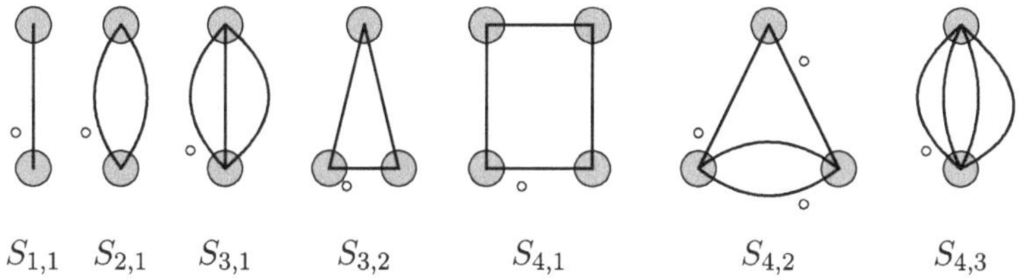

Figure 3.2.1: Non-separable outer-planar maps of size 1–4.

Example 3. Non-separable outer-planar maps. In Liu YP [33], it was shown that the enufunction of non-separable outer-planar rooted maps by size $f_{\text{nsp}} = f_{\text{nsp}}(z)$ is a solution of equation (3.2.7).

Figures 3.2.1 and 3.2.2 show all such maps of size from 1 through 5 where $S_{n,i}$ is the ith equivalent class of size n.

Hollows near the edges distinguish the non-equivalent classes, or rooted manners. Thus,

$$\begin{aligned} 1(S_{1,1}) &\quad \text{for size 1,} \\ 1(S_{2,1}) &\quad \text{for size 2,} \\ 1(S_{3,1}) + 1(S_{3,2}) = 2 &\quad \text{for size 3,} \\ 1(S_{4,1}) + 3(S_{3,2}) + 1(S_{4,3}) = 5 &\quad \text{for size 4,} \end{aligned}$$

and

$$1(S_{5,1}) + 3(S_{5,2}) + 3(S_{5,3}) + 4(S_{5,4}) + 2(S_{5,5}) + 1(S_{5,6}) = 14 \quad \text{for size 5.}$$

To be compared with (3.2.9) we have

$$\begin{aligned} &\partial_z^1 f_{\text{nsop}} = 1 = \partial_z^1 f; \quad \partial_z^2 f_{\text{nsop}} = 1 = \partial_z^2 f; \\ &\partial_z^3 f_{\text{nsop}} = 2 = \partial_z^3 f; \quad \partial_z^4 f_{\text{nsop}} = 5 = \partial_z^4 f; \\ &\partial_z^5 f_{\text{nsop}} = 14 = \partial_z^5 f. \end{aligned}$$

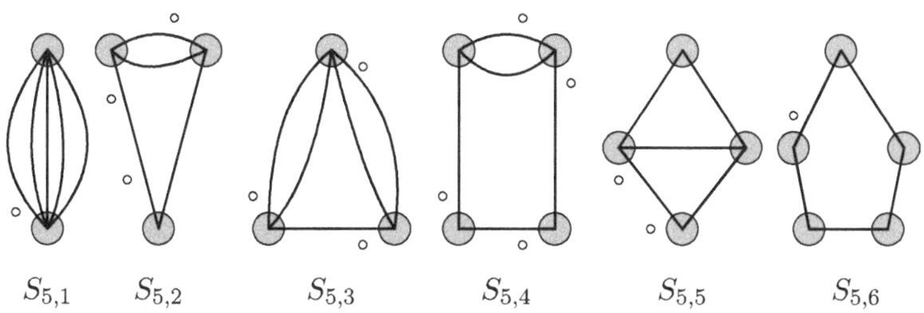

Figure 3.2.2: Non-separable outer-planar maps of size 5.

3.3 At least two coefficients variable

If an equation of function has at least two coefficients variable, then it is called a *levity variform*.

In this section, only quadratic equations with all three coefficients variable are considered and are called *total variform*.

This equation is of the form

$$\begin{cases} a(1+z)f^2 - b(1+z)f + cz = 0; \\ f|_{z=0} = d, \end{cases} \tag{3.3.1}$$

where $a, b, c, d \in \mathcal{R}_+$, $abc > 0$. This is equation (3) in Introduction.

Let $F_i = \partial_z^i f$ and $F_i^{[2]} = \partial_z^i f^2$, $i \geq 0$, then

$$F_i^{[2]} = \sum_{j=0}^{i} F_j F_{i-j}. \tag{3.3.2}$$

On the basis of equation (3.3.1), because of

$$bf = a(1+z)f^2 - bzf + cz, \tag{3.3.3}$$

the equation system

$$\begin{cases} bF_0 = aF_0^{[2]}, & \text{when } n = 0; \\ bF_1 = c + aF_1^{[2]} + aF_0^{[2]} - bF_0, & n = 1; \\ b(F_n) = a(F_n^{[2]} + F_{n-1}^{[2]}) + bF_{n-1}, & n \geq 2, \end{cases} \tag{3.3.4}$$

is easily found for $F_0 = d$.

Let $\hat{F}_{n-1}^{[2]} = F_n^{[2]} - 2F_0F_n$, then from (3.3.2), we have

$$\hat{F}_{n-1}^{[2]} = \sum_{i=1}^{n-1} F_i F_{n-i}. \tag{3.3.5}$$

Lemma 3.3.1. *Equation* (3.3.1) *is equivalent to the following equation system:*

$$\begin{cases} bF_0 = aF_0^{[2]}, & \textit{when } n = 0; \\ (b - 2aF_0)F_1 = c, & \textit{when } n = 1; \\ (b - 2aF_0)F_n = a(\hat{F}_{n-1}^{[2]} + F_{n-1}^{[2]}) - bF_{n-1}, & \textit{when } n \geq 2, \end{cases} \tag{3.3.6}$$

for $F_0 = d$ *and* $f \in \mathcal{R}\{z\}$.

Proof. By a series of transformations which are equivalent in $\mathcal{R}\{z\}$, equation (3.3.6) is obtained from equation (3.3.1). □

On the basis of equation (3.3.6), a number of conditions on a, b, c and d for the consistence of equation (3.3.5) and hence equation (3.3.1) are found.

Observation 3.3.2. *For $abc > 0$ with all of $a, b, c \in \mathbb{Z}_+$, the solution of equation* (3.3.6) *satisfies $d(b - ad) = 0$.*

Proof. From the first equation of equation (3.3.6), the initial condition leads to $d = ad^2$. This is the conclusion. □

This observation enables us to discuss equation (3.3.5) in two cases: (i) $d = 0$ and (ii) $d = b/a$.

Let $F_n^{[2]} = 2F_0F_n + \cdots + 2F_{i-1}F_{n-i+1} + \hat{F}_{n-i}^{[2]}$ where

$$\hat{F}_{n-i}^{[2]} = \sum_{j=i}^{n-i} F_iF_{n-i} \tag{3.3.7}$$

for $1 \le i \le \lfloor n/2 \rfloor$ and $n \ge 1$; then equation (3.3.6) becomes

$$\begin{cases} aF_0^{[2]} - bF_0 = 0, & \text{when } n = 0; \\ F_1 = \dfrac{c}{b - 2aF_0}, & \text{when } n = 1; \\ F_n = \dfrac{a(\hat{F}_{n-1}^{[2]} + F_{n-1}^{[2]}) - bF_{n-1}}{b - 2aF_0}, & \text{when } n \ge 2. \end{cases} \tag{3.3.8}$$

The condition $abc > 0$ in equation (3.3.1) is for avoiding unnecessary cumbersome variants.

First, for $d = 0$, equation (3.3.8) becomes

$$\begin{cases} F_0 = 0, & \text{when } n = 0; \\ F_1 = \dfrac{c}{b}, & \text{when } n = 1; \\ F_n = \dfrac{a(\hat{F}_{n-1}^{[2]} + F_{n-1}^{[2]}) - bF_{n-1}}{b}, & \text{when } n \ge 2, \end{cases} \tag{3.3.9}$$

for $n \ge 1$.

Theorem 3.3.3. *Equation* (3.3.9) *is well-defined on $\mathcal{R}\{z\}$ for $abc > 0$, $a, b, c \in \mathbb{Z}_+$.*

Proof. It is well-known that we have F_0 and F_1 for $n = 0$ and 1. For $n \ge 2$, according to the principle of induction on n, the assumption of all F_l for $0 \le l \le m-1$ known is made. From the assumption, all of $\hat{F}_{n-1}^{[2]}$, $F_{n-1}^{[2]}$ and F_{n-1} are known. Therefore, F_4 is clear from the third equation of equation (3.3.9). The uniqueness of the solution is determined only by the initial condition. □

This theorem enables us to induce a number of corollaries.

Corollary 3.3.4. *The solution of equation* (3.3.9) *holds for all coefficients non-negative integers if, and only if, b is a common factor of a and c.*

Proof. The reason is the fact that, for positive integers a, b and c, both c/b and a/b are integers if, and only if, b is a common factor of a and c on the basis of (3.3.8) and (3.3.9). □

When $a = b = c = 1$, equation (3.3.9) becomes

$$\begin{cases} F_0 = 0, & \text{when } n = 0; \\ F_1 = 1, & \text{when } n = 1; \\ F_n = (\hat{F}^{[2]}_{n-2} + F^{[2]}_{n-1}) + F_{n-1}, & \text{when } n \geq 2, \end{cases} \tag{3.3.10}$$

for $n \geq 0$.

Corollary 3.3.5. *The solution of equation* (3.3.10) *holds for all coefficients non-negative integers.*

Proof. This is a direct result of Theorem 3.3.3 and Corollary 3.3.4 in the case of $a = b = c = 1$. □

Then, for $d = b/a$, equation (3.3.8) becomes

$$\begin{cases} F_0 = \frac{b}{a}, & \text{when } n = 0; \\ F_1 = -\frac{c}{b}, & \text{when } n = 1; \\ F_n = F_{n-1} - \frac{a}{b}(\hat{F}^{[2]}_{n-1} + F^{[2]}_{n-1}), & \text{when } n \geq 2, \end{cases} \tag{3.3.11}$$

for $n \geq 0$.

Theorem 3.3.6. *Equation* (3.3.11) *is well-defined on* $\mathcal{R}\{z\}$ *for* $abc > 0$, $a, b, c \in \mathbb{Z}_+$.

Proof. Similar to the proof of Theorem 3.3.3. □

Let f and f' be the functions in $\mathcal{R}\{z\}$ determined by, respectively, F_n in (3.3.9) and F'_n in (3.3.11) for $n \geq 0$.

Theorem 3.3.7. *For integers* $n \geq 0$, $F_0 = F'_0 - b/a$ *and* $F_n = -F'_n$ *for* $n \geq 1$.

Proof. We consider equations (3.3.9) and (3.3.11), which are known to be true for $n = 0$ and 1. We proceed by induction on $n \geq 1$. Assume that, for any positive integer $l \leq l \leq n-1$, $F_l + F'_l = 0$, to see what happens for $l = n$. Because of

$$\begin{aligned} F_n + F'_n &= \frac{a}{b}(\hat{F}^{[2]}_{n-1} + F^{[2]}_{n-1}) - F_{n-1} + F'_{n-1} \\ &\quad - \frac{a}{b}(\hat{F'}^{[2]}_{n-1} + F'^{[2]}_{n-1}) \\ &= \frac{a}{b}(\hat{F}^{[2]}_{n-1} - \hat{F'}^{[2]}_{n-1}) + 2F'_{n-1} \\ &\quad + \frac{a}{b}(F^{[2]}_{n-1} - F'^{[2]}_{n-1}) \end{aligned}$$

$$
\begin{aligned}
&= 2F'_{n-1} + \frac{a}{b}(-2F'_0 F'_{n-1}) \\
&= 2F'_{n-1} - 2F'_{n-1} \\
&= 0,
\end{aligned}
$$

the conclusion is obtained. □

This theorem tells us that $f' = 1 - f$.

When $a = b = c = 1$, equation (3.3.11) becomes

$$
\begin{cases}
F_0 = 1, & \text{when } n = 0; \\
F_1 = -1, & \text{when } n = 1; \\
F_n = F_{n-1} - (\hat{F}^{[2]}_{n-1} + F^{[2]}_{n-1}), & \text{when } n \geq 2,
\end{cases}
\tag{3.3.12}
$$

for $n \geq 0$.

Theorem 3.3.8. *Equation* (3.3.12) *is well-defined on* $\mathcal{R}\{z\}$ *for* $abc > 0$, $a, b, c \in \mathbb{Z}_+$.

Proof. This is a result of Theorem 3.3.6 in the case of $a = b = c = 1$. □

Corollary 3.3.9. *The solution of equation* (3.3.12) *holds for all coefficients as non-positive integers except only for* $F_0 = 1$.

Proof. This is a result of Theorem 3.3.7 and Corollary 3.3.5. □

Let h and h' be the functions determined, respectively, by $H_n = F_n$ in (3.3.10) and $H'_n = F_n$ in (3.3.12).

Corollary 3.3.10. $h' = 1 - h$.

Proof. This is the special case of $a = b = c = 1$ in Theorem 3.3.7. □

On the basis of Corollary 3.3.5, some examples are shown to have applications in combinatorics.

Example 1. Classification of non-separable outer-planar simple rooted maps by size. Let $f_{\text{nsops}} \in \mathcal{R}\{z\}$ be the enufunction for enumerating non-isomorphic classes of non-separable outer-planar simple rooted maps with size n as parameter. That implies $F_n = \partial_z^n f_{\text{nsopt}}$, $n \geq 0$, in (3.3.10).

By employing (3.3.10), one can see that

$$F_1 = 1, \quad F_2 = 0, \quad F_3 = F_4 = 1, \quad F_5 = 3, \quad F_6 = 6 \quad \text{and} \quad F_7 = 15.$$

See Figure 3.3.2 and Figure 3.3.3, where we refer to

$$
\begin{aligned}
F_1 \text{ by } 1P_{1,1}, &\quad F_3 \text{ by } 1P_{3,1}, \\
F_4 \text{ by } 1P_{4,1}, &\quad F_5 \text{ by } 1P_{5,1} + 2P_{5,2}, \\
F_6 \text{ by } &5P_{6,1} + 1P_{6,2} \quad \text{and} \\
F_7 \text{ by } 1P_{7,1} &+ 6P_{7,2} + 3P_{7,3} + 4P_{7,6}.
\end{aligned}
$$

Example 2. Classification of non-isomorphic non-separable outer-planar rooted triangulations by the valency m of root-face. Because it is well-known that a non-separable outer-planar simple map of size n is a non-separable outer-planar triangulation of root-face valency m if, and only if, $n = 2m - 3$, we have

$$\partial_z^m f_{\text{nsotr}} = \partial_z^{2m-3} f_{\text{nsops}}|_{\text{root-face valency } m}$$

where $f_{\text{nsotr}} \in \mathcal{R}\{z\}$ is the enufunction of non-isomorphic non-separable outer-planar rooted triangulations by the valency m of the root-face.

Only $n = 3, 5$ and 7 are available for, respectively, $m = 3, 4$ and 5. Then $\partial_z^3 f_{\text{notr}} = 1$, $\partial_z^4 f_{\text{notr}} = 2$ and $\partial_z^5 f_{\text{notr}} = 4$ are represented by, respectively, $1P_{3,1}$, $2P_{5,2}$ and $4P_{7,4}$ in Figure 3.3.2 and Figure 3.3.3.

Example 3. Isomorphic classification of planted trinary trees. A *planted trinary tree* is a trinary tree of end vertex rooted type. A *trinary tree* is a tree with all vertices of valency 3 except for *ends* (of valency 1).

According to the out-duality, a map is a planted trinary tree if, and only if, its out-dual is a non-separable outer-planar triangulation. The number of end vertices of a planted trinary tree is the valency of the root-face of the outer-planar triangulation.

Because the number of end vertices in a planted trinary tree is 2 less than the number of its 3-valent vertices, a trinary tree of order $n \geq 4$ has $m = (n + 2)/2$ end vertices.

Let $f_{\text{ptri-t}} \in \mathcal{R}\{z\}$ be the enufunction of planted trinary trees by the number s of 3-valent vertices as parameter, then

$$\partial_z^s f_{\text{ptri-t}} = \partial_z^{2s-3} f_{\text{nsops}}|_{\text{root-face valency } s}$$

where $s \geq 1$.

In Figure 3.3.1, one can see the number of isomorphic classes of planted trinary trees for $s = 1, 2$ and 3 as $(1T_{1,1}) + (2T_{2,1}) + (3T_{3,1})$.

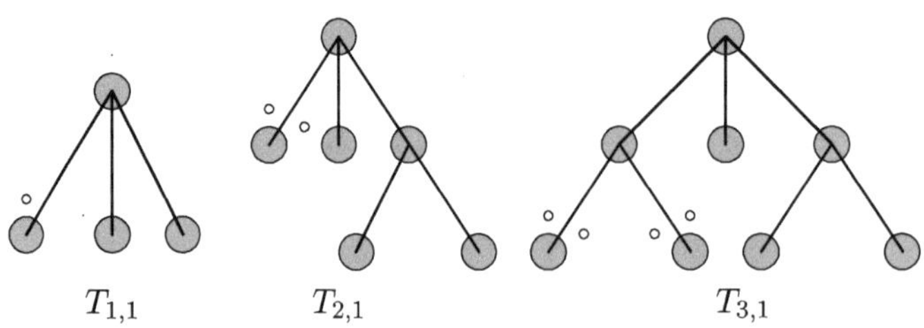

Figure 3.3.1: Planted trinary trees with 1–3 trivalent vertices.

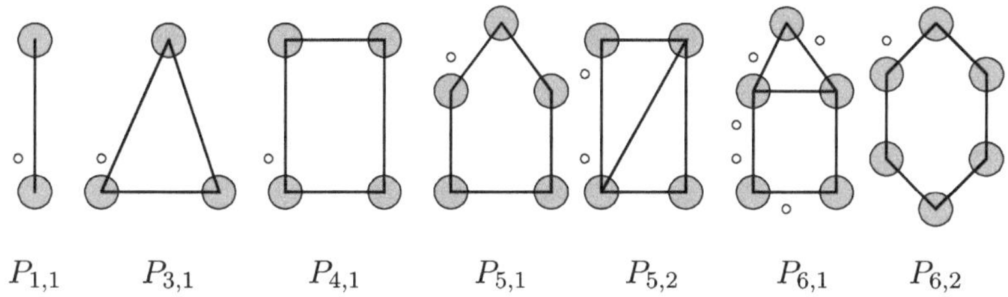

Figure 3.3.2: Non-separable outer-planar simple rooted maps of size 0–6.

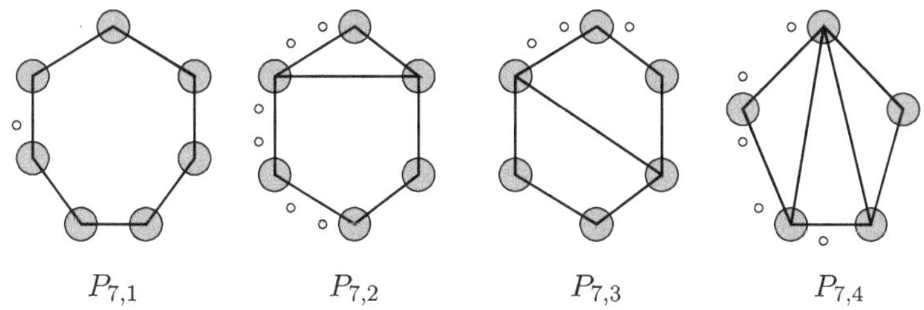

Figure 3.3.3: Non-separable outer-planar simple rooted maps of size 7.

Example 4. Isomorphic classification of non-separable outer-planar simple rooted maps including the vertex map. By the substitution $f = g - 1$ in equation (3.3.1) with $a = b = c = 1$ and $d = 0$, one can see the equation about g is

$$\begin{cases}(1+z)g^2 - 3(1+z)g + 3z + 2 = 0;\\ g|_{z=0} = 1.\end{cases} \tag{3.3.13}$$

The solution of equation (3.3.13) is $g = f + 1$ where f is determined by $F_m = \partial_z^m f$, $m \geq 0$, which are shown in (3.3.10).

3.4 Model of triangulations

Consider the equation (Liu YP [51], p. 70)

$$\begin{cases}f^3 + \dfrac{1-z}{z^2}f^2 + \dfrac{z-2}{z}f + 1 = 0;\\ f|_{z=0} = 0,\end{cases} \tag{3.4.1}$$

for $f \in \mathcal{R}\{z\}$ with all coefficients in $\mathbb{Z}_+$. This is equation (4) in Introduction I when $a = b = c = 1$ and $d = 0$ this being meaningful in the classification of outer-planar triangulations.

For convenience, by the cancelation law, equation (3.4.1) is transformed into

$$\begin{cases} 2zf = z^2f^3 + (1-z)f^2 + z^2f + z^2; \\ f|_{z=0} = 0. \end{cases} \tag{3.4.2}$$

Because of $f \in \mathcal{R}\{z\}$, f is determined by $F_m = \partial_z^m f$ for $n \geq 0$, and hence f^2, f^3 by, respectively,

$$\begin{cases} F_m^{[2]} = \sum_{i=0}^{m} F_i F_{m-i}, \\ F_M^{[3]} = \sum_{i=0}^{m} F_i F_{m-i}^{[2]}, \end{cases} \tag{3.4.3}$$

for $n \geq 0$.

By the initial condition

$$\begin{cases} F_0^{[2]} = F_1^{[2]} = 0; \\ F_0^{[3]} = F_1^{[3]} = F_2^{[3]} = 0. \end{cases} \tag{3.4.4}$$

Then (3.4.3) becomes

$$\begin{cases} F_m^{[2]} = \sum_{i=1}^{m-1} F_i F_{m-i}, & \text{for } m \geq 2; \\ F_m^{[3]} = \sum_{i=1}^{m-2} F_i F_{m-i}^{[2]}, & \text{for } m \geq 3, \end{cases} \tag{3.4.5}$$

for $n \geq 0$.

On the basis of equation (3.4.2), by transforming in $\mathcal{R}\{z\}$, it is seen that the equation system

$$\begin{cases} z^0 : F_0^{[2]} = 0 \\ \quad \text{(for constant term);} \\ z^1 : 2F_0 = F_1^{[2]} - F_0^{[2]} \\ \quad \text{(for first term);} \\ z^2 : 2F_1 = F_0^{[3]} + F_2^{[2]} - F_1^{[2]} + F_0 + 1 \\ \quad \text{(for second term);} \\ z^3 : 2F_2 = F_1^{[3]} + F_3^{[2]} - F_2^{[2]} + F_1 \\ \quad \text{(for third term);} \\ z^4 : 2F_3 = F_2^{[3]} + F_4^{[2]} - F_3^{[2]} + F_2 \\ \quad \text{(for third term);} \\ z^{n+1} : 2F_n = F_{n-1}^{[3]} + F_{n+1}^{[2]} - F_n^{[2]} + F_{n-1} \\ \quad \text{(for } n+1\text{-st term, } n \geq 4\text{),} \end{cases} \tag{3.4.6}$$

is equivalent to equation (3.4.2) and hence to equation (3.4.1)

For z^0, $F_0^{[2]} = 0 \Rightarrow F_0^2 = 0 \Rightarrow$

$$F_0 = 0. \tag{3.4.7}$$

This is the initial condition.

For z^1, because of $F_0 = 0$ and hence $F_1^{[2]} = 0$, $2F_0 = F_1^{[2]} - F_0^{[2]}$ is an identity.

For z^2, from $F_0 = F_1^{[2]} = F_0^{[3]} = 0$, $2F_1 = F_0^{[3]} + F_2^{[2]} - F_1^{[2]} + F_0 + 1 \Rightarrow 2F_1 = F_2^{[2]} + 1 \Rightarrow$

$$(F_1 - 1)^2 = 1 \Rightarrow F_1 = 1. \tag{3.4.8}$$

For z^3, from $F_1^{[3]} = 3F_0^2 F_1 = 0$, $F_3^{[2]} = 2F_1F_2 = 2F_2$ and $F_2^{[2]} = F_1^2 = 1$, $2F_2 = F_1^{[3]} + F_3^{[2]} - F_2^{[2]} + F_1 = 2F_2 \Rightarrow$ an identity.

For z^4, from $F_2^{[3]} = 0$, $F_4^{[2]} = 2F_3 + F_2^2$, $F_3^{[2]} = 2F_2$,

$$\begin{aligned} 2F_3 = F_2^{[3]} + F_4^{[2]} - F_3^{[2]} + F_2 &\Rightarrow 2F_3 = 2F_3 + F_2^2 - 2F_2 + F_2 \\ &\Rightarrow F_2^2 - F_2 = F_2(F_2 - 1) = 0. \end{aligned}$$

Because of $F_2 = 1$ not being available, for $F_n \in \mathbb{Z}_+$ for all $n \geq 2$, this enables us to have

$$F_2 = 0. \tag{3.4.9}$$

Observation 3.4.1. *If $F_2 = 1$, then $F_3 \notin \mathbb{Z}_+$.*

Proof. From $F_0 = 0$, $F_1 = 1$ and $F_2 = 1$, we have the equation

$$\begin{aligned} 2F_4 &= F_3^{[3]} + F_5^{[2]} - F_4^{[2]} + F_3 \Rightarrow \\ 2F_4 &= 1 + (2F_4 + 2F_3) - (2F_3 + 1) + F_3 \Rightarrow \\ 0 &= 1 + 2F_3 - 2F_3 - 1 \Rightarrow F_3 = 0. \end{aligned}$$

We proceed addressing F_4. By $F_1 = F_2 = 1$ and $F_3 = 0$,

$$\begin{aligned} 2F_5 &= F_4^{[3]} + F_6^{[2]} - F_5^{[2]} + F_4 \Rightarrow \\ 2F_5 &= 3F_1^2F_2 + (2F_1F_5 + 2F_2F_4 + F_3^2) - (2F_1F_4 + 2F_2F_3) + F_4 \Rightarrow \\ 2F_5 &= 3 + (2F_5 + 2F_4) - (2F_4) + F_4 \\ &\Rightarrow F_4 + 3 = 0. \end{aligned}$$

However, $F_4 = -3 \notin \mathbb{Z}_+$. □

This enables us to choose $F_2 = 0$ instead of $F_2 = 1$.

For z^5, from $F_3^{[3]} = F_1^3 = 1$,

$$\begin{aligned} F_5^{[2]} &= 2F_1F_4 + 2F_2F_3 = 2F_4, \\ F_4^{[2]} &= 2F_1F_3 + F_2^2 = 2F_3, \end{aligned}$$

$$\begin{aligned}2F_4 &= F_3^{[3]} + F_5^{[2]} - F_4^{[2]} + F_3\\ &\Rightarrow 2F_4 = 1 + 2F_4 - 2F_3 + F_3\\ &\Rightarrow 0 = 1 - F_3 \Rightarrow\\ F_3 &= 1.\end{aligned} \tag{3.4.10}$$

Now we consider $z^{n+2}(n \geq 4)$. Because of $F_0 = 0$, $F_1 = 1$ and $F_2 = 0$, we have

$$\begin{aligned}F_n^{[3]} &= F_0 F_n^{[2]} + \sum_{i=1}^{n-1} F_i F_{n-i}^{[2]}\\ &= \sum_{1\leq i,j,k\leq n-1} F_i F_j F_k,\\ F_{n+1}^{[2]} &= 2F_0F_{n+1} + 2F_1F_n + 2F_2F_{n-1} \sum_{i=3}^{n-2} F_iF_{n+1-i}\\ &= 2F_n + \sum_{i=2}^{n-1} F_iF_{n+1-i},\end{aligned}$$

and, by $F_2 = 0$,

$$\begin{aligned}F_{n+2}^{[2]} &= 2F_1F_{n+1} + 2F_2F_n + \sum_{i=3}^{n-1} F_iF_{n+2-i}\\ &= 2F_{n+1} + \sum_{i=3}^{n-1} F_iF_{n+2-i}.\end{aligned}$$

Thus, $2F_{n+1} = F_n^{[3]} + F_{n+2}^{[2]} - F_{n+1}^{[2]} + F_n \Rightarrow$

$$\begin{aligned}0 = \sum_{\substack{1\leq i,j,k\leq n-1\\ i+j+k=n}} F_iF_jF_k + \sum_{i=3}^{n-1} F_iF_{n+2-i} - F_n - \sum_{i=2}^{n-1} F_iF_{n+1-i} \Rightarrow\\ F_n = \Sigma_{n-1}^{(1)} + \Sigma_{n-1}^{(2)}\end{aligned} \tag{3.4.11}$$

where, by $F_2 = 0$,

$$\begin{cases}\Sigma_{n-1}^{(1)} = \sum\limits_{\substack{1\leq i,j,k\leq n-1\\ i+j+k=n}} F_iF_jF_k;\\ \Sigma_{n-1}^{(2)} = \sum\limits_{i=3}^{n-1} F_iF_{n+2-i} - \sum\limits_{i=3}^{n-2} F_iF_{n+1-i}.\end{cases} \tag{3.4.12}$$

Observation 3.4.2. *For any integer $n \geq 3$, $\Sigma_{n-1}^{(2)} \geq 0$.*

Proof. Because of $F_{n+2-i} - F_{n+1-i} \geq 0$,

$$\begin{aligned}\sum_{i=3}^{n-1} F_i F_{n+2-i} - \sum_{i=3}^{n-2} F_i F_{n+1-i} &= F_{n-1}F_3 + \sum_{i=3}^{n-2} F_i F_{n+2-i} - \sum_{i=3}^{n-2} F_i F_{n+1-i} \\ &= F_{n-1} + \sum_{i=3}^{n-2} F_i (F_{n+2-i} - F_{n+1-i}) \\ &\geq 0.\end{aligned}$$

From (3.4.12), this is the conclusion. □

In the proof of Observation 3.4.2, all $F_n \in \mathbb{Z}_+$ and $F_n \geq F_{n-1}$ for $n \geq 3$ are known to be true by induction on n from $F_0, F_1, F_2 \in \mathbb{Z}_+$.

Observation 3.4.3. *For any integer $n \geq 3$, $F_n \in \mathbb{Z}_+$ is determined only by F_l, $0 \leq l \leq n-1$, on the basis of $F_0 = 0$ and $F_2 = 0$.*

Proof. On the basis of Observation 3.4.2, having checked $\Sigma_{n-1}^{(1)}$ and $\Sigma_{n-1}^{(2)}$, the conclusion is drawn. □

Theorem 3.4.4. *Equation* (3.4.6), *and hence equation* (3.4.1), *is well-defined for all $F_n (n \geq 0) \in \mathbb{Z}_+$ if, and only if, $F_2 = 0$.*

Proof. This results from the procedure for evaluating a solution of equation (3.4.6) mentioned above. □

This theorem enables us to complete the procedure for getting the solution only from the initial conditions $F_0 = 0$ and $F_2 = 0$.

Theorem 3.4.5. *The solution of equation* (3.4.6), *and hence equation* (3.4.1), *is in the form of a sum with all terms positive,*

$$F_n = \begin{cases} 0, & \text{when } n = 0; \\ 1, & \text{when } n = 1; \\ 0, & \text{when } n = 2; \\ \Sigma_{n-1}^{(1)} + \Sigma_{n-1}^{(2)}, & \text{when } n \geq 3, \end{cases} \tag{3.4.13}$$

where $\Sigma_{n-1}^{(1)}$ and $\Sigma_{n-1}^{(2)}$ are seen in (3.4.12).

Proof. On the basis of Observation 3.4.2 and Theorem 3.4.4, the conclusion is drawn. □

Although (3.4.13) is particularly suitable for us to go further on the other two stages: efficientization and intelligentization, an explicision in the form of a sum with all terms positive is also found.

Theorem 3.4.6. *In* (3.4.13), *for integer* $n \geq 0$,

$$F_n = \begin{cases} 0, & \textit{when } n = 0 \textit{ and } n = 2; \\ 1, & \textit{when } n = 1; \\ \sum\limits_{\lceil \frac{n+1}{2} \rceil \leq l \leq n-1} \binom{n-1}{l}\binom{3l-n-3}{l-1}, & \textit{when } n \geq 3. \end{cases} \tag{3.4.14}$$

Proof. Because

$$\Sigma_{n-1}^{(1)} + \Sigma_{n-1}^{(2)} = \sum_{\lceil \frac{n+1}{2} \rceil \leq l \leq n-1} \binom{n-1}{l}\binom{3l-n-3}{l-1}$$

for $n \geq 3$ in (3.4.13), from Theorem 3.4.5, the conclusion is drawn. □

Some applications are shown of usage of the equation discussed in this section.

Example 1. Classification of outer-planar rooted triangulations by the valency of rooted face. In Liu YP [51] (Section 3.1), one can see that the solution of equation (3.4.1) is $f = zg_{\text{ot}}$ where g_{ot} is the enufunction of outer-planar rooted triangulations with the valency of the rooted face as parameter, *i. e.*, $\partial_z^n g_{\text{ot}} = \partial_z^{n+1} f$ for $n \geq 0$.

In Figures 3.4.1–3.4.3, one can see the rooted non-isomorphic classes of outer-planar triangulations with the root-face valency at most 7. As a result of (3.4.14) in

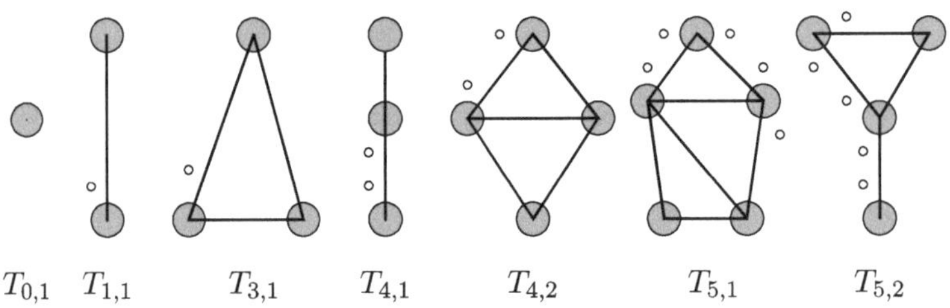

Figure 3.4.1: Classification of outer-planar triangulations of root-face valency 0–5.

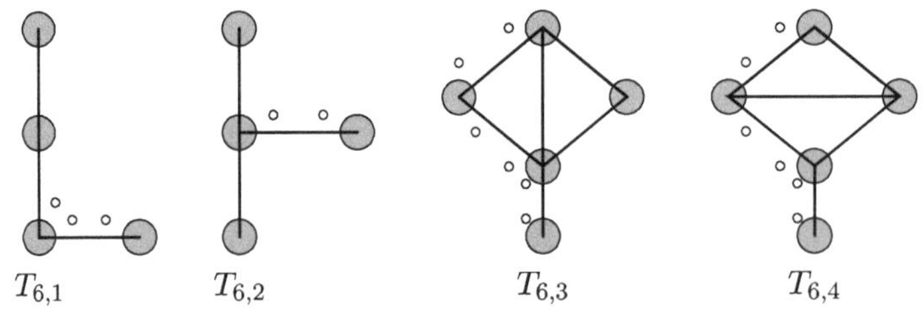

Figure 3.4.2: Classification of outer-planar triangulations of root-face valency 6 (I).

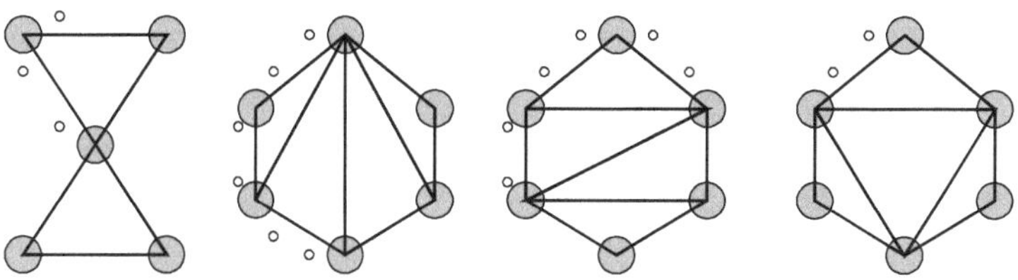

Figure 3.4.3: Classification of outer-planar triangulations of root-face valency 6 (II).

Theorem 3.4.6

$$\partial_z^0 g_{\rm ot} = \partial_z^1 f = 1, \quad \partial_z^1 g_{\rm ot} = \partial_z^2 f = 0, \quad \partial_z^2 g_{\rm ot} = \partial_z^3 f = 1,$$
$$\partial_z^3 g_{\rm ot} = \partial_z^4 f = 1, \quad \partial_z^4 g_{\rm ot} = \partial_z^5 f = 4, \quad \partial_z^5 g_{\rm ot} = \partial_z^6 f = 10,$$
$$\partial_z^6 g_{\rm ot} = \partial_z^7 f = 34.$$

Example 2. From Figure 3.4.1 through Figure 3.4.3, one might also see the rooted non-isomorphic non-separable outer-planar triangulations contained in outer-planar triangulations, and hence those with given value of the root-face valency. This task will be completed in the next chapter as corollaries.

Here, the rooted non-isomorphic classes of non-separable outer-planar triangulations with a number of inner faces of at most 4 are provided as

$$(1T_{3,1}) + (2T_{4,2}) + (5T_{5,1}) + (6T_{6,6} + 6T_{6,7} + 2T_{6,8}).$$

On account of the relation between the root-face valency and the number of inner faces in a non-separable outer-planar triangulation, where $T_{3,1}$; $T_{4,2}$; $T_{5,1}$ and $T_{6,i}$, $6 \le i \le 8$, we have those with the number of inner faces, respectively, 1; 2; 3 and 4.

As a matter of fact, if the vertex map is arranged as a non-separable outer-planar triangulation without inner face in degeneracy, then the enufunction $f_{\rm not} = f_{\rm not}(z)$ of non-separable outer-planar rooted triangulations with the number of inner faces as parameter is a solution of the equation

$$\begin{cases} zf^2 - f + 1 = 0; \\ f_{z=0} = 1. \end{cases} \tag{3.4.15}$$

This again goes back to $t_{\rm bint}$ in Section 3.1.

3.5 Model of quadrangulations

Consider the equation

$$\begin{cases} zf^3 - 3zf^2 + (3z-1)f + 1 = 0; \\ f|_{z=0} = 1, \end{cases} \tag{3.5.1}$$

where $f \in \mathcal{R}\{z\}$ with all coefficients in $\mathbb{Z}_+$. This is equation (5) in Introduction when $a = b = c = d = 1$ because of it is meaningful in the classification of non-separable outer-planar quadrangulations.

For any integer $n \geq 0$, let $F_n = \partial_z^n f$ be the coefficient of the term with z^n in $f \in \mathcal{R}\{z\}$, then the coefficients $F_n^{[2]}$ and $F_n^{[3]}$ of z^n in, respectively, f^2 and f^3 have the form

$$\begin{cases} F_n^{[2]} = \sum_{i=0}^{n} F_i F_{n-i}; \\ F_n^{[3]} = \sum_{\substack{i+j+k=n \\ 0 \leq i,j,k \leq n}} F_i F_j F_k = \sum_{i=0}^{n} F_i F_{n-i}^{[2]}, \end{cases} \tag{3.5.2}$$

for $n \geq 0$.

Theorem 3.5.1. *Equation* (3.5.1) *is equivalent to the equation system as*

$$\begin{cases} -F_0 + 1 = 0; & \textit{when } n = 0; \\ F_{n-1}^{[3]} - 3F_{n-1}^{[2]} + 3F_{n-1} - F_n = 0, & \textit{when } n \geq 1, \end{cases} \tag{3.5.3}$$

for $F_n \in \mathcal{R}\{z\}$, $n \geq 0$.

Proof. By employing the cancelation law on $\mathcal{R}\{z\}$, for z^0, equation (3.5.1) leads to the equation

$$-F_0 + 1 = 0.$$

This is the first equation of equation (3.5.3) for the initial condition.

Then, for any integer $n \geq 2$, the coefficient of the term involving z_n, equation (3.5.1), leads to the equation

$$F_{n-1}^{[3]} - 3F_{n-1}^{[2]} + 3F_{n-1} - F_n = 0.$$

This is the other equation in the system equation (3.5.3) for all $n \geq 1$. Since all transformations used above are equivalences on $\mathcal{R}\{z\}$, the conclusion is drawn. □

On the basis of this theorem, we are allowed to concentrate on the equation system (3.5.3) instead of equation (3.5.1).

Let $(F_0, F_1, F_2, \ldots, F_n, \ldots)$ be a solution of equation (3.5.3). Because $F_0 = 1$, which is the initial condition determined by the first equation of equation (3.5.3), F_n for $n \geq 1$ can be extracted by the following procedure.

From $F_0 = 1$, it is seen that

$$F_0^{[2]} = 1 \quad \text{and} \quad F_0^{[3]} = 1. \tag{3.5.4}$$

When $n = 1$, $F_0^{[3]} - 3F_0^{[2]} + 3F_0 - F_1 = 0 \Rightarrow 1 - F_1 = 0 \Rightarrow$

$$F_1 = 1. \tag{3.5.5}$$

Then $F_1^{[2]} = 2$ and $F_1^{[3]} = 3$.

When $n = 2$, $F_1^{[3]} - 3F_1^{[2]} + 3F_1 - F_2 = 0 \Rightarrow 0 - F_2 = 0 \Rightarrow$

$$F_2 = 0. \tag{3.5.6}$$

Then $F_2^{[2]} = 1$ and $F_2^{[3]} = 3$.

When $n = 3$, $F_2^{[3]} - 3F_2^{[2]} + 3F_2 - F_3 = 0 \Rightarrow 0 - F_3 = 0 \Rightarrow$

$$F_3 = 0. \tag{3.5.7}$$

Then $F_3^{[2]} = 0$ and $F_3^{[3]} = 3$.

When $n = 4$, $F_3^{[3]} - 3F_3^{[2]} + 3F_3 - F_4 = 0 \Rightarrow 3 - F_4 = 0 \Rightarrow$

$$F_4 = 3. \tag{3.5.8}$$

Then $F_3^{[2]} = 0$ and $F_3^{[3]} = 3$.

When $n \geq 4$, $F_{n-1}^{[3]} - 3F_{n-1}^{[2]} + 3F_{n-1} - F_n = 0 \Rightarrow$

$$F_n = F_{n-1}^{[3]} + 3F_{n-1} - 3F_{n-1}^{[2]}. \tag{3.5.9}$$

Observation 3.5.2. *If $n = 2(\mathrm{mod}\ 3)$ or $0(\mathrm{mod}\ 3)$, then $F_n = 0$ for $n \geq 1$.*

Proof. In a way to sketch the proof by the meaning of F_n in non-separable outer-planar quadrangulations without inner-face number $n = 0(\mathrm{mod}\ 3)$ or $2(\mathrm{mod}\ 3)$ shown in Example 1 afterward in this section. □

Observation 3.5.3. *For any integer $n \geq 0$, $F_n^{[3]} + 3F_n - 3F_n^{[2]} \geq 0$.*

Proof. We proceed by induction on n. The cases for $n \leq 3$ are known. For $n \geq 4$, suppose $F_{n-1}^{[3]} + 3F_{n-1} - 3F_{n-1}^{[2]} \geq 0$ is known, we prove $F_n^{[3]} + 3F_n - 3F_n^{[2]} \geq 0$. According to Observation 3.5.2, we are allowed to consider $n = 1(\mathrm{mod}\ 3)$. From $F_0 = F_0^{[2]} = F_0^{[3]} = 1$,

$$F_n^{[3]} = F_{n-1}^{[3]} + 3F_n \quad \text{and} \quad F_n^{[2]} = F_{n-1}^{[2]} + 2F_n.$$

Because of

$$\begin{aligned} F_n^{[3]} + 3F_n - 3F_n^{[2]} &= F_{n-1}^{[3]} - 3F_{n-1}^{[2]}, \quad \text{by Observation, 3.5.2,} \\ &= F_{n-1}^{[3]} + 3F_{n-1} - 3F_{n-1}^{[2]}, \quad \text{by induction hypothesis,} \\ &\geq 0. \end{aligned}$$

This is the conclusion. □

The two observations above enable us to establish our theorems.

Theorem 3.5.4. *Equation (3.5.1) is well-defined in $\mathcal{R}\{z\}$.*

Proof. Our proof is on the basis of Theorem 3.5.1, Observation 3.5.2 and Observation 3.5.3. On account of the procedure shown in (3.5.4)–(3.5.8), F_n is only determined by $F_l, 0 \leq l \leq n-1$, for any integer $n \geq 1$. The conclusion is drawn. □

This theorem enables us only to discuss equation (3.5.3) for equation (3.5.1) in what follows.

Theorem 3.5.5. *The solution of equation* (3.5.1) *is determined by* $F_n = \partial_z^n f$, $n \geq 0$, *in the form*

$$F_n = \begin{cases} 1, & \text{when } n = 0; \\ 0, & \text{when } n = 0(\text{mod } 3) \text{ or } 1(\text{mod } 3); \\ \Sigma_{n-1} & \text{otherwise,} \end{cases} \tag{3.5.10}$$

where

$$\Sigma_{n-1} = F_n^{[3]} + 3F_n - 3F_n^{[2]}$$

for $n = 1(\text{mod } 3)$, $n \geq 1$.

Proof. For $n = 0$, the case follows from the initial condition. For $n = 0(\text{mod } 3)$ or $1(\text{mod } 3)$, the case follows from Observation 3.5.2. The last case is done from (3.5.9). □

This theorem suggests us to introduce a substitute as

$$f = 1 + zg \tag{3.5.11}$$

in $\mathcal{R}\{z\}$ to transform equation (3.5.1) into

$$\begin{cases} z^3 g^3 - 1 = g; \\ g|_{z=0} = 1. \end{cases} \tag{3.5.12}$$

Theorem 3.5.6. *Equation* (3.5.12) *is well-defied in* $\mathcal{R}\{z\}$.

Proof. Let $x = z^3$ where $h = g|_{z^3=x}$, then equation (3.5.12) becomes

$$\begin{cases} xh^3 - 1 = h; \\ h|_{x=0} = 1. \end{cases} \tag{3.5.13}$$

We work on the basis of $H_n = \partial_x^n h$ for $n \geq 0$. When $n = 0$, we have the equation $H_0 = 1$. This is the initial condition. Then, for $n \geq 1$, we proceed by induction on n. On account of the equation, $H_n = H_{n-1}^{[3]}$. Assume, for $0 \leq s \leq n-1$, that $H_s = H_{s-1}^{[3]}$. We prove $H_n = H_{n-1}^{[3]}$. Because $H_{n-1}^{[3]}$ is only dependent on H_s for $n-1 \geq s \geq 0$, H_n is determined. Because h is well-defined for the equation, g is well-defined for equation (3.5.12). □

Corollary 3.5.7. *Equation* (3.5.13) *is well-defined. Its solution is in the form of a summation-free explicision, for* $n \geq 0$,

$$H_n = \begin{cases} 1, & \text{when } n = 0; \\ \dfrac{1}{2n+1}\dbinom{3n}{n}, & \text{when } n \geq 1. \end{cases} \tag{3.5.14}$$

Proof. We work on the basis of $H_n = H_{n-1}^{[3]}$ from $H_0 = 1$ for $n \geq 0$. We proceed by induction on n, and the conclusion can be drawn. □

This corollary enables us to deduce an explicision of the solution of equation (3.5.12) and hence equation (3.5.1).

Theorem 3.5.8. *The solution of equation* (3.5.12) *is determined by* $G_n = \partial_z^n g$ *for* $n \geq 0$ *in the form*

$$G_n = \begin{cases} 1, & \text{when } n = 0; \\ \left.\begin{array}{ll} 0, & n = 1(\text{mod } 3) \text{ and } 2(\text{mod } 3); \\ G_{n-3}^{[3]}, & n = 0(\text{mod } 3) \end{array}\right\}, \\ \quad \text{when } n \geq 1. \end{cases} \tag{3.5.15}$$

Proof. This is a result of Theorem 3.5.6. □

Furthermore, a summation-free explicision of G_n can also be found.

Theorem 3.5.9. *The solution of equation* (3.5.12) *has a summation-free explicision in the form*

$$G_n = \begin{cases} 1, & \text{when } n = 0 \text{ and } 1; \\ \dfrac{1}{2s+1}\dbinom{3s}{s}, & \text{when } n = 2s+1,\ s \geq 1; \\ 0, & \text{otherwise}. \end{cases} \tag{3.5.16}$$

Proof. This is a result of Theorem 3.5.6 and its Corollary 3.5.7. □

Example 1. Classification of non-separable outer-planar quadrangulations by inner size. By *inner size* is meant the number of *inner edges*, or say, those that are not on the root-face boundary. Let q be the enufunction for counting non-isomorphic classes of non-separable outer-planar rooted quadrangulations by inner size. From non-separability and outer planarity, the least one is a quadrangle without inner edge.

On the basis of the quadrangle, any non-separable outer-planar rooted quadrangulation can be done by adding three non-root edges, while one inner edge and one inner face (quadrangle!) are produced step by step. If s is the inner size of a quadrangulation, then it has inner size $s+1$, number of non-rooted edges $3s$ and size $3s+1$. Thus, for $s \geq 0$,

$$\partial_z^{3s+4} f = \partial_z^{3s+3} g = \partial_z^s q. \tag{3.5.17}$$

In Figure 3.5.1, one can see the rooted non-isomorphic classes $\partial_z^s q$ for $0 \leq s \leq 3$:

$$(1T_{0,1}) + (3T_{1,1}) + (8T_{2,1} + 4T_{2,2}) + (20T_{3,1} + 10T_{3,2} + 10T_{3,3} + 5T_{3,4} + 10T_{3,5}).$$

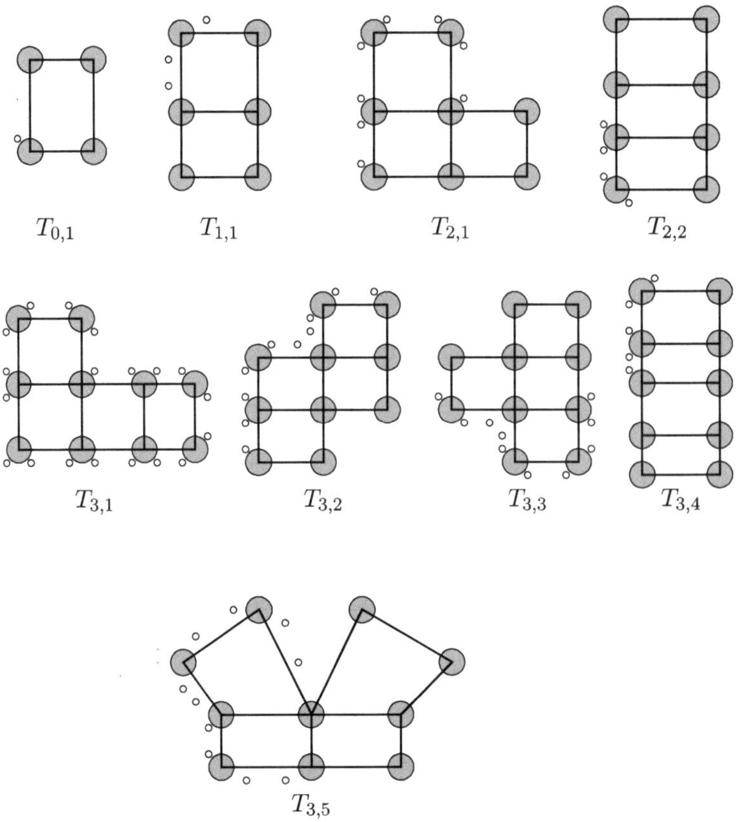

Figure 3.5.1: Classification of non-separable outer-planar quadrangulations.

Example 2. Isomorphic classes of planted quadrary tree by the number ends. A *quadrary tree* is a tree whose vertices of valencies number either 1 or 4. On the basis of Example 1, because the outer dual of a quadrary tree is a non-separable outer-planar quadrangulation, from a bijection between their isomorphic classes, the corresponding task can be done.

3.6 General model

Consider the equation

$$\begin{cases} zf^4 - (1-z)f^3 + (1-3z)f^2 + 3zf - z = 0; \\ f|_{z=0} = 1, \end{cases} \tag{3.6.1}$$

for $f \in \mathcal{R}\{z\}$. This is equation (6) in Introduction when $a = b = c = d = 1$ because it is meaningful in the classification of general outer-planar maps.

This equation is extracted from an investigation of the classification of general outer-planar maps within rooted isomorphism. One may refer to Liu YP ([51], p. 100)

Because of $f|_{z=0} = 1$, for $n \geq 0$, let

$$\begin{cases} \partial_z^n f^i (= F_n^{[i]}) = \sum_{k=0}^{i} F_k F_{i-k}^{[i-1]}, & i \geq 2; \\ \partial_z^n f = F_n (= F_n^{[1]}). \end{cases} \tag{3.6.2}$$

On the extension of integral domain $\mathcal{R}\{z\}$, equation (3.6.1) is equivalently transformed as

$$\begin{aligned} -F_0^{[3]} + F_0^{[2]} = 0 \Longrightarrow\ & F_0^{[2]}(-F_0 + 1) = 0, \\ & \text{by } f|_{z=0} = 1, \\ \Longrightarrow\ & F_0 = 1, \\ & F_0^{[i]} = 1 \quad (2 \leq i \leq 4); \end{aligned}$$

$$\begin{aligned} & F_0^{[4]} - F_1^{[3]} + F_0^{[3]} + F_1^{[2]} - 3F_0^{[2]} + 3F_0 - 1 = 0 \\ & \Longrightarrow 1 - 3F_1 + 1 + 2F_1 - 3 + 3 - 1 = 0 \\ & \Longrightarrow -F_1 + 1 = 0 \\ & \Longrightarrow F_1 = 1, \\ & \quad\ F_1^{[2]} = 2, \quad F_1^{[3]} = 3, \quad F_1^{[4]} = 4; \end{aligned}$$

$$\begin{aligned} & F_{n-1}^{[4]} - F_n^{[3]} + F_{n-1}^{[3]} + F_n^{[2]} - 3F_{n-1}^{[2]} + 3F_{n-1} = 0 \\ & \Longrightarrow F_{n-1}^{[4]} - \left(3F_n + \sum_{k=1}^{n-1} F_k F_{n-k}^{[2]} \right) + F_{n-1}^{[3]} \\ & \quad + \left(2F_n + \sum_{k=1}^{n-1} F_k F_{n-k} \right) - 3F_{n-1}^{[2]} + 3F_{n-1} = 0. \end{aligned}$$

Thus, we have

$$\begin{aligned} F_n &= F_{n-1}^{[4]} - \sum_{k=1}^{n-1} F_k (F_{n-k} + F_{n-k}^{[2]}) \\ &\quad + F_{n-1}^{[3]} + \sum_{k=1}^{n-1} F_k F_{n-k} - 3F_{n-1}^{[2]} + 3F_{n-1} \\ &= \left(F_{n-1}^{[4]} - \sum_{k=1}^{n-1} F_k F_{n-k}^{[2]} \right) \\ &\quad + (F_{n-1}^{[3]} + 3F_{n-1} - 3F_{n-1}^{[2]}) \\ &\quad (= F_{i;0 \leq i \leq n-1} = F_{\leq n-1}), \quad n \geq 2. \end{aligned} \tag{3.6.3}$$

Because of $F_0 = 1$, $2F_0F_{n-1} = 2F_{n-1}$ in $F^{[2]}_{n-1}$, equation (3.6.3) becomes

$$\begin{cases} F_n = \Sigma^{(1)}_{n-1} + \Sigma^{(2)}_{n-1}, & n \geq 2; \\ F_0 = F_1 = 1, \end{cases} \tag{3.6.4}$$

where

$$\begin{cases} \Sigma^{(1)}_{n-1} = F^{[4]}_{n-1} - \sum_{k=1}^{n-1} F_k F^{[2]}_{n-k}; \\ \Sigma^{(2)}_{n-1} = F^{[3]}_{n-1} - 3F_{n-1} - 3\sum_{i=1}^{n-2} F_i F_{n-1-i}. \end{cases} \tag{3.6.5}$$

Lemma 3.6.1. *Equation* (3.6.1) *for* $f \in \mathcal{R}\{z\}$ *is equivalent to equations* (3.6.3) *and hence to equation* (3.6.4) *for* $\mathbf{F} \in \mathcal{R}^{\infty}\{z\}$ *where* $\mathbf{F} = (F_0, F_1, F_2, \ldots)$.

Proof. Because all transformations in the process from equation (3.6.1) through equations (3.6.3) are equivalent, the conclusion is drawn. □

This lemma enables us only to consider the system of equations (3.6.3), or equivalently equation (3.6.4), instead of equation (3.6.1).

Theorem 3.6.2. *In* $\mathcal{R}^{\infty}\{z\}$, *the system of equations* (3.6.4) *and hence equation* (3.6.1) *in* $\mathcal{R}\{z\}$ *has, and is the only one to have, a solution.*

Proof. On the basis of the equivalence between (3.6.1) and (3.6.4), for any integer $n \geq 1$, $F_n \in \mathcal{R}$ is uniquely derived from $F_0 = 1$ by (3.6.4). Then the conclusion is directly drawn. □

Now, one might like to investigate some constructions of the solution for evaluating a compact expression.

Observation 3.6.3. *For any integer* $n \geq 3$,

$$F^{[4]}_{n-1} \geq \sum_{k=2}^{n-1} F_k F^{[2]}_{n-k}. \tag{3.6.6}$$

Proof. We proceed by induction on n, and the conclusion can be drawn. □

Similarly, another inequality can also be obtained.

Observation 3.6.4. *For any integer* $n \geq 3$,

$$F^{[3]}_{n-1} \geq 3F_{n-1} + 3\sum_{i=1}^{n-2} F_i F_{n-1-i}. \tag{3.6.7}$$

Proof. We proceed by induction on n, and the conclusion can be drawn. □

Theorem 3.6.5. *The solution of equations* (3.6.4) *and hence equation* (3.6.1) *in* $\mathcal{R}\{z\}$ *is of the form*

$$F_n = \begin{cases} F_0 = 1; & \textit{when } n = 0; \\ \Sigma_{n-1}^{(1)} + \Sigma_{n-1}^{(2)}, & \textit{when } n \geq 1, \end{cases} \tag{3.6.8}$$

where $\Sigma_{n-1}^{(1)}$ *and* $\Sigma_{n-1}^{(2)}$, *both non-negative, are shown in* (3.6.5).

Proof. By considering the procedure shown in the beginning of this section, it is seen that F_n for $n \geq 0$ determines a solution of equations (3.6.4) and hence equation (3.6.1) in $\mathcal{R}\{z\}$. Its integrality and non-negativity are shown in Observation 3.6.3 and Observation 3.6.4. This is the conclusion. □

Now, one might like to seek some constructions of the solution for evaluating a compact expression.

Example 1. Classification of general outer-planar quadrangulations within a rooted isomorphism by size. From the quadrangularity of the inner faces, no self-loop occurs. From the outer planarity, no multi-edge occurs. Hence, all quadrangulations considered here are simple (by no means seen without difficulty!).

In Liu YP [51] (pp. 99–100), one can see the proof that the solution f_{oq} of (3.6.1) is the enufunction of general outer-planar rooted quadrangulations with size as the parameter.

By (3.6.4), the solution of equation (3.6.1) was calculated as

$$\begin{aligned} f_{\mathrm{oq}}(z) &= 1 + z + 2z^2 + 5z^3 + 15z^4 + 48z^5 + 160z^3 \\ &\quad + 552z^7 + 1953z^8 + \cdots . \end{aligned}$$

For example, the coefficient 1953 of z^8 is meant that all general outer-planar quadrangulations of size 8 have 1953 rooted isomorphic classes. Because trees are all outer planar without inner face, from (3.1.6),

$$\partial_z^n t_{\mathrm{root}} = \frac{(2n)!}{n!(n+1)!},$$

then the general trees of size 8 have

$$\frac{(16)!}{8!9!} = 13 \times 11 \times 10 = 1430$$

rooted isomorphic classes. Hence, general outer-planar quadrangulations with at least one inner face have 1953 − 1430 = 523 rooted isomorphic classes as shown in Figure 3.6.1–Figure 3.6.6 where we have $Q_{1,i}$, $1 \leq i \leq 30$, general outer-planar rooted quadrangulations with one inner face, and $Q_{2,i}$, $1 \leq i \leq 3$, general outer-planar rooted quadrangulations with two inner faces. A hollow represents the location of a root.

There being no hollow in a figure represents that all incident pairs {end,side} (or {semi-edge,semi-side}) on the boundary of the outer face have a hollow.

In Figure 3.6.1, rooted isomorphic classes are provided as

$$12Q_{1,1} + 12Q_{1,2} + 24Q_{1,3} + 24Q_{1,4} + 12Q_{1,5} + 12Q_{1,6};$$

altogether 96 rooted isomorphic classes.

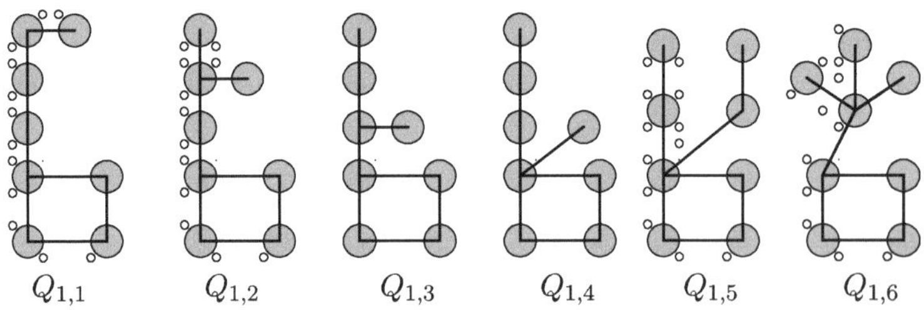

Figure 3.6.1: Classification of general outer-planar quadrangulations: $Q_{1,1}$–$Q_{1,6}$.

In Figure 3.6.2, one can see

$$24Q_{1,7} + 24Q_{1,8} + 12Q_{1,9} + 12Q_{1,10} + 24Q_{1,11} + 12Q_{1,12},$$

altogether 108 rooted isomorphic classes.

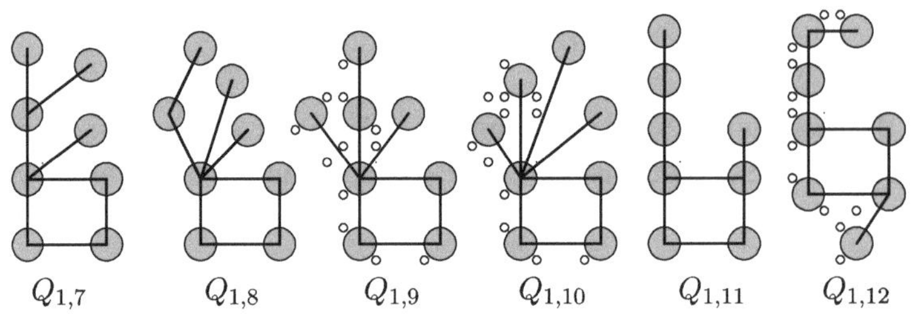

Figure 3.6.2: Classes of general outer-planar quadrangulations: $Q_{1,7}$–$Q_{1,12}$.

In Figure 3.6.3, is shown

$$24Q_{1,13} + 12Q_{1,14} + 24Q_{1,15} + 24Q_{1,16} + 12Q_{1,17} + 6Q_{1,18},$$

altogether 102 rooted isomorphic classes.

In Figure 3.6.4, it is shown that

$$24Q_{1,19} + 12Q_{1,20} + 6Q_{1,21} + 12Q_{1,22} + 24Q_{1,23} + 24Q_{1,24}$$

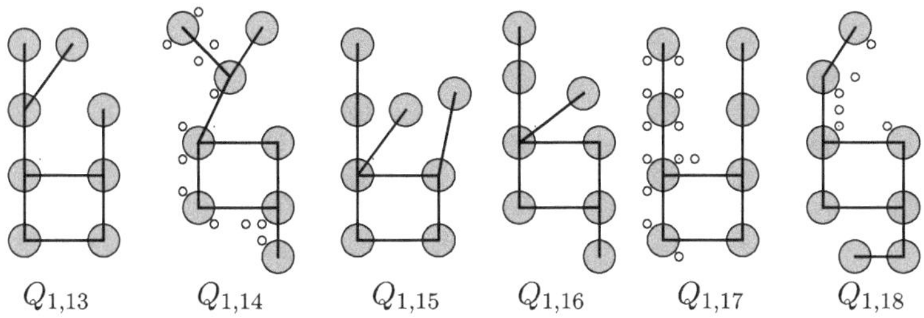

Figure 3.6.3: Classes of general outer-planar quadrangulations: $Q_{1,13}$–$Q_{1,18}$.

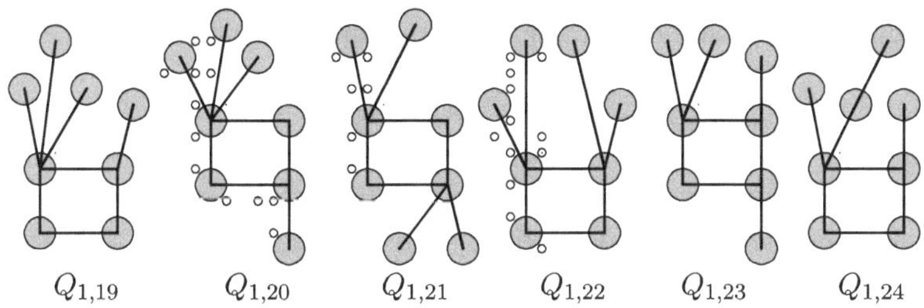

Figure 3.6.4: Classes of general outer-planar quadrangulations: $Q_{1,19}$–$Q_{1,24}$.

has altogether 102 rooted isomorphic classes.

In Figure 3.6.5, it is seen that

$$24Q_{1,25} + 12Q_{1,26} + 24Q_{1,27} + 12Q_{1,28} + 12Q_{1,29} + 3Q_{1,30}$$

has, altogether, 87 rooted isomorphic classes.

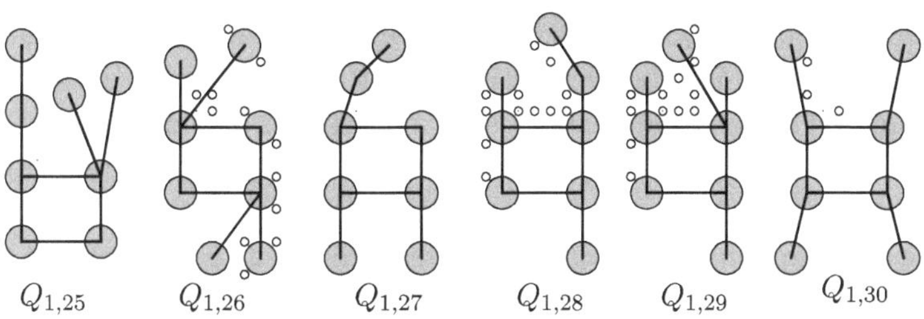

Figure 3.6.5: Classes of general outer-planar quadrangulations: $Q_{1,15}$–$Q_{1,30}$.

In Figure 3.6.6, it is seen that

$$16Q_{2,1} + 8Q_{2,2} + 4Q_{1,27}$$

has, altogether, $16 + 8 + 4 = 28$ rooted isomorphic classes.

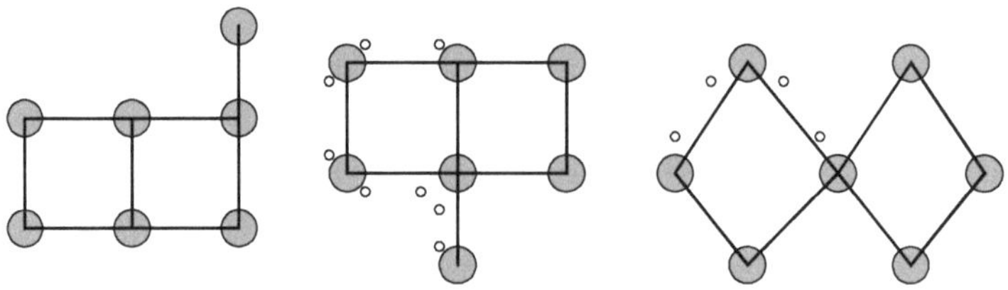

Figure 3.6.6: Classes of general outer-planar quadrangulations: $Q_{2,1}$–$Q_{2,3}$.

Since general outer-planar quadrangulations with one inner face have

$$96 + 108 + 102 + 102 + 87 = 495$$

classes and general outer-planar quadrangulations with two inner faces have 28 classes, general outer-planar quadrangulations with at least one inner face have $28 + 495 = 523$ classes. By considering 1430 classes of general outer-planar quadrangulations without inner face, $1930 + 523 = 1953$ classes are done.

3.7 Notes

3.7.1. From Example 1 and Example 2 in Section 3.2, one can see that, for arbitrarily given size, the number of isomorphic classes for planted trees is equal to lei petal bundles. Is there a bijection between them? The bijection is known to be the planar dual between them.

3.7.2. Similarly, from Example 2 and Example 3 in Section 3.2, for size given, planted trees and non-separable outer-planar rooted maps have the same number of isomorphic classes. Is there a bijection between their isomorphic classes? This question remained unanswered yet in the literature. Here, such a bijection is provided.

For a non-separable outer-planar map (including the *link map*, a map with a single edge which is a link, as a degenerate case), the map obtained by arranging each edge on the outer face boundary and each inner face as, respectively, an end vertex and a inner vertex, two vertices adjacent as their corresponding faces adjacent or one incident to the other, is called its *outer dual*. The root-edge is chosen in a corresponding way. It can be shown that the outer dual of a non-separate outer-planar rooted map is a planted plane tree. Because of the symmetry, the outer dual of a planted plane tree is a non-separable outer-planar rooted map. This is a bijection, or a 1–1 correspondence. The reader is referred to Liu YP [44], or for some details to Liu YP [56].

3.7.3. By duality and outer duality, problem 5.1 and problem 5.2 in Liu YP [33] have been solved. This shows that the equations considered being well-defined stimulates the discoveries of some laws, rules and theorems to reflect certain inner relationships among some distinguished objects characterized by those equations.

3.7.4. In Section 3.4, by substituting $f = zg$ into equation (3.4.1), we have

$$\begin{cases} z^3g^3 + (1-z)g^2 + (z-2)g + 1 = 0; \\ g|_{z=0} = 1. \end{cases} \tag{3.7.1}$$

This is equation (11.4) about Φ in Liu YP [32]. Although an explicision is found, the form is rather complicated as a multiple sum of alternative terms. In Example 1 of Section 3.4, g_{ot} is g in (3.7.1). However, the solution of equation (3.4.1) determined by (3.4.14) is a sum of positive terms evaluated only via transformations in the extension of integral domain $\mathcal{R}\{z\}$. This is an answer of problem 11.1. Although this result is equivalent to that in Dong FM-Liu YP [8], the procedure looks simpler.

3.7.5. Because of the emphasis on recursion in the form of a sum with all terms positive, the details for extracting a compact explicision is often omitted without description for the reader. However, as soon as it is clear that an equation is well-defined, how to transform the solution in the form of an explicision with all coefficients as a sum of all terms positive, and further only one term, or as we say, a summation-free form, is still an indispensable problem!

4 Equations of function with several variables

4.1 Elimination of variables

Consider the equation

$$\begin{cases} axy^2f^2 + b(x-1)f + c(x-1) = 0; \\ f|_{x=0,y=0} = d, \end{cases} \tag{4.1.1}$$

for $f \in \mathcal{R}\{x,y\}$ and $a,b,c,d \in \mathbb{Z}_+$. This is equation (7) in Introduction when $b = 1$.

When $a = b = c = d = 1$, one can see that the result of equation (4.1.1) is just the equation that appeared in Liu YP [13] (equation (5.3), 1984), for enumerating plane tree-like maps with the size and the outer face valency as two parameters.

Although this equation has two variables, x and y, we notice that, by the substitution

$$z = \frac{xy^2}{1-x}, \tag{4.1.2}$$

equation (4.1.1) becomes

$$\begin{cases} azf^2 - bf - c = 0; \\ f|_{z=0} = d, \end{cases} \tag{4.1.3}$$

for $f \in \mathcal{R}\{z\}$. This is an equation with only one variable z, *i. e.*, equation (3.1.1).

Observation 4.1.1. *Equation* (4.1.1) *on* $\mathcal{R}\{x,y\}$ *is equivalent to equation* (4.1.3) *on* $\mathcal{R}\{z\}$.

Proof. Because of

$$\partial_x^s z^l = y^{2l}\binom{s-1}{l-1} = y^{2l}\frac{(s-1)!}{(l-1)!(s-l)!} \tag{4.1.4}$$

for $l \geq 1$, we have $z \in \mathcal{R}\{x,y\}$ and hence $\mathcal{R}\{z\} = \mathcal{R}\{x,y\}$. □

This observation enables us to consider equation (4.1.3) instead of equation (4.1.1).

Theorem 4.1.2. *Equation* (4.1.1) *is well-defined if, and only if,* $c = bd$ *and* $abd \neq 0$.

Proof. This is a result of Observation 4.1.1 and Theorem 3.1.1. □

From this theorem, we are allowed to only discuss

$$\begin{cases} azf^2 - bf - bd = 0; \\ f|_{z=0} = d, \end{cases} \tag{4.1.5}$$

for $a,b,d \in \mathbb{Z}_+$ and $abd > 0$.

https://doi.org/10.1515/9783110625837-004

Theorem 4.1.3. *The solution of equation* (4.1.5) *is determined by*

$$\partial_z^m f = \left(\frac{a}{b}\right)^m \frac{d^{m+1}}{2m+1}\binom{2m+1}{m} \tag{4.1.6}$$

for any integer $m \geq 0$.

Proof. Because of equation (4.1.5), same as equation (3.1.5), Theorem 3.1.2 results in (4.1.6). □

On the basis of (4.1.2), (4.1.4) and Theorem 4.1.3, one might think of an explicision of $\partial_{x,y}^{m,n} f$ in the solution of equation (4.1.1) for $m, n \geq 0$.

Theorem 4.1.4. *The solution f of equation* (4.1.1) *is determined by the following explicisions:*

$$\partial_{x,y}^{m,n} f = \begin{cases} d, & \text{when } m = n = 0; \\ \left(\frac{a}{b}\right)^m \frac{d^{m+1}}{2m+1}\binom{2m+1}{m}\binom{m-1}{k-1}, & \\ & \text{when } n = 2k \text{ and } m \geq k \geq 1; \\ 0, & \text{otherwise,} \end{cases} \tag{4.1.7}$$

for $m, n \geq 0$.

Proof. By (4.1.6),

$$\begin{aligned} \partial_{x,y}^{m,n} f &= \left(\frac{a}{b}\right)^m \frac{d^{m+1}}{2m+1}\binom{2m+1}{m}\partial_{x,y}^{m,n} z^m, \quad \text{by (4.1.4)}, \\ &= \left(\frac{a}{b}\right)^m \frac{d^{m+1}}{2m+1}\binom{2m+1}{m}\binom{m-1}{k-1}, \end{aligned}$$

the conclusion is drawn. □

The explanation above shows how to eliminate a variable in an equation with at least two variables by introducing a substitution of variables.

In what follows, two other methods are introduced for eliminating a variable. One is to fix a variable at a constant number. The other is to consider the coefficients of terms in the unknown function expressed by the variable one hopes to be eliminate. Only the former is discussed in this section because of the latter being met everywhere for each equation considered.

In Dong FM-Liu YP [8], one can find the equation

$$\begin{cases} \dfrac{xyf}{x-f} - (1+y)f + x^2 y = 0; \\ f_{x=0,y=0} = 0. \end{cases} \tag{4.1.8}$$

This is equation (8) in Introduction when $a = c = 1$ and $d = 0$ because it is meaningful in a classification of non-separable simple outer planar maps.

By the cancelation law, equation (4.1.8) is transformed to

$$\begin{cases} xf = (1+y)f^2 - x^2yf + x^3y; \\ f_{x=0,y=0} = 0, \end{cases} \tag{4.1.9}$$

for $f \in \mathcal{R}\{x,y\}$.

Because of equivalence, we are allowed to discuss equation (4.1.9) instead of equation (4.1.8).

Let $f \in \mathcal{R}\{x,y\}$ be a solution of equation (4.1.9); then f is determined by $F_{m,n} = \partial_{x,y}^{m,n} f$ for $m, n \geq 0$. Furthermore, let $F_{*,m} = \partial_x^m f$ and $F_{*,n} = \partial_y^n f$, then

$$\begin{cases} F_{*,n} = \sum\limits_{m\geq 0} F_{m,n} \in \mathcal{R}\{x\}, & \text{for } n \geq 0; \\ F_{*,m} = \sum\limits_{n\geq 0} F_{m,n} \in \mathcal{R}\{y\}, & \text{for } m \geq 0, \end{cases} \tag{4.1.10}$$

and hence f is determined by $F_{*,n}(n \geq 0)$, or by $F_{*,m}(m \geq 0)$ as well.

Now, $F_{*,n}$ for all $n \geq 0$ are chosen to determine f as follows:

$$\begin{aligned} y^0 : xF_{*,0} &= F_{*,0}^{[2]} + F_{*,-1}^{[2]} - x^2F_{*,-1}, \quad \text{by no sense of } F_{*,-1}, \\ &= F_{*,0}^{[2]} = F_{*,0}^2. \end{aligned}$$

By the initial condition, we have

$$F_{*,0} = 0 \Rightarrow F_{*,0}^{[2]} = 0 \text{ and } F_{*,1}^{[2]} = 0. \tag{4.1.11}$$

For $F_{*,1}$,

$$\begin{aligned} y^1 : xF_{*,1} &= F_{*,1}^{[2]} + F_{*,0}^{[2]} - x^2F_{*,0} + x^3 \quad \text{by (4.1.11)}, \\ &= x^3. \end{aligned}$$

By the cancelation law for x,

$$F_{*,1} = x^2 \Rightarrow F_{*,1}^{[2]} = 0 \text{ and } F_{*,2}^{[2]} = x^4. \tag{4.1.12}$$

For $F_{*,2}$, we have

$$\begin{aligned} y^2 : xF_{*,2} &= F_{*,2}^{[2]} + F_{*,1}^{[2]} - x^2F_{*,1} \quad \text{by (4.1.12)}, \\ &= x^4 - x^2x^2. \end{aligned}$$

Hence,

$$F_{*,2} = 0 \Rightarrow F_{*,2}^{[2]} = x^4 \text{ and } F_{*,3}^{[2]} = 0. \tag{4.1.13}$$

For $F_{*,3}$,

$$\begin{aligned} y^3 : xF_{*,3} &= F^{[2]}_{*,3} + F^{[2]}_{*,2} - x^2 F_{*,2} \quad \text{by (4.1.13)},\\ &= x^4. \end{aligned}$$

By the cancelation law for x,

$$F_{*,3} = x^3 \Rightarrow F^{[2]}_{*,3} = 0 \text{ and } F^{[2]}_{*,4} = 2x^5. \tag{4.1.14}$$

For $F_{*,4}$,

$$\begin{aligned} y^4 : xF_{*,4} &= F^{[2]}_{*,4} + F^{[2]}_{*,3} - x^2 F_{*,3} \quad \text{by (4.1.14)},\\ &= 2x^5 - x^2x^3 = x^5. \end{aligned}$$

By the cancelation law for x,

$$F_{*,4} = x^4 \Rightarrow F^{[2]}_{*,4} = 2x^5 \text{ and } F^{[2]}_{*,5} = 2x^6. \tag{4.1.15}$$

For $F_{*,5}$,

$$\begin{aligned} y^5 : xF_{*,5} &= F^{[2]}_{*,5} + F^{[2]}_{*,4} - x^2 F_{*,4} \quad \text{by (4.1.15)},\\ &= 2x^6 + 2x^5 - x^2x^4 = x^6 + 2x^5. \end{aligned}$$

By the cancelation law for x,

$$F_{*,5} = x^5 + 2x^4 \Rightarrow F^{[2]}_{*,5} = 2x^6 \text{ and } F^{[2]}_{*,6} = 2x^7 + 5x^6. \tag{4.1.16}$$

For $F_{*,n}$, $n \geq 6$, we have

$$F_{*,n} = \frac{1}{x}(F^{[2]}_{*,n} + F^{[2]}_{*,n-1}) - xF_{*,n-1}. \tag{4.1.17}$$

Theorem 4.1.5. *Equation* (4.1.9), *and hence equation* (4.1.8), *is well-defined on* $\mathcal{R}\{x, y\}$.

Proof. On the basis of (4.1.11)–(4.1.16), when $0 \leq n \leq 5$, $F_{*,n}$ are determined from $F_{*,0} = 0$ (the initial condition). For $n \geq 6$, if all $F_{*,k}$ are determined for $k \leq n-1$, then from (4.1.17), $F_{*,n}$ is determined by all $F_{*,k}$ for $k \leq n-1$. Hence, the conclusion is drawn. □

Combinatorial structures of the solution of equation (4.1.9) have to be investigated on all $F_{*,n}$, $n \geq 0$, for any given integer $m \geq 3$, because they are known for $m \leq 2$ from (4.1.11)–(4.1.13).

Observation 4.1.6. *For any integer $m \geq 3$ given, if $n \geq 2m-2$ or $n \leq m-1$, then $F_{*,n} = 0$, else $F_{*,n} \nleq 0$.*

Proof. We proceed by induction on m. For $0 \leq m \leq 5$, from (4.1.11)–(4.1.16), the conclusion are checked. In general, for $m \geq 6$, whenever all $F_{k.n}$ for $k \leq m-1$ are known, from (4.1.17), $F_{m,n}$ is deduced to obey the conclusion. □

In fact, the observation can be shown for non-separable outer planar maps. From Theorem 4.1.5, the conclusion of this observation is easily obtained.

Observation 4.1.7. *For any integer $m \geq 3$, if $m \leq n \leq 2m-3$, then $F^{[2]}_{*,n} - x^2 F_{*,n-1} \geq 0$.*

Proof. On the basis of Observation 4.1.6, for $m \geq 3$, only when $m \leq n \leq 2m-3$, $F_{*,n}$ is allowed to be non-zero. Because of $F_{*,n} \geq 0$,

$$F^{[2]}_{*,n} \geq 2F_{*,1}F_{*,n-1} = 2x^2 F_{*,n-1} \geq x^2 F_{*,n-1}.$$

This is the conclusion. □

Because of $F_{*,0} = 0$, this observation enables us to write

$$\Sigma^{04.1}_{1(n-1)} = \frac{F^{[2]}_{*,n} - x^2 F_{*,n-1}}{x} = xF_{*,n-1} + \frac{1}{x}\sum_{i=2}^{n-2} F_{*,i}F_{*,n-i}, \tag{4.1.18}$$

which only depends on $F_{*,k}$, $0 \leq k \leq n-1$.

Observation 4.1.8. *For any integer $m \geq 3$, if $m \leq n \leq 3m-5$, then*

$$x | (F^{[2]}_{*,n} + F^{[2]}_{*,n-1}), \tag{4.1.19}$$

i. e., x is a factor of $F^{[2]}_{*,n} + F^{[2]}_{*,n-1}$.

Proof. Because of any non-zero term in $F^{[2]}_{*,n}$ and $F^{[2]}_{*,n-1}$ with a factor x^2, the conclusion is true. □

On the basis of this observation, it is well-known that all coefficients of $\Sigma^{04.1}_{1(n-1)}$ given by (4.1.19) are all in $\mathbb{Z}_+$.

Theorem 4.1.9. *The solutionf of equation (4.1.9) and hence equation (4.1.8) determined by $F_{*,n}$, $n \geq 0$, has the form*

$$F_{*,n} = \begin{cases} 0, & \text{when } n = 0; \\ x^2, & \text{when } n = 1; \\ 0, & \text{when } n = 2; \\ x^3, & \text{when } n = 3; \\ x(F_{*,n-1} + 2F_{*,n-2}) & \\ \quad + \sum_{i=3}^{n-3} F_{*,i}\left(\frac{F_{*,n-i} + x^2 F_{*,n-1-i}}{x}\right), & \\ \quad \text{when } n \geq 4. & \end{cases} \tag{4.1.20}$$

Proof. The cases of $0 \le n \le 2$ are clear from (4.1.11)–(4.1.13). The general case of $n \ge 3$ is based on (4.1.14)–(4.1.17) and Observation 4.1.7. □

This theorem enables us to evaluate the solution of equation (4.1.9), and hence equation (4.1.8), directly from a recursion in the form of a sum with all terms positive.

Theorem 4.1.10. *The solution f of equation* (4.1.9) *and hence equation* (4.1.8) *is determined by $\partial^{(m,n)}_{(x,y)} f = F_{m,n}$ for $1 \le m \le n \le 2m-3$, which is in a summation-free explicision:*

$$\partial^{(m,n)}_{(x,y)} f = \begin{cases} 1, & \text{when } m = 2 \text{ and } n = 1; \\ \frac{1}{n}\binom{n}{m-1}\binom{m-3}{n-m}, & \text{when } 3 \le m \le n \le 2m-3. \end{cases} \tag{4.1.21}$$

Proof. On the basis of Theorem 4.1.9, we proceed by induction on $m \ge 1$, and the conclusion can in principle be drawn. □

In equation (4.1.9), by $x = 1$, let $h = f|_{x=1} \in \mathcal{R}\{y\}$, then h satisfies the equation

$$\begin{cases} (1+y)h^2 - (1+y)h + y = 0; \\ h|_{y=0} = 0. \end{cases} \tag{4.1.22}$$

This is an equation in total variform shown in Section 3.3 when $a = b = c = 1$ and $d = 0$.

Theorem 4.1.11. *The solution h of equation* (4.1.22) *determined by $\partial^n_y h = H_n$ for $n \ge 1$, which is in an explicision in the form of a sum with all terms positive:*

$$\partial^n_y h = \begin{cases} 1, & \text{when } n = 1; \\ 0, & \text{when } n = 2; \\ \sum_{m=3}^{n} \frac{1}{n}\binom{n}{m-1}\binom{m-3}{n-m}, & \text{when } 9 \ge n \ge 3; \\ \sum_{m=\lceil (n-3)/2 \rceil}^{n} \frac{1}{n}\binom{n}{m-1}\binom{m-3}{n-m}, & \text{when } n \ge 10. \end{cases} \tag{4.1.23}$$

Proof. Because of $H_n = F_{*,n}|_{x=1}$, from (4.1.21), (4.1.23) is easily obtained. □

4.2 Linear forms

Consider the equation

$$\begin{cases} f = x^2 y + \dfrac{xy}{1-xy}\left(\dfrac{x}{1-x}h - \dfrac{1}{1-x}f\right) \\ f|_{x=0,y=0} = 0, \end{cases} \tag{4.2.1}$$

where $f \in \mathcal{R}\{x,y\}$ and $h = f(1,y) \in \mathcal{R}\{y\}$.

This is equation (9) in Introduction when $a = c = 1$ and $d = 0$, it being meaningful in a classification for restriction to outer planar maps.

For convenience, it is transformed on $\mathcal{R}\{x,y\}$ into its equivalent

$$\left(1+\frac{xy}{(1-xy)(1-x)}\right)f = x^2y + \frac{x^2y}{(1-xy)(1-x)}h. \tag{4.2.2}$$

For any integer $n \geq 0$, let

$$\partial_y^n h = H_n \quad \text{and} \quad \partial_y^n f = F_{*,n} \tag{4.2.3}$$

where

$$F_{*,n} = \sum_{m\geq 0} F_{m,n}x^m, \quad F_{m,n} = \partial_{(x,y)}^{(m,n)} f \quad \text{and} \quad H_n = F_{*,n}|_{x=1}. \tag{4.2.4}$$

By employing the cancelation law on $\mathcal{R}\{x,y\}$, equation (4.2.2) is transformed into

$$(1-x)f = (1-x)x^2y - (x-x^2)x^2y^2 + x^2y(h-f), \tag{4.2.5}$$

and we obtain

$$y^0: \ (1-x)F_{*,0} = 0 \implies F_{*,0} = 0 \text{ and } H_0 = 0, \tag{4.2.6}$$

$$\begin{aligned} y^1 &: (1-x)F_{*,1} = (1-x)x^2 + x^2(H_0 - F_{*,0}) \\ &\implies (1-x)F_{*,1} = (1-x)x^2 \\ &\implies F_{*,1} = x^2 \text{ and } H_1 = 1, \end{aligned} \tag{4.2.7}$$

$$\begin{aligned} y^2 &: (1-x)F_{*,2} = -(x-x^2)x^2 + x^2(H_1 - F_{*,1}) \\ &\implies (1-x)F_{*,2} = -(1-x)x^3 + x^2(1-x^2) \\ &\implies F_{*,2} = x^2 \text{ and } H_1 = 1, \end{aligned} \tag{4.2.8}$$

in general, for $n \geq 3$,

$$y^n: \ (1-x)F_{*,n} = x^2(H_{n-1} - F_{*,n-1}). \tag{4.2.9}$$

Lemma 4.2.1. *Equation (4.2.2) on $\mathcal{R}\{x,y\}$ is equivalent to the equation system*

$$\begin{cases} F_{*,1} = x^2 \Rightarrow H_1 = 1, & \text{when } n = 1; \\ F_{*,1} = x^2 \Rightarrow H_1 = 1, & \text{when } n = 2; \\ F_{*,n} = xF_{*,n} + x^2(H_{n-1} - F_{*,n-1}). & \\ \quad \Rightarrow H_n = F_{*,n}|x = 1, & \text{when } n \geq 3, \end{cases} \tag{4.2.10}$$

from $F_{+,0} = 0 \Rightarrow H_0 = 0$ on $\mathcal{R}\{x\}^{\infty}$.

Proof. Because all transformations based on the cancelation law that are an equivalence on $\mathcal{R}\{x,y\}$, the conclusion is drawn. □

This lemma enables us discuss a solution of the equations of system (4.2.10) instead of equation (4.2.2), and hence equation (4.2.1).

Theorem 4.2.2. *Equation* (4.2.2), *and hence equation* (4.2.1), *is well-defined on* $\mathcal{R}\{x,y\}$.

Proof. We proceed by induction on $n \geq 0$. When $n = 0$, the result is known from the initial condition. When $n = 1$ and 2, the results are from (4.2.7) and (4.2.8). In general, (4.2.10) leads to the conclusion. □

Because we aim at an expression in the form of a sum with all terms positive, relative constructions of the solution have to be investigated in correspondence.

Observation 4.2.3. *For any integer* $n \geq 1$, $F_{*,n}$ *has a factor* x^2.

Proof. We proceed by induction on n. From (4.2.7) and (4.2.8), the conclusion for $n = 1$ and 2 is true. For $n \geq 3$, assume that $x^2|F_{*,n-1}$. We prove $x^2|F_{*,n}$. Because of

$$\begin{aligned}(1-x)F_{*,n} &= x^2(H_{n-1} - F_{*,n-1}), \quad \text{by the assumption,}\\ &= x^2 \sum_{m\geq 2} F_{m,n-1}(1-x)\left(\sum_{i=0}^{m-1} x^i\right),\end{aligned}$$

by the cancelation law, we have

$$F_{*,n} = x^2 \sum_{m\geq 2} F_{m,n-1}\left(\sum_{i=0}^{m-1} x^i\right). \tag{4.2.11}$$

Therefore, $x^2|F_{*,n}$. □

This observation tells us that, for any integer $n \geq 1$, the minimum degree of x in $F_{*,n}$ is not less than 2.

Observation 4.2.4. *For any integer* $n \geq 3$, $H_n - F_{8,n}$ *has a factor* $1 - x$.

Proof. This follows from the proof of Observation 4.2.3. □

Observation 4.2.5. *For any integer* $n \geq 3$, *if* $m \geq n$, *then* $F_{m,n} = 0$.

Proof. We proceed by induction on n, on the basis of (4.2.10). □

This observation shows that, for any integer $n \geq 3$, $F_{*,n}$ is a polynomial of x with degree at most n.

Observation 4.2.6. *For any integer* $n \geq 3$, $F_{m,n} \in \mathcal{Z}_+$.

Proof. We proceed by induction on n, on the basis of (4.2.10). □

The observations above claim that all $F_{*,n}$ for $n \geq 3$ are polynomials of x with minimum degree not less than 2, the maximum degree is not greater than n, and all coefficients are non-negative.

Observation 4.2.7. *For any integer $n \geq 3$,*

$$\frac{H_{n-1} - F_{*,n-1}}{1-x} \geq 0. \tag{4.2.12}$$

Proof. This is a result of Observation 4.2.6. □

The observations described above are helpful as they clarify the structure of the solution of equation (4.2.1).

Theorem 4.2.8. *The solution of equation* (4.2.2), *and hence equation* (4.2.1), *determined by $F_{*,n}$ for $n \geq 1$, is a recursion in the form of a sum of finite non-negative terms:*

$$F_{*,n} = \begin{cases} x^2, & \text{when } n = 1; \\ x^2, & \text{when } n = 2; \\ x^2 \sum_{j=0}^{n-2} (\sum_{i=j+1}^{n-1} F_{i,n-1}) x^j, & \text{when } n \geq 3. \end{cases} \tag{4.2.13}$$

Proof. From (4.2.10), it is only necessary to consider the case of $n \geq 3$. On the basis of Observations 4.2.5 and 4.2.7, the conclusion is drawn. □

Although equation (4.2.2) is linear about $f \in \mathcal{R}\{x,y\}$, one has another function $h = f|_{x=1} \in \mathcal{R}\{y\}$. This function h cannot be determined directly by taking $x = 1$ on the equation. A new parameter $\xi \in \mathcal{R}\{y\}$ is introduced satisfying the relations

$$\begin{cases} 1 + \dfrac{\xi y}{(1-\xi y)(1-\xi)} = 0; \\ \xi^2 y + \dfrac{\xi^2 y}{(1-\xi y)(1-\xi)} h = 0, \end{cases} \tag{4.2.14}$$

called a *characteristic equation* of equation (4.2.2).

There is an equivalence between equation (4.2.14) and

$$\begin{cases} \xi = 1 + \xi^2 y; \\ h = \xi y. \end{cases} \tag{4.2.15}$$

From $\xi \in \mathcal{R}\{y\}$, by the first relation in (4.2.15), we have

$$\partial_y^n \xi = \begin{cases} 1, & \text{when } n = 0; \\ \sum_{i=0}^{n-1} (\partial_y^i \xi)(\partial_y^{n-1-i} \xi), & \text{when } n \geq 1, \end{cases} \tag{4.2.16}$$

for $n \geq 0$.

Then, by the second relation in (4.2.15), we have

$$\partial_y^n h = \begin{cases} 0, & \text{when } n = 0; \\ \partial_y^{n-1}\xi, & \text{when } n \geq 1, \end{cases} \tag{4.2.17}$$

for $n \geq 0$ where $\partial_y^{n-1}\xi$ is given by (4.2.16) for $n \geq 1$.

Theorem 4.2.9. *The function h in equation* (4.2.2) *is determined by* H_n, $n \geq 0$, *in a summation-free explicision:*

$$H_n = \begin{cases} 0, & \textit{when } n = 0; \\ \frac{(2n-2)!}{n!(n-1)!}, & \textit{when } n \geq 1. \end{cases} \tag{4.2.18}$$

Proof. Because equation (4.2.15) is the same as equation (3.2.7), from (3.2.9) and (4.2.8), (4.2.18) is obtained. □

Further, for $m, n \geq 0$ and $f \in \mathcal{R}\{x, y\}$ being the solution of equation (4.2.2) and hence equation (4.2.2), a summation-free explicision of $\partial_{x,y}^{m,n} f$ can also obtained.

Theorem 4.2.10. *Let f be the solution of equation* (4.2.2), *hence equation* (4.2.1), *and* $F_{m,n} = \partial_{x,y}^{m,n} f$ *for* $m, n \geq 0$. *Then* $F_{m,n}$ *has the form*

$$F_{m,n} = \begin{cases} 1, & \textit{when } m = 2 \textit{ and } n = 1; \\ \frac{m-1}{n-1}\binom{2n-m-2}{n-2}, & \textit{when } 2 \leq m \leq n \textit{ and } n \geq 2; \\ 0, & \textit{otherwise.} \end{cases} \tag{4.2.19}$$

Proof. Method 1 is by induction on the basis of (4.2.13). Method 2 is by induction on the basis of (4.2.15) and (4.2.16). □

Example 1. We study an isomorphic classification of restrict outer planar rooted maps with size and valency of root-vertex. A outer planar map is called *restrict* if there is no loop in the root-loop and nor after a contraction of the root-loop.

Consider the equation in Liu YP [45] (pp. 90–92), or [48] (pp. 69–70),

$$\begin{cases} f = 1 + x^2 y f + \dfrac{xy}{1-x}(h - xf); \\ f|_{x=0,y=0} = 1, \end{cases} \tag{4.2.20}$$

where $h = f_{x=1}$.

Or equivalently, its normal has the form of

$$\left(1 - x^2 y + \frac{x^2 y}{1-x}\right) f = 1 + xy\frac{h}{1-x}. \tag{4.2.21}$$

By the cancelation law,

$$(1 - x + x^3 y) f = 1 - x + xyh. \tag{4.2.22}$$

Theorem 4.2.11. *Equation* (4.2.22), *or equation* (4.2.21), *is well-defined in* $\mathcal{R}\{x,y\}$.

Proof. The proof is similar to the proof of Theorem 4.2.2. □

Because of

$$\sum_{m\geq 1} F_{m,n-1} \sum_{i=0}^{m+1} x^{i+1} = \sum_{m=1}^{3}\left(\sum_{j=1}^{2(n-1)} F_{j,n-1}\right)x^m + \sum_{m=4}^{2n}\left(\sum_{j=m-2}^{2(n-1)} F_{j,n-1}\right)x^m,$$

from (4.2.21),

$$F_{m,n} = \begin{cases} 0, & \text{when } m = 0; \\ \sum_{j=1}^{2(n-1)} F_{j,n-1}, & \text{when } 1 \leq m \leq 3; \\ \sum_{j=m-2}^{2(n-1)} F_{j,n-1}, & \text{when } 4 \leq m \leq 2n \\ 0, & \text{when } m \geq 2n+1. \end{cases} \tag{4.2.23}$$

By (4.2.23), for $n = 1, 2$ and 3,

$$\begin{aligned} &F_{\star,1} = x + x^2; \quad F_{\star,2} = 2x + 2x^2 + 2x^3 + x^4; \\ &F_{\star,3} = 7x + 7x^2 + 7x^3 + 5x^4 + 3x^5 + x^6. \end{aligned} \tag{4.2.24}$$

On this basis, Figures 4.2.1–4.2.6 show the classification of such maps with size 3 and less. In these figures, $T_{m,n(i)}$ stands for the ith map with size n and root-vertex valency m. In each figure, a small hollow indicates the location of the root.

From (4.2.24), it is seen that restrict outer planar maps of size 1 have two possibilities: root-vertex valency: 1 and 2. They are $(F_{1,1}, F_{2,1}) = (1,1)$. Figure 4.2.1 shows the two classes:

$$F_{1,1} = 1 \implies 1T_{1,1(1)}; \quad F_{2,1} = 1 \implies 1T_{2,1(1)}.$$

From (4.2.23), it is seen that restricting to outer planar maps of size 2 leaves four possibilities of root-vertex valencies m: $1 \leq m \leq 4$. They are

$$(F_{1,2}, F_{2,2}, F_{3,2}, F_{4,2}) = (2,2,2,1).$$

$T_{1,1(1)}$ $T_{2,1(1)}$

Figure 4.2.1: Classes of restrict outer planar maps with size 1 and valencies 1–2.

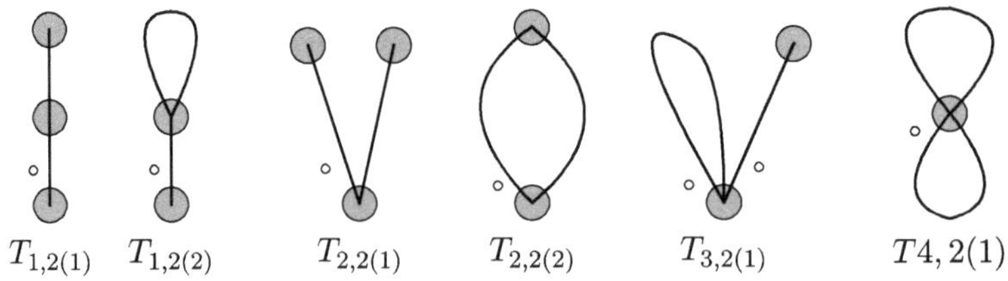

Figure 4.2.2: Classification of restrict outer planar maps of size 2 and valencies 1–4.

Figure 4.2.2 shows the classes. They are

$$F_{1,2} = 2 \implies 1T_{1,2(1)} + 1T_{1,2(2)};$$
$$F_{2,2} = 2 \implies 1T_{2,2(1)} + 1T_{2,2(2)};$$
$$F_{3,2} = 2 \implies 2T_{3,2(1)};$$
$$F_{4,2} = 1 \implies 1T_{4,2(1)}.$$

From (4.2.23), it is seen that restricting to outer planar maps of size 3 and root-vertex valencies m, $1 \le m \le 6$, we have

$$(F_{1,3}, F_{2,3}, F_{3,3}, F_{4,3}, F_{5,3}, F_{6,3}) = (7,7,7,5,3,1).$$

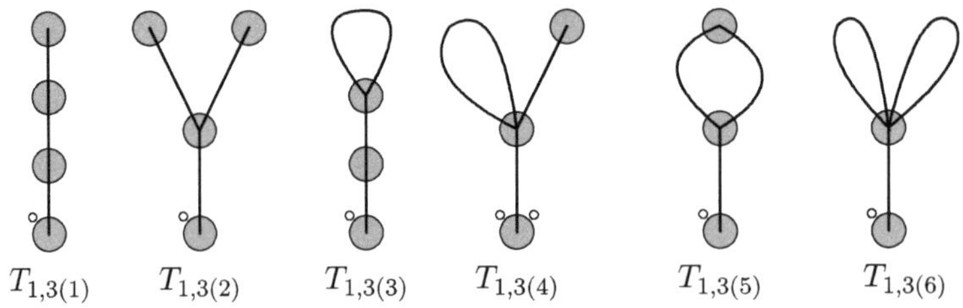

Figure 4.2.3: Classification of restrict outer planar maps of size 3 and valency 1.

Figure 4.2.3 provides the case of valency 1:

$$F_{1,3} = 7 \implies 1T_{1,3(1)} + 1T_{1,3(2)} + 1T_{1,3(3)} + 2T_{1,3(4)} + 1T_{1,3(5)} + 1T_{1,3(6)}.$$

Figure 4.2.4 provides such classes in two cases:

$$F_{2,3} = 7 \implies 2T_{2,3(1)} + 1T_{2,3(2)} + 1T_{2,3(3)} + 1T_{2,3(4)} + 2T_{2,3(5)}.$$

Figures 4.2.5 and 4.2.6 provide the classes for root-vertex valencies, respectively, 3 and 4–6:

$$F_{3,3} = 7 \implies 2T_{3,3(1)} + 2T_{3,3(2)} + 2T_{3,3(3)} + 1T_{3,3(4)}$$

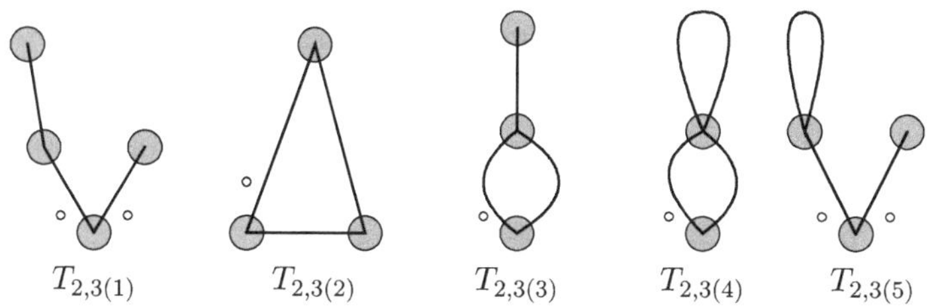

Figure 4.2.4: Classification of restrict outer planar maps of size 3 and valency 2.

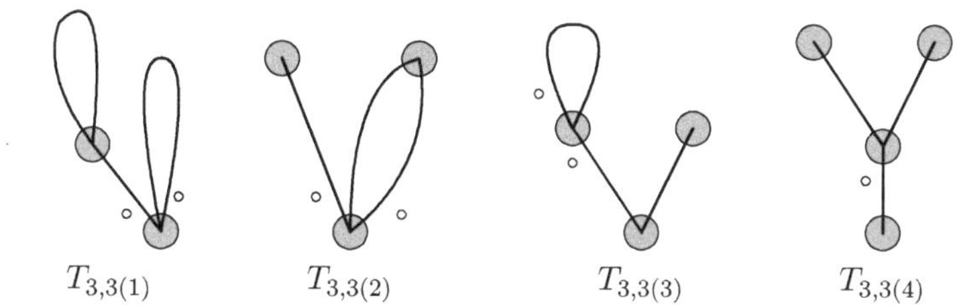

Figure 4.2.5: Classification of restrict outer planar maps of size 3 and valency 3.

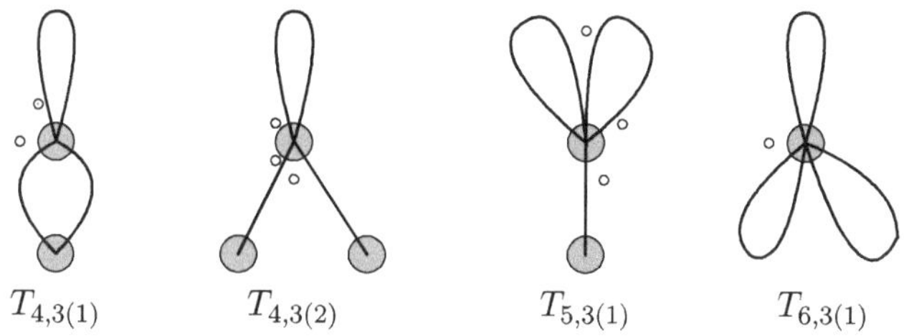

Figure 4.2.6: Classification of restrict outer planar maps of size 3 and valencies 4–6.

and

$$\begin{aligned} F_{4,3} &= 5 \implies 2T_{4,3(1)} + 3T_{4,3(2)}; \\ F_{5,3} &= 3 \implies 3T_{5,3(1)}; \\ F_{6,3} &= 1 \implies 1T_{6,3(1)}. \end{aligned}$$

One might like to see if it is possible to get an explicision of the solution $f \in \mathcal{R}\{x, y\}$ of equation (4.2.20). Because $h \in \mathcal{R}\{y\}$ is unknown as well, a parameter $\xi \in \mathcal{R}\{y\}$ instead of x has to be considered to satisfy the characteristic equation as in (4.2.14),

$$\begin{cases} 1 - \xi + \xi^3 y = 0; \\ h = \xi^2, \end{cases} \tag{4.2.25}$$

from the equivalence (4.2.22) of equation (4.2.20).

Let ξ be determined by $P_i = \partial_y^i \xi$ for $i \geq 0$. From equation (4.2.25), we have

$$\begin{array}{llll}
y^0 & P_0 = 1 \Rightarrow P_0 = 1 & \Rightarrow & P_0^{[2]} = 1, P_0^{[3]} = 1; \\
y^1 & P_1 = P_0^{[3]} \Rightarrow P_0 = 1 & \Rightarrow & P_1^{[2]} = 2, P_1^{[3]} = 3; \\
y^2 & P_2 = P_1^{[3]} \Rightarrow P_2 = 3 & \Rightarrow & P_2^{[2]} = 7, P_2^{[3]} = 12; \\
y^3 & P_2 = P_2^{[3]} \Rightarrow P_3 = 12 & \Rightarrow & P_3^{[2]} = 30, P_3^{[3]} = 55; \\
y^n(n \geq 4) & \begin{cases} P_n = P_{n-1}^{[3]} \Rightarrow \\ P_3 = \sum_{i=0}^{n-1} P_i P_{n-1-i}^{[2]} \end{cases} & \Rightarrow & \begin{cases} P_n^{[2]} = \sum_{i=0}^{n-1} P_i P_{n-1-i}, \\ P_n^{[3]} = \sum_{i=0}^{n} P_i P_{n-i}^{[2]}. \end{cases}
\end{array} \tag{4.2.26}$$

Let h be determined by H_n for $n \geq 0$. Because of $h = \xi^2$ in (4.2.25), we have

$$H_n = \begin{cases} 1, & \text{when } n = 0; \\ \sum_{i=0}^{n} P_i P_{n-i}, & \text{when } n \geq 1. \end{cases} \tag{4.2.27}$$

Theorem 4.2.12. *Let $f_{\mathrm{rop}} = f(x,y)$ be the solution of equation (4.2.20) and $h_{\mathrm{rop}} = f_{\mathrm{rop}}|_{x=1} = h(y)$, then we have*

$$\partial_y^n h_{\mathrm{rop}} = \begin{cases} 1, & \text{when } n = 0; \\ \frac{2(3n+1)!}{n!(2n+2)!}, & \text{when } n \geq 1, \end{cases} \tag{4.2.28}$$

and

$$\partial_{(x,y)}^{(m,n)} f_{\mathrm{rop}} = \begin{cases} 1, & \text{when } m = 0, n = 0; \\ \sum_{k=\lceil m/2 \rceil}^{m} \frac{3k-m}{n-k} \binom{3n-m-1}{n-k-1} \binom{k}{m-k}, & \text{when } n \geq 1, 1 \leq m \leq 2n; \end{cases} \tag{4.2.29}$$

Proof. Because of $h_{\mathrm{rop}} = h$ and hence $\partial_y^n h_{\mathrm{rop}} = H_n$ in (4.2.27), (4.2.28) can be done by induction on n. On the basis of (4.2.23), (4.2.29) can be done by induction on n. □

Corollary 4.2.13. *For integer $n \geq 1$, we have the identity*

$$\frac{2(3n+1)!}{n!(2n+2)!} = \sum_{\substack{\lceil m/2 \rceil \leq k \leq m \\ 1 \leq m \leq 2n}} \frac{3k-m}{n-k} \binom{3n-m-1}{n-k-1} \binom{k}{m-k}. \tag{4.2.30}$$

Proof. It can be directly proved by induction on n as a result of (4.2.25) and (4.2.26). □

This corollary suggests us to determine the function h of one variable first if the direct determination of the function f of two variables needs some sophistication in solving the equation for $f \in \mathcal{R}\{x,y\}$ involving $h = f|_{x=1} \in \mathcal{R}\{y\}$.

4.3 Quadratic forms

Consider the equation about $f \in \mathcal{R}\{x, y\}$ as

$$\begin{cases} x^2y(1-x^2)f^2 - (1 - x^2 + x^2y)f + (1 - x^2) + x^2yh = 0; \\ f|_{x=0,y=0} = 1, \end{cases} \tag{4.3.1}$$

where $h = f(1, y) \in \mathcal{R}\{y\}$. This is equation (10) in Introduction when $a = b = c = d = 1$ because of it is meaningful in a classification of Eulerian planar maps.

For convenience, its equivalence

$$f = 1 + \frac{x^2y}{1-x^2}(h - f) + x^2yf^2 \tag{4.3.2}$$

is used.

Because of the occurrences of x always in the form of x^2, by the substitution $z = x^2$, equation (4.3.2) becomes

$$f = 1 + \frac{zy}{1-z}(h - f) + zyf^2. \tag{4.3.3}$$

Let $F_{\star,n} = \partial_y^n f$ and $H_n = \partial_y^n h$, then we have

$$H_n = F_{\star,n}|_{z=1} = \sum_{m\geq 0} F_{m,n} \tag{4.3.4}$$

where $F_{m,n} = \partial_{(z,y)}^{(m,n)} f$, $m, n \geq 0$.

Writing $F_{\star,n}^{[2]} = \partial_y^n f^2$,

$$F_{\star,n}^{[2]} = \sum_{i=0}^{n} F_{\star,i} F_{\star,n-i}, \tag{4.3.5}$$

by (4.3.3),

$$F_{\star,n} = \begin{cases} 1, & \text{when } n = 0; \\ zF_{\star,n-1}^{[2]} + \frac{z(H_{n-1} - F_{\star,n-1})}{1-z}, & \text{when } n \geq 2. \end{cases}$$

From (4.3.4), it is seen that

$$\begin{aligned} H_{n-1} - F_{\star,n-1} &= \sum_{m\geq 0} (F_{m,n-1} - F_{m,n-1}z^m) \\ &= (1-z) \sum_{m\geq 1} F_{m,n-1}\left(\sum_{i=0}^{m-1} z^i\right) \\ &= (1-z) \sum_{i\geq 0}\left(\sum_{m\geq i+1} F_{m,n-1}\right) z^i. \end{aligned}$$

Therefore,

$$F_{\star,n} = \begin{cases} 1, & \text{when } n = 0; \\ z(F_{\star,n-1}^{[2]} + \sum_{i\geq 0}(\sum_{m\geq i+1} F_{m,n-1})z^i), & \text{when } n \geq 1. \end{cases} \tag{4.3.6}$$

Observation 4.3.1. *For any integer $n \geq 1$, $F_{*,n}$ is determined by all $F_{*,k}$, $0 \leq k \leq n-1$.*

Proof. From (4.3.6), it is seen that for the whole right hand side of the relation involving $F_{*,n}$ there is only dependence on $F_{*,k}$, $0 \leq k \leq n-1$. This implies the conclusion. □

On the basis of this observation, we are allowed to establish the qualitative theory of equation (4.3.1).

Theorem 4.3.2. *Equation (4.3.1) on $\mathcal{R}\{z, y\}$ is well-defined.*

Proof. On the basis of Observation 4.3.1, all $F_{*,n}$ can be determined from the initial condition on $\mathcal{R}\{x, y\}$. Hence, equation (4.3.1) has, and is the only one to have, a solution on $\mathcal{R}\{x, y\}$. □

In order to evaluate the solution $f \in \mathcal{R}\{x, y\}$ of equation (4.3.1), its relative structures have be investigated.

Lemma 4.3.3. *For any integer $n \geq 0$, $F_{\star,n}$ is a polynomial of z with degree of z at most n.*

Proof. We proceed by induction on n. Because of $F_{\star,0} = 1$, the conclusion is true when $n = 0$. For general $n \geq 1$, assume the truth for k, $n = k \geq 1$. By (4.3.6),

$$F_{\star,k+1} = z\Big(F^{[2]}_{\star,k} + \sum\Big).$$

From the assumption, $F^{[2]}_{\star,k}$ is a polynomial of z with the degree at most k. Since the degree of $\sum$ is less than that of $F^{[2]}_{\star,k}$, the degree of $zF^{[2]}_{\star,k}$ is at most $k+1$. Hence, for $n = k+1$, the degree of $F_{*,n}$ is at most n. This is the conclusion. □

This lemma enables us to get the solution of equation (4.3.1) in a recursion as a sum of finite terms.

Lemma 4.3.4. *For any integer $n \geq 0$, $F_{\star,n}$ has all coefficients of terms in $\mathbb{Z}_+$.*

Proof. We proceed by induction on n. Because of only addition and multiplication being used for getting $F_{*,n}$ from $F_{*,k}$, $0 \leq k \leq n-1$, the conclusion is drawn. □

This lemma enables us to see that all coefficients of the terms in $F_{*,n}$ are positive.

Theorem 4.3.5. *The solution $f \in \mathcal{R}\{x, y\}$ of equation (4.3.1) determined by $F_{*,n} = \partial_y^n f$ for $n \geq 0$ as a recursion in form, as a sum with all terms positive, is*

$$F_{\star,n} = \begin{cases} 1, & \textit{when } n = 0; \\ z(\sum_{i=0}^{n-1} F_{\star,i}F_{\star,n-i} + \sum_{\substack{i+1\leq m\leq n-1 \\ 0\leq i\leq n-1}} F_{m,n-1}z^i), & \textit{when } n \geq 1. \end{cases} \tag{4.3.7}$$

Proof. On the basis of the two lemmas, by (4.3.5), (4.3.6) becomes (4.3.7). □

Example 1. Isomorphic classification of planar Eulerian rooted maps with size and root-vertex valency are arbitrarily given. By (4.3.7),

$$\begin{aligned}
F_{\star,1} &= z(F_{\star,0}^2 + 0) = z(1+0) = z,\\
F_{\star,2} &= z(2F_{\star,0}F_{\star,1} + F_{1,1}) = z(2z+1)\\
&= z + 2z^2,\\
F_{\star,3} &= z\Big(2F_{\star,0}F_{\star,2} + F_{\star,1}^2 + \sum_{\substack{i+1\le m\le 2\\ 0\le i\le 1}} F_{m,2}x^i\Big)\\
&= z(2z + 5z^2 + 3 + 2z)\\
&= 3z + 4z^2 + 5z^3,\\
F_{\star,4} &= z\Big(2F_{\star,0}F_{\star,3} + 2F_{\star,0}F_{\star,3} + \sum_{\substack{i+1\le m\le 3=0\\ 0\le i\le 2}} F_{m,3}x^i\Big)\\
&= z(6z + 10z^2 + 14z^3 + 12 + 9z + 5z^2)\\
&= 12z + 15z^2 + 15z^3 + 14z^4.
\end{aligned}$$

We might state that the solutions of equation (4.3.1), denoted by $f_{\mathrm{pE}} = f(x,y) \in \mathcal{R}\{x,y\}$ and $h_{\mathrm{pE}} = f_{\mathrm{pE}}|_{x=1} = h(y) \in \mathcal{R}\{y\}$, are, respectively,

$$\begin{aligned}
f_{\mathrm{pE}} &= 1 + (z)y + (z + 2z^2)y^2 + (3z + 4z^2 + 5z^3)y^3\\
&\quad + (15z^2 + 15z^3 + 14z^4)y^4 + \cdots
\end{aligned}$$

and

$$h_{\mathrm{pE}} = 1 + (1)y + (3)y^2 + (12)y^3 + (56)y^4 + \cdots$$

where $z = x^2$.

In Figures 4.3.1–4.3.5, $aT_{m,n}$ stands for a figure which is a map of size n and root-vertex valency m with a root-isomorphic classes, $2 \le m \le 8$ and $1 \le n \le 4$.

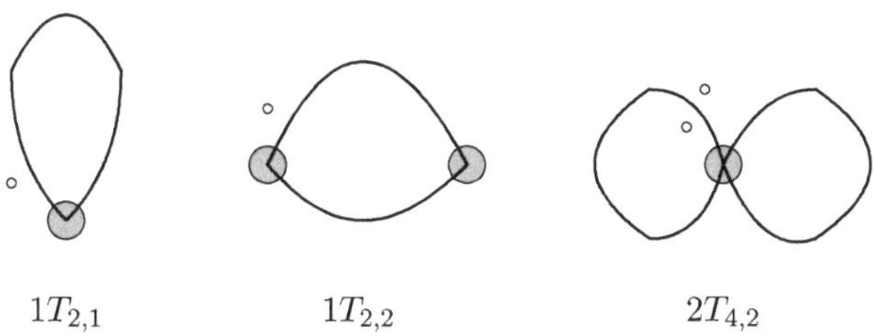

Figure 4.3.1: Classes of planar Euler root-maps of size 1–2.

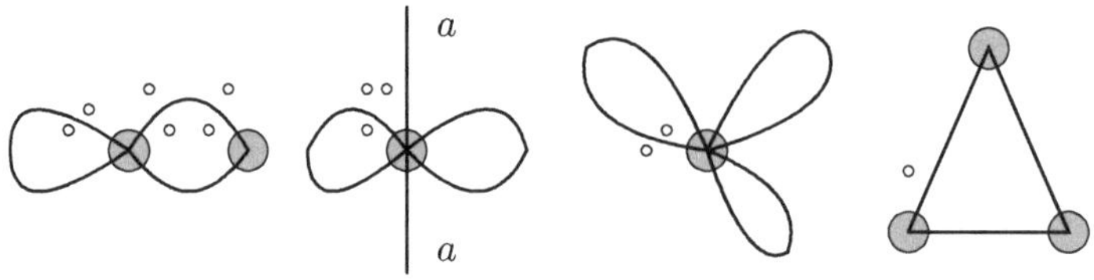

$2T_{2,3} + 4T_{4,3}$ $3T_{6,3}$ $2T_{6,3}$ $1T_{2,3}$

Figure 4.3.2: Classes of planar Euler root-maps of size 3.

From Figures 4.3.1 through 4.3.2, it is seen that planar Euler root-maps of size 1 have 1 class, *i. e.*, $1T_{2,1}$, the coefficient $F_{\star,1} = x^2$ of the term y in f_{pE}; planar Euler root-maps of size 2 have 3 classes, *i. e.*, $1T_{2,2}+2T_{4,2}$, the coefficient $F_{\star,2} = x^2+2x^4$ of term y^2 in f_{pE} and the planar Euler root-maps have 3 classes, *i. e.*, $(2T_{2,3}+4T_{4,3})+(3T_{6,3})+(2T_{6,3})+(1T_{2,3}) = 3T_{2,3} + 4T_{4,3} + 5T_{6,3}$, the coefficient $F_{\star,4} = 3x^2 + 4x^4 + 5x^6$ of the term y^4 in f_{pE}.

From Figs. 4.3.3 through 4.3.5, it is seen that planar Euler root-maps of size 4 have 56 classes, *i. e.*, $(2T_{2,3} + 4T_{4,3}) + (3T_{6,3}) + (2T_{6,3}) + (1T_{2,3}) = 3T_{2,3} + 4T_{4,3} + 5T_{6,3}$, the coefficient $F_{\star,4} = 3x^2 + 4x^4 + 5x^6$ of the term y^4 in f_{pE}.

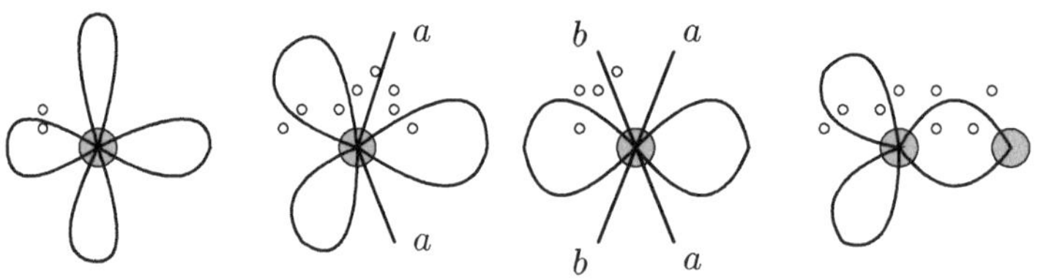

$2T_{8,4}$ $8T_{8,4}$ $4T_{8,4}$ $6T_{8,4} + 2T_{2,4}$

Figure 4.3.3: Classes of planar Euler root-maps of size 4 I.

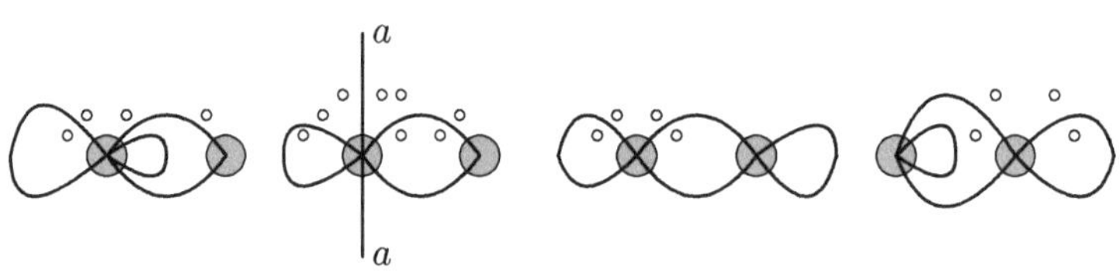

$3T_{6,4} + 1T_{2,4}$ $6T_{6,4} + 2T_{2,4}$ $4T_{4,4}$ $4T_{4,4}$

Figure 4.3.4: Classes of planar Euler root-maps of size 4 II.

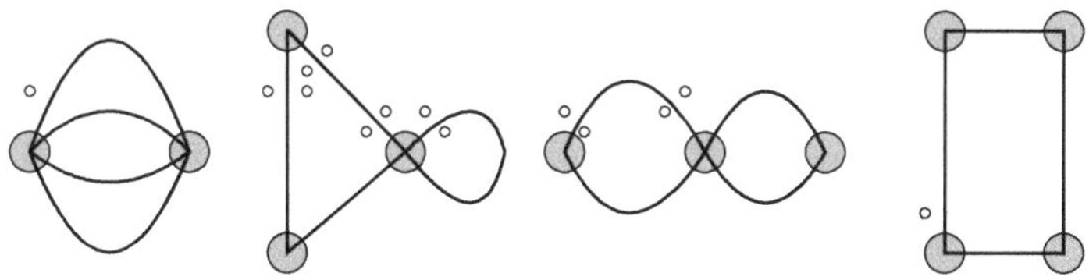

$1T_{4,4}$ $4T_{2,4} + 4T_{4,4}$ $2T_{2,4} + 2T_{4,4}$ $1T_{2,4}$

Figure 4.3.5: Classes of planar Euler root-maps of size 4 III.

Example 2. Consider the equation about $f \in \mathcal{R}\{x, y\}$ as

$$\begin{cases} x^2y(1-x)f^2 - (1-x+x^2y)f + xyh + (1-x) = 0; \\ f|_{x=0,y=0} = 1, \end{cases} \tag{4.3.8}$$

where $h = f(1, y) \in \mathcal{R}\{y\}$.

For convenience, its equivalent form

$$f = 1 + \frac{xy}{1-x}(h - xf) + x^2yf^2 \tag{4.3.9}$$

is adopted. Similarly, a qualitative theory is established and for equation (4.3.8) its solution is evaluated for getting a recursion in the form of a sum with all terms positive.

4.4 Forms of degree not less three

Consider the equation about $f \in \mathcal{R}\{x, y\}$ as

$$\begin{cases} f = x^2y + \dfrac{x^2y(f-h)}{x^2(1+h)^2 - (1+f)^2}; \\ f|_{y=0} = 0 \ (\Rightarrow \ h_{y=0} = 0), \end{cases} \tag{4.4.1}$$

where $h = f(1, y) \in \mathcal{R}\{y\}$. This is equation (11) in Introduction when $a = c = 1$ and $d = 0$ because it is meaningful in a classification for non-separable Euterian planar maps

Since only x^2 appears in the equation, let $z = x^2$, equation (4.4.1) becomes

$$\begin{cases} f = zy + \dfrac{zy(f-h)}{z(1+h)^2 - (1+f)^2}; \\ f|_{y=0} = 0 \ (\Rightarrow \ h_{y=0} = 0), \end{cases} \tag{4.4.2}$$

where $h = f(1, y)$.

For convenience, equation (4.4.2) is equivalently transformed into

$$\begin{cases} (f - zy)(z(1+h)^2 - (1+f)^2) = zy(f-h); \\ f|_{y=0} = 0 (\Rightarrow h_{y=0} = 0). \end{cases} \tag{4.4.3}$$

For any integer $n \geq 0$, let

$$F_{*,n}^{[i]} = \begin{cases} \partial_y^n f = [f]_n = F_{*,n}, & \text{when } i = 1; \\ \partial_y^n f^i = [f^i]_n = [f^{[i]}]_n, & \text{when } i \geq 2. \end{cases} \tag{4.4.4}$$

As a matter of fact, only $i = 2$ is used here. Because of

$$F_{*,n}^{[2]} = \sum_{j=0}^{n} F_{*,n-i} F_{*,i}, \quad n \geq 0, \tag{4.4.5}$$

and because of $h = f|_{z=1}$, for any integer $n \geq 0$,

$$\begin{cases} H_n = \partial_y^n h = F_{*,n}|_{z=1}; \\ H_n^{[2]} = \sum_{j=0}^{n} H_{n-i} H_i, \end{cases} \tag{4.4.6}$$

our aim is to determine all functions $F_{*,n}$ with one variable z for $n \geq 0$.

For convenience in the usage of equation (4.4.3), we notice that

$$[1+f]_n = \partial_y^n (1+f) = \begin{cases} 1 + F_{*,0}, & \text{when } n = 0; \\ F_{*,n}, & \text{when } n \geq 1, \end{cases} \tag{4.4.7}$$

and

$$[1+f]_n^{[2]} = \partial_y^n (1+f)^2 = \begin{cases} (1 + F_{*,0})^2, & \text{when } n = 0; \\ 2F_{*,n} + F_{*,n}^{[2]}, & \text{when } n \geq 1. \end{cases} \tag{4.4.8}$$

From (4.4.6), we have

$$[1+h]_n = \partial_y^n (1+h) = \begin{cases} 1 + H_0, & \text{when } n = 0; \\ H_n, & \text{when } n \geq 1, \end{cases} \tag{4.4.9}$$

and

$$[1+h]_n^{[2]} = \partial_y^n (1+h)^2 = \begin{cases} (1 + H_0)^2, & \text{when } n = 0; \\ 2H_n + H_n^{[2]}, & \text{when } n \geq 1. \end{cases} \tag{4.4.10}$$

Because of

$$\begin{aligned} [(f - zy)(z(1+h)^2 - (1+f)^2)]_0 &= [f - zy]_0 [z(1+h)^2 - (1+f)^2]_0 \\ &= [f]_0 (z[(1+h)^2]_0) - [(1+f)^2]_0, \quad \text{by (4.4.10) and (4.4.8)}, \\ &= F_{*,0}(z(1+H_0)^2 - (1+F_{*,0})^2) \end{aligned}$$

and

$$[zy(f-h)]_0 = 0.$$

From equation (4.4.3), we have

$$F_0\big(z(1+H_0)^2 - (1+F_0)^2\big) = 0. \tag{4.4.11}$$

It is seen that the initial condition of equation (4.4.3) is satisfied, *i. e.*,

$$F_{*,0} = 0 \Longrightarrow \begin{cases} H_0 = 0 \Longrightarrow H_0^{[2]} = 0 \\ F_{*,0}^{[2]} = 0. \end{cases} \tag{4.4.12}$$

For any integer $n \geq 1$, because of

$$\begin{aligned}
&[(f-zy)(z(1+h)^2-(1+f)^2)]_n \\
&\quad = \sum_{i=0}^{n} [f-zy]_i [z(1+h)^2-(1+f)^2]_{n-i}, \quad \text{by (4.4.12)}, \\
&\quad = \begin{cases} (F_{*,1}-z)[z(1+h)^2-(1+f)^2]_0 = (F_{*,1}-z)(z-1), \\ \quad \text{when } n=1; \\ F_{*,n}(z-1) + \sum_{i=1}^{n-1} F_{*,n-i}(z[1+h]_i^{[2]} - [1+f]_i^{[2]}), \\ \quad \text{when } n \geq 2 \end{cases}
\end{aligned}$$

and

$$[zy(f-h)]_n = z(F_{n-1} - H_{n-1}),$$

from equation (4.4.3), we have

$$\begin{cases} (F_{*,1}-z)(z-1) = z(F_0 - H_0) \Longrightarrow F_{*,1} - z = 0 \\ \quad \Longrightarrow F_1 = z, \quad \text{when } n = 1; \\ F_{*,2}(z-1) = z(F_1 - H_1) \Longrightarrow F_{*,2}(z-1) = z(z-1) \\ \quad \Longrightarrow F_2 = z, \quad \text{when } n = 2; \\ F_{*,n}(z-1) = z(F_{*,n-1} - H_{n-1}) + \sum_{i=1}^{n-2} F_{*,n-i}([1+f]_i^{[2]} - z[1+h]_i^{[2]}) \\ \quad \Longrightarrow F_{*,n} = z\dfrac{F_{*,n-1} - H_{n-1}}{z-1} + \sum_{i=1}^{n-2} F_{*,n-i}\dfrac{[1+f]_i^{[2]} - z[1+h]_i^{[2]}}{z-1}, \\ \quad \text{when } n \geq 3. \end{cases} \tag{4.4.13}$$

Theorem 4.4.1. *Equation* (4.4.2) *is well-defined on* $\mathcal{R}\{z,y\}$.

Proof. From (4.4.4) and (4.4.10), all $F_{*,n}$, $n \geq 2$, are determined by F_i, $0 \leq i \leq n-1$. It is easy to see that $F_{*,n} \in \mathcal{R}\{z\} \subseteq \mathcal{R}\{z,y\}$. Therefore, $f \in \mathcal{R}\{z,y\}$ is a solution of equation (4.4.3). Furthermore, $F_{*,n}$, $n \geq 1$, is determined by the value of F_0. From the initial condition of equation (4.4.3), this solution is the only one. □

Assume that $F_{*,n}$ is a polynomial of z with degree m_n, *i. e.*, it has the form

$$F_{*,n} = \sum_{j=0}^{m_n} F_{j,n} z^j \tag{4.4.14}$$

where $F_{j,n} \in \mathcal{R}$, $0 \le j \le m_n$, $n \ge 1$.

From (4.4.6),

$$\begin{aligned} F_{n-1} - H_{n-1} &= \sum_{j=1}^{m_{n-1}} F_{j,n-1}(z^j - 1) \\ &= (z-1) \sum_{j=1}^{m_{n-1}} F_{j,n-1}(1 + z + \cdots + z^{j-1}) \\ &= (z-1) \sum_{k=0}^{m_{n-1}-1} \left(\sum_{j=k+1}^{m_{n-1}} F_{j,n-1} \right) z^k. \end{aligned} \tag{4.4.15}$$

On one hand, from (4.4.4) and (4.4.6),

$$\begin{aligned} F_{*,i}^{[2]} - zH_i^{[2]} &= \sum_{j=1}^{i} (F_{*,j}F_{*,i-j} - zH_jH_{i-j}) \\ &= z(z-1) \sum_{j=1}^{i} \sum_{t=0}^{m_j+m_{i-j}-1} \left(\sum_{s=t+1}^{m_j+m_{i-j}} \Lambda_{j,i}^{(s)} \right) z^t, \end{aligned}$$

where

$$\Lambda_{j,i}^{(s)} = \begin{cases} 0, & \text{when } 0 \le s < 2; \\ \sum_{\substack{k+l=s \\ 1\le k\le m_j \\ 1\le l\le m_{i-j}}} F_{k,j}F_{l,i-j}, & \text{when } s \ge 2, \end{cases} \tag{4.4.16}$$

and

$$\begin{aligned} F_{*,i} - zH_i &= \sum_{j=1}^{m_i} F_{j,i}(z^j - z) \\ &= z(z-1) \sum_{j=2}^{m_i} F_{j,i}(1 + z + \cdots + z^{j-2}) \\ &= z(z-1) \sum_{t=0}^{m_i-2} \left(\sum_{j=t+2}^{m_i} F_{j,i} \right) z^t. \end{aligned}$$

From (4.4.8) and (4.4.10),

$$\begin{aligned} [1+f]_i^{[2]} - z[1+h]_i^{[2]} &= 2(F_{*,i} - zH_i) + (F_{*,i}^{[2]} - zH_i^{[2]}) \\ &= z(z-1) \left(2 \sum_{t=0}^{m_i-2} \left(\sum_{j=t+2}^{m_i} F_{j,i} \right) z^t + \sum_{j=1}^{i} \sum_{t=0}^{m_j+m_{i-j}-1} \left(\sum_{s=t+1}^{m_j+m_{i-j}} \Lambda_{j,i}^{(s)} \right) z^t \right). \end{aligned} \tag{4.4.17}$$

By (4.4.15) and (4.4.17), (4.4.13) leads to

$$F_{*,n} = \begin{cases} 0, & \text{when } n = 0; \\ z, & \text{when } n = 1, 2, 3; \\ z \sum_{k=0}^{m_{n-1}-1} (\sum_{j=k+1}^{m_{n-1}} F_{j,n-1}) z^k \\ \quad + z \sum_{i=1}^{n-2} F_{n-i} (2 \sum_{t=0}^{m_i - 2} (\sum_{j=t+2}^{m_i} F_{j,i}) z^t \\ \quad + \sum_{j=1}^{i} \sum_{t=0}^{m_j + m_{i-j} - 1} A_{j,i} z^t), \\ \quad \text{when } n \geq 4, \end{cases} \tag{4.4.18}$$

where

$$A_{j,i} = \sum_{s=t+1}^{m_j + m_{i-j}} \Lambda_{j,i}^{(s)}, \tag{4.4.19}$$

and $\Lambda_{j,i}^{(s)}$ is given by (4.4.16).

From (4.4.18) and (4.4.19), it is seen that $F_{*,n}$, $n \geq 0$, provides an expression of the solution $f \in \mathcal{R}\{z, y\}$ of equation (4.4.2).

Lemma 4.4.2. *For any integer $n \geq 2$, $F_{*,n}$ is a polynomial of z with degree $\lfloor n/2 \rfloor$.*

Proof. For a polynomial P of z, denote by $\mu(P)$ the degree of P, we show $m_n = \mu(F_{*,n}) = \lfloor n/2 \rfloor$. From (4.4.5),

$$\mu(F_{*,n}^{[2]}) = \max_{0 \leq i \leq n} (m_{n-i} + m_i). \tag{4.4.20}$$

It is easily seen that when $n = 2$ and 3, the conclusion is checked to be true. We proceed by induction on $n \geq 4$, assume for any $0 \leq i \leq n-1$, $\mu(F_{*,i}) = m_i = \lfloor i/2 \rfloor$, to prove $m_n = \lfloor n/2 \rfloor$.

From (4.4.5), it is seen that $\mu(F_{*,n}^{[2]}) \leq \max\{\mu(F_{n-i}) + \mu(F_{*,i}) \mid 0 \leq i \leq n\} = \lfloor n/2 \rfloor$. Because of $F_{*,0} = 0$, $\lfloor n/2 \rfloor = \max\{\mu(F_{*,n-i}) + \mu(F_{*,i}) \mid 1 \leq i \leq n-1\} = \lfloor n/2 \rfloor$. By the assumption,

$$\lfloor n/2 \rfloor \leq \max_{1 \leq i \leq n-1} \{\mu(F_{*,n-i}) + \mu(F_{*,i}) \leq \lfloor (n-i)/2 \rfloor + \lfloor i/2 \rfloor\} \leq \lfloor n/2 \rfloor.$$

On the other hand, for $n \geq 4$, from the induction assumption, by (4.4.17), we have

$$\begin{aligned} \mu(F_{*,n}) &= \mu\left(\frac{[1+f]_{n-1}^{[2]} - z[1+h]_{n-1}^{[2]}}{z-1} \right) = 1 + \mu(F_{*,n-2}^{[2]}) \\ &= 1 + \lfloor (n-1)/2 \rfloor = 1 + \lfloor n/2 \rfloor - 1 = \lfloor n/2 \rfloor. \end{aligned}$$

From (4.4.13), the conclusion is drawn. □

On the basis of this lemma, in (4.4.18) and (4.4.19), all m_i, $0 \leq i \leq n$, can be replaced by $\lfloor i/2 \rfloor$.

Lemma 4.4.3. *For any integer $n \geq 2$, all coefficients of polynomial $F_{*,n}$ are non-negative integers.*

Proof. From (4.4.18) and (4.4.19), by induction on n, the conclusion can be drawn. □

Theorem 4.4.4. *The solution $\hat{f}$ of equation (4.4.1) determined by $\hat{F}_{*,n}$ for $n \geq 0$ has a recursion in the form of a sum with all terms positive,*

$$\hat{F}_{*,n} = \begin{cases} 0, & \text{when } n = 0; \\ x^2, & \text{when } n = 1,2,3; \\ x^2 \sum_{k=0}^{m_{n-1}-1} (\sum_{j=k+1}^{\lfloor (n-1)/2 \rfloor} \hat{F}_{j,n-1}) x^{2k} \\ \quad + x^2 \sum_{i=1}^{n-2} \hat{F}_{*,n-i} (2 \sum_{t=0}^{\lfloor (i-2)/2 \rfloor} (\sum_{j=t+2}^{\lfloor i/2 \rfloor} \hat{F}_{j,i}) x^{2t} \\ \quad + \sum_{j=1}^{i} \sum_{t=0}^{\lfloor j/2 \rfloor + \lfloor (i-j)/2 \rfloor - 1} \hat{A}_{j,i} x^{2t}), \\ \qquad \text{when } n \geq 4, \end{cases} \tag{4.4.21}$$

where

$$\hat{A}_{j,i} = \sum_{s=t+1}^{\lfloor j/2 \rfloor + \lfloor (i-j)/2 \rfloor} \hat{\Lambda}_{j,i}^{(s)} \tag{4.4.22}$$

and

$$\Lambda_{j,i}^{(s)} = \begin{cases} 0, & \text{when } s = 0 \text{ and } 1; \\ \sum_{\substack{k+l=s \\ 1 \leq k \leq \lfloor j/2 \rfloor \\ 1 \leq l \leq \lfloor (i-j)/2 \rfloor}} \hat{F}_{k,j} \hat{F}_{l,i-j}, & \text{when } s \geq 2. \end{cases} \tag{4.4.23}$$

Proof. By considering the relationship between the solution f of equation (4.4.2) and the solution $\hat{f}$ of equation (4.4.1) and Lemma 4.4.2, from (4.4.18) and Lemma 4.4.3, the conclusion is drawn. □

Example 1. Isomorphic classification of non-separable Euler planar rooted maps by size and root-vertex valency. As a matter of fact, in (4.4.21), $\hat{F}_{m,n}$ is the number of isomorphic classes of non-separable Euler planar rooted maps with size n and root-vertex valency $2m$. Figures 4.4.1–4.4.3 provide, respectively, the cases for sizes 1–3, 4–5 and 6. For instance, from Figure 4.4.3, it is seen that

$$\begin{aligned} &(1T_{2,6}) + (2T_{4,6}) + (2T_{2,6} + 2T_{4,6}) + (2T_{2,6} + 4T_{4,6}) + (1T_{2,6} + 4T_{4,6}) + (T_{6,6}) \\ &\quad = 6T_{2,6} + 12T_{4,6} + 1T_{6,6}. \end{aligned}$$

We might state that in non-separable Euler planar rooted maps of size 6, the root-vertex valencies have three possibilities: 2, 4 and 6. They have, respectively, 6, 12 and 1 isomorphic classes.

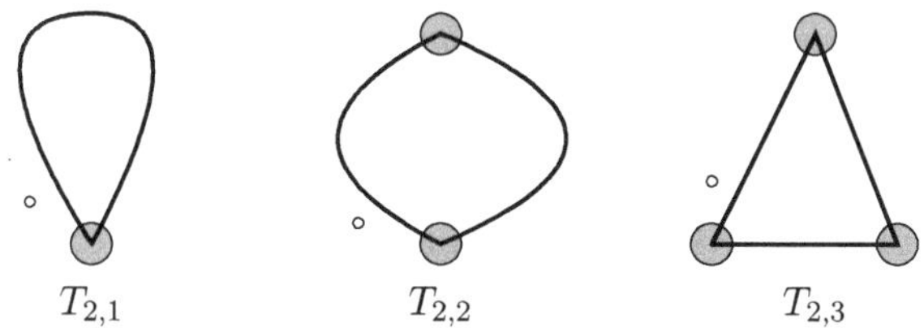

Figure 4.4.1: Non-separable Euler planar rooted maps of sizes 1–3.

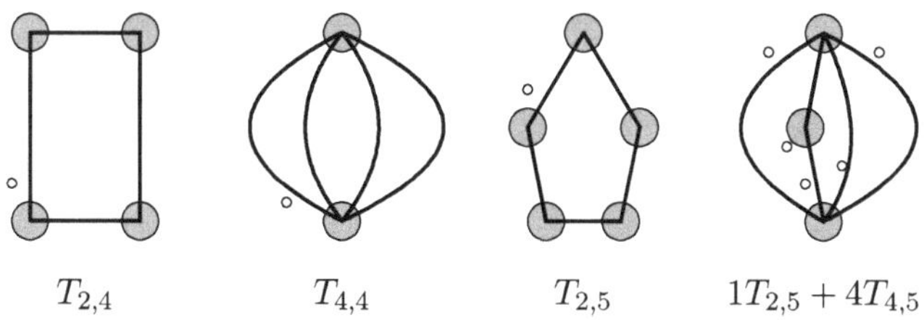

Figure 4.4.2: Non-separable Euler planar rooted maps of sizes 4–5.

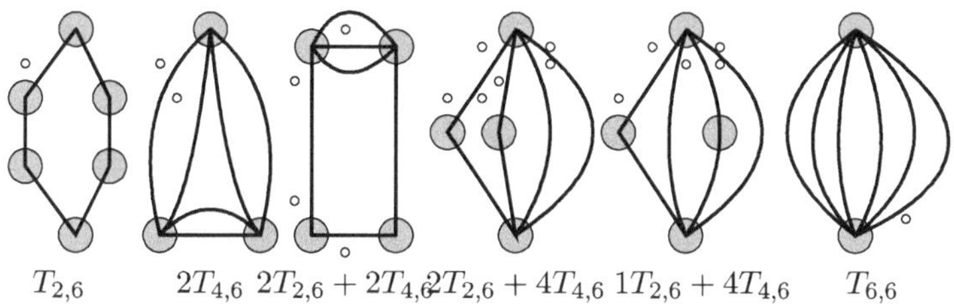

Figure 4.4.3: Non-separable Euler planar rooted maps of size 6.

4.5 Notes

4.5.1. In solving an equation of a function with several variables, a universal principle is available to find a way to reduce variables, and/or decrease its degree so that the equation is transformed into one with less variables, or a system of equations with less degree. How could we seek a proper manner for employing the principle? This depends on the level of understanding the structures of the equation considered.

Most equations encountered in this book still involve some parts of the unknown or undeterminate. In order to extract the solution needed for the equation as an explicit expression (or abbreviated, as explicision), via a certain number of characteristic curves excluding the whole unknown, to evaluate parts of the unknown such that the equation is transformed into an ordinary one without any part of the unknown. The Tutte quadratic method, or multi-root method, is an example only for quadratic equations. See Tutte WT [80–84].

However, in this book, all equations considered, or relevant, are addressed without usage of this method. In consequence, the solutions with its companion parts are evaluated at the same time.

4.5.2. A linear equation of an unknown with a companion part in variables comes from the research of root-isomorphic classification for outer planar maps; see Liu YP [23–26]. Based on these is the method of *characteristic curves*, as shown by (4.2.14) in 4.2. It is different from Tutte's.

4.5.3. If f in equation (4.1.5) is replaced by

$$g = \frac{f - x^2y}{x}, \quad i.\ e., \quad f = xg - (xy)^2, \tag{4.5.1}$$

then we have

$$xg = \frac{(xy)g + (xy)^2}{1 - (xy) - g} - (xy)g - (xy)^2. \tag{4.5.2}$$

By substituting $z = xy$, we have

$$\begin{cases} (x + z)g^2 - (x - xz - 2z^2)g + z^2 = 0; \\ g|_{x=0,z=0}. \end{cases} \tag{4.5.3}$$

Theorem 4.5.1. *Equation* (4.5.3) *is well-defined in* $\mathcal{R}\{x, z\}$. *Its solution* g, *determined by* $G_n = \partial_z^n g (n \geq 0) \in \mathcal{R}\{x\}$, *is of the form of a sum with terms positive,*

$$xG_n = \begin{cases} 0, & \text{when } 0 \leq n \leq 2; \\ 0, & \text{when } n = 3; \\ x \sum_{i=3}^{n-3} G_{*,i}G_{*,n-i} + \sum_{i=3}^{n-4} G_{*,i}G_{*,n-1-i} + xG_{*,n-1} + 2G_{*,n-2}, \\ \quad \text{when } n \geq 4. \end{cases} \tag{4.5.4}$$

Proof. The first conclusion is from Theorem 4.4.1. The second conclusion is drawn similarly to the proof of Theorem 4.4.4. □

4.5.4. In Liu YP [51] (equation (8.1.15), p. 211) a very early version in [13] (equation (3.8), 1984), one might find the equation for $f \in \mathcal{R}\{x, y\}$,

$$\begin{cases} x^2y(1 - x)f^2 - (1 - x + x^2y)f + xyh + (1 - x) = 0; \\ f|_{x=0,y=0} = 1, \end{cases} \tag{4.5.5}$$

where $h = f(1, y) \in \mathcal{R}\{y\}$.

By transforming equation (4.5.5) into its equivalent form,

$$\begin{cases} f = 1 + \dfrac{xy}{1 - x}(h - xf) + x^2yf^2; \\ f|_{y=0 \Rightarrow x=0} = 1, \end{cases} \tag{4.5.6}$$

similar to Theorems 4.3.2 and 4.3.5, that equation (4.5.6) is well-defined and its solution has the form of a finite sum with all terms positive can also be seen. In fact, the solution is determined by $F_{*,n}$ $(n \geq 1)$, which is a polynomial of x with degree not greater than $2n$ and all coefficients in $\mathcal{Z}_+$.

4.5.5. From the decomposition principle presented by Liu YP [21] (1986), equation (4.4.1) can easily be derived. In Liu YP [40] (1992), an explicision of its solution in summation-free form is also provided. Equation (1) in Liu YP [42] (1992) is just equation (4.4.23).

4.5.6. From (4.4.21), by substituting $z = x^2$, a solution of equation (4.4.1) is done in $\mathcal{R}\{x, y\}$. Although recursion is very convenient for efficientization and intelligentization, explicision is necessary for use of computers. For convenience, an explicision of an implicit function is also necessary. We need to get an explicision of the solution f of equation (4.4.1) determined by $F_{*,n}$ $(n \geq 0)$ without the knowledge of $F_{*,i}$ $(0 \leq i \leq n-1)$.

Let $g = f + 1$ and $l = h + 1$, for getting an equivalence of equation (4.4.1),

$$\begin{cases} g^3 - (1 + x^2y)g^2 + x^2(y - l^2)g + x^2(l^2 - yl + x^2yl^2) = 0; \\ g|_{x=y=0} = 1, \end{cases} \tag{4.5.7}$$

where $l = g(1, y) \in \mathcal{R}\{y\}$.

Because equation (4.4.21) is of degree three with a part l of the unknown f, certain complications are involved in the evaluation. In order to determine l, a parameter $q = q(y) \in \mathcal{R}\{x, y\}$ is introduced satisfying the relation

$$q - 1 = l(yq^2) \tag{4.5.8}$$

and we have the equation

$$\begin{cases} x^2y(x^2 - 1)p^2 + (x^2y - x^2 + 1)p + x^2 - 1 - x^2yq = 0; \\ p|_{x=y=0} = 1, \end{cases} \tag{4.5.9}$$

to determine $q = p(1, y)$.

On $\mathcal{R}\{x, y\}$, whenever the substitution $z = x^2$ is employed, equation (4.5.9) becomes

$$\begin{cases} zy(z - 1)f^2 + (zy - z + 1)f + z - 1 - zyq = 0; \\ f|_{x=y=0} = 1, \end{cases} \tag{4.5.10}$$

where $q = f(1, y)$.

In Section 4.3, the solution of equation (4.3.3) is shown by (4.3.6). The case of $z = 1$ is just q.

5 Difference equations

5.1 With straight difference of one variable

Consider the equation for $f \in \mathcal{R}\{x\}$

$$\begin{cases} f(1 - xy\delta_{1,x}(xf)) = 1; \\ f|_{y=0 \Rightarrow x=0} = 1, \end{cases} \tag{5.1.1}$$

where $\delta_{1,x}(xf) = ((xf)_{x=1} - (xf))/(1-x)$ is the straight difference of xf between 1 and x. This is equation (12) in Introduction when $a = c = d = 1$ it being meaningful in the classification for loopless planar maps.

Since only one variable x is involved, the equation is said to be in *straight difference form of one variable*.

For any integer $n \geq 0$, let $F_{*,n} = [f]_n = \partial_y^n f$, then

$$\partial_y^n f(1 - xy\delta_{1,x}(xf)) = \sum_{i=0}^{n} F_{*,i}[1 - xy\delta_{1,x}(xf)]_{n-i}. \tag{5.1.2}$$

Write $h = f|_{x=1}$ and $H_n = [h]_n = \partial_y^n h$, $n \geq 0$, then

$$[1 - xy\delta_{1,x}(xf)]_j = \left[1 - xy\frac{h - xf}{1-x}\right]_j. \tag{5.1.3}$$

When $j = 0$,

$$[1 - xy\delta_{1,x}(xf)]_0 = 1. \tag{5.1.4}$$

When $j \geq 1$,

$$\begin{aligned} [1 - xy\delta_{1,x}(xf)]_j &= -\left[xy\frac{h - xf}{1-x}\right]_j \\ &= -\frac{x}{1-x}[h - xf]_{j-1} \\ &= -\frac{x}{1-x}(H_{j-1} - xF_{*,j-1}). \end{aligned} \tag{5.1.5}$$

Lemma 5.1.1. *Equation* (5.1.1) *for* $f \in \mathcal{R}\{x, y\}$ *is equivalent to the equation system*

$$\begin{cases} \partial_y^0(f(1 - xy\delta_{1,x}(xf))) = 1, & n = 0; \\ \partial_y^n(f(1 - xy\delta_{1,x}(xf))) = 0, & n \geq 1m \end{cases} \tag{5.1.6}$$

for $\{F_{*,n} = \partial_y^n f \mid n \geq 0\} \subseteq \mathcal{R}\{x\}$.

https://doi.org/10.1515/9783110625837-005

Proof. Because of

$$\partial_y^0(f(1 - xy\delta_{1,x}(xf))) = [f(1 - xy\delta_{1,x}(xf))]_0$$

and (5.1.2),

$$\begin{aligned}[f(1 - xy\delta_{1,x}(xf))]_0 &= F_{*,0}[1 - xy\delta_{1,x}(xf)]_0, \quad (5.1.2),\\ &= F_{0,0}, \quad \text{from the initial condition,}\\ &= 1.\end{aligned} \tag{5.1.7}$$

This is the case of $n = 0$ in (5.1.6).

For any integer $n \geq 1$, from (5.1.2) and (5.1.6),

$$\begin{aligned}\partial_y^n(f(1 - xy\delta_{1,x}(xf))) &= [f(1 - xy\delta_{1,x}(xf))]_n, \quad \text{from (5.1.2)},\\ &= \sum_{i=0}^{n} F_i[1 - xy\delta_{1,x}(xf)]_{n-i}, \quad \text{from (5.1.4) and (5.1.5)},\\ &= F_{*,n} - \frac{x}{1-x}\sum_{i=0}^{n-1} F_{*,i}(H_{n-i-1} - xF_{*,n-i-1})\\ &= 0.\end{aligned}$$

This is the general case of equation (5.1.1), *i. e.*,

$$F_{*,n} = \frac{x}{1-x}\sum_{i=0}^{n-1} F_{*,i}(H_{n-i-1} - xF_{*,n-i-1}). \tag{5.1.8}$$

Therefore, the conclusion is drawn. □

Observation 5.1.2. *For integer* $j \geq 0$, $(1 - x)|(H_j - xF_{*,j})$.

Proof. For $j \geq 0$, it is seen that $H_j = F_{0,j} + F_{1,j} + \cdots + F_{i,j} + \cdots$ where $F_{*,j} = F_{0,j} + F_{1,j}x + \cdots + F_{i,j}x^i + \cdots$. When $i = j = 0$, because of $F_{0,0} - xF_{0,0} = 1 - x$, the conclusion is true. For $i, j \geq 1$, because of

$$F_{i,j} - xF_{i,j}x^i = F_{i,j}(1 - x^{i+1}) = F_{i,j}(1 - x)\left(\sum_{l=0}^{i} x^l\right),$$

we have $(1 - x)|(F_{i,j} - xF_{i,j}x^i)$. From

$$H_j = \sum_{i\geq 0} F_{i,j}$$

we have $(1 - x)|(H_j - xF_{*,j})$. This is the conclusion. □

This observation enables us to see that, for any integer $n \geq 0$,

$$\frac{H_n - xF_{*,n}}{1-x} \in \mathcal{R}\{x\}$$

is a polynomial of x.

Observation 5.1.3. *For integer $j \geq 0$, $F_{*,j}$ is a polynomial of x with degree j.*

Proof. When $n = 0$, $F_{*,0} = \delta_{*,0}F_{0,0} = 1$ is a polynomial of x with degree $0 = n$. By induction, for $n \geq 1$, assume for any $0 \leq k \leq n-1$, $F_{*,k}$ is a polynomial of x with degree k; we prove $F_{*,n}$ is a polynomial of x with degree n. Let $d_F = d(F)$ be the degree of F and write $d_k = d(F_{*,k})$. From (5.1.8), we have

$$\begin{aligned} d(F_{*,j}) &= 1 + d\left(\sum_{i=0}^{j-1} F_{*,i}\frac{H_{j-i-1} - xF_{*,j-i-1}}{1-x}\right) \\ &= 1 + d(F_{*,j-1}), \quad \text{by induction assumption,} \\ &= 1 + (j-1) = j. \end{aligned} \tag{5.1.9}$$

This is the conclusion. □

This observation enables us only to discuss $F_{m,n}$ for $0 \leq m \leq n$ whenever $n \geq 1$ is given. Moreover, for $n \geq 1$, all $F_{*,n}$ are seen without a constant term not zero. So, it is only necessary to consider $n \geq 1$.

Theorem 5.1.4. *Equation* (5.1.1) *is well-defined in $\mathcal{R}\{x, y\}$.*

Proof. Because of a solution f of equation (5.1.1) is determined by $F_{*,n}$ for $n \geq 0$ in Lemma 5.1.1, (5.1.7) and (5.1.8), one is led to the conclusion. □

Because there is no constant in $F_{*,n}$ for $n \geq 1$, (5.1.9) enables us to write

$$\begin{cases} F_{*,n} = F_{1,n}x + F_{2,n}x^2 + \cdots + F_{n,n}x^n, ; \\ H_n = F_{1,n} + F_{2,n} + \cdots + F_{n,n}, \end{cases} \tag{5.1.10}$$

for $\{F_{j,n} \mid n \geq j \geq 1\} \subseteq \mathcal{R}$.

For any integer $j \geq 0$, from (5.1.10),

$$\begin{aligned} H_j - xF_{*,j} &= \sum_{l=1}^{j} F_{l,j}(1 - x^{j+1}) \\ &= (1-x)\sum_{l=1}^{j} F_{l,j}\left(\sum_{s=0}^{j} x^s\right) \\ &= (1-x)\left(H_j + \sum_{\substack{l \leq s \leq j \\ 1 \leq l \leq j}} F_{s,j}x^l\right), \end{aligned}$$

and hence

$$\frac{H_j - xF_{*,j}}{1-x} = H_j + \sum_{\substack{l \leq s \leq j \\ 1 \leq l \leq j}} F_{s,j}x^l. \tag{5.1.11}$$

By substituting (5.1.11) into (5.1.8),

$$F_{*,n} = x \sum_{i=0}^{n-1} F_{*,i}\Big(H_{n-i-1} + \sum_{\substack{l\le s\le n-i-1\\ 1\le l\le n-i-1}} F_{s,n-i-1}x^l\Big). \tag{5.1.12}$$

From (5.1.7) and (5.1.12),

$$F_{*,n} = \begin{cases} 1, & \text{when } n = 0;\\ x\sum_{i=0}^{n-1} F_{*,i}(H_{n-i-1} + \sum_{l=1}^{n-i-1} \Lambda_{l,n-i-1}x^l), & \text{from } F_{0,k} = 0\ (k \ge 1),\\ = x(F_{*,n-1} + \sum_{\substack{0\le l\le n-i-1\\ 0\le i\le n-2}} F_{*,i}\Lambda_{l,n-i-1}x^l), & \text{when } n \ge 1, \end{cases} \tag{5.1.13}$$

where

$$\Lambda_{l,n-i-1} = \sum_{s=l}^{n-i-1} F_{s,n-i-1}. \tag{5.1.14}$$

Theorem 5.1.5. *The solution f of equation* (5.1.1) *on* $\mathcal{R}\{x,y\}$ *is determined by* $F_{*,n} = \partial_y^n f \in \mathcal{R}\{x\}$ *for* $n \ge 0$ *i in the form of a sum with finite terms all positive as*

$$F_{*,n} = \begin{cases} 1, & \textit{when } n = 0;\\ x(F_{*,n-1} + \sum_{\substack{0\le l\le n-i-1\\ 0\le i\le n-2}} F_{*,i}\Lambda_{l,n-i-1}x^l), & \textit{when } n \ge 1, \end{cases} \tag{5.1.15}$$

where $\Lambda_{l,n-i-1}$ *is given in* (5.1.14).

Proof. This is a result of (5.1.13) and (5.1.14). □

Example 1. Classification of loopless planar rooted maps by size and root-vertex valency. In Liu YP [36], one finds the equation

$$f = 1 + \frac{xy\delta_{1,x}(xf)}{1 - xy\delta_{1,x}(xf)}. \tag{5.1.16}$$

One of its solutions is just the enufunction of root-isomorphic classes of loopless planar maps with size and root-vertex valency as two parameters. However, attention should be paid to the fact that u, v, h and h_1 in [36] are, respectively, x, yf and h here.

Because of the existence of $(1 - xy\delta_{1,x}(xf))^{-1}$ on $\mathcal{R}\{x,y\}$, equation (5.1.16) is equivalent to equation (5.1.1). See Figure 5.1.1 for sizes between 0 and 2.

In Figure 5.1.2, the distinct root-isomorphic classes of loopless planar maps with size 3 are provided by

$$\begin{aligned}(T_{1,3} + T_{3,3}) &+ (T_{1,3} + 2T_{2,3}) + (T_{1,3} + 2T_{2,3} + 3T_{3,3}) + (T_{3,3}) + (T_{2,3})\\ &= (1 + 1 + 1)T_{1,3} + (2 + 2 + 1)T_{2,3} + (1 + 3 + 1)T_{3,3}\\ &= 3T_{1,3} + 5T_{2,3} + 5T_{3,3}.\end{aligned}$$

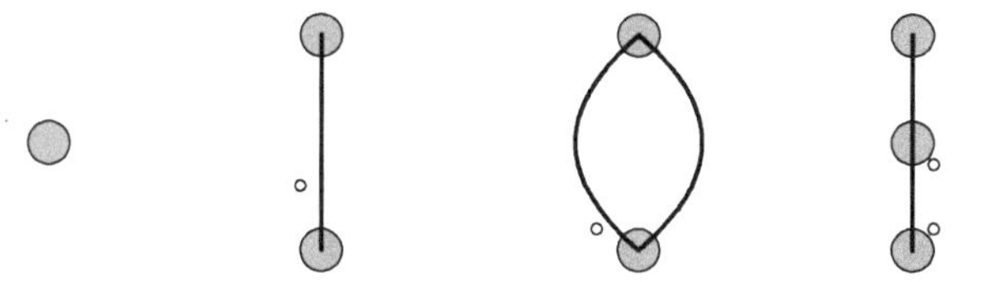

$T_{0,0}$ $T_{1,1}$ $T_{2,2}$ $T_{1,2} + T_{2,2}$

Figure 5.1.1: Classes of loopless planar rooted maps with sizes: 0–2.

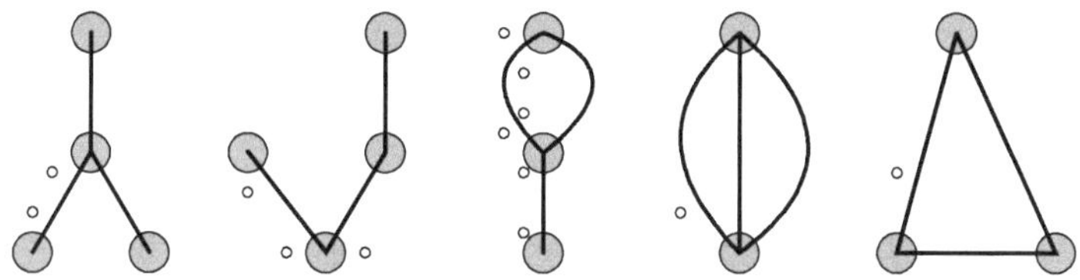

$T_{1,3} + T_{3,3}$ $T_{1,3} + 2T_{2,3}$ $T_{1,3} + 2T_{2,3} + 3T_{3,3}$ $T_{3,3}$ $T_{2,3}$

Figure 5.1.2: Classes of loopless planar rooted maps with size 3.

In Figure 5.1.3, the distinct root-isomorphic classes of loopless planar maps with size 4 are provided by

$$\begin{aligned}
&(T_{1,4} + 3T_{2,4}) + (3T_{1,4} + 2T_{2,4} + 3T_{3,4}) + (T_{1,4} + T_{4,4}) \\
&\quad + (T_{1,4} + 4T_{2,4} + 3T_{3,4}) + (T_{1,4} + 4T_{2,4} + 3T_{3,4}) \\
&\quad = (1 + 3 + 1 + 1 + 1)T_{1,4} + (2 + 2 + 4 + 4)T_{2,4} \\
&\quad + (3 + 3 + 3)T_{3,4} + T_{4,4} \\
&\quad = 7T_{1,4} + 13T_{2,4} + 9T_{3,4} + T_{4,4}.
\end{aligned}$$

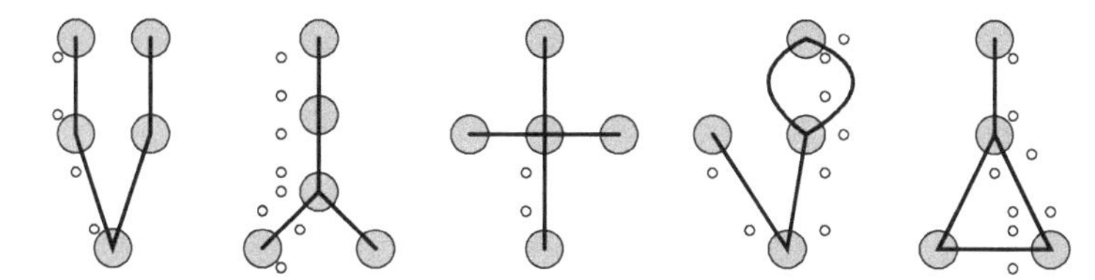

$T_{1,4} + 3T_{2,4}$ $3T_{1,4} + 2T_{2,4} + 3T_{3,4}$ $T_{1,4} + T_{4,4}$ $T_{1,4} + 4T_{2,4} + 3T_{3,4}$ $T_{1,4} + 4T_{2,4} + 3T_{3,4}$

Figure 5.1.3: Classes of loopless planar rooted maps with size 4 I.

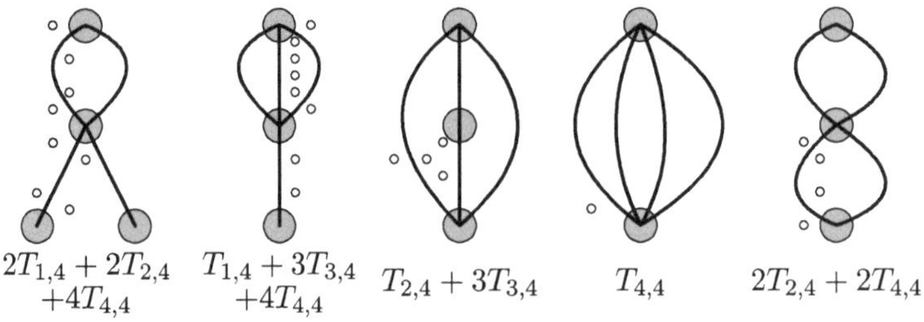

Figure 5.1.4: Classes of loopless planar rooted maps with size 4 II.

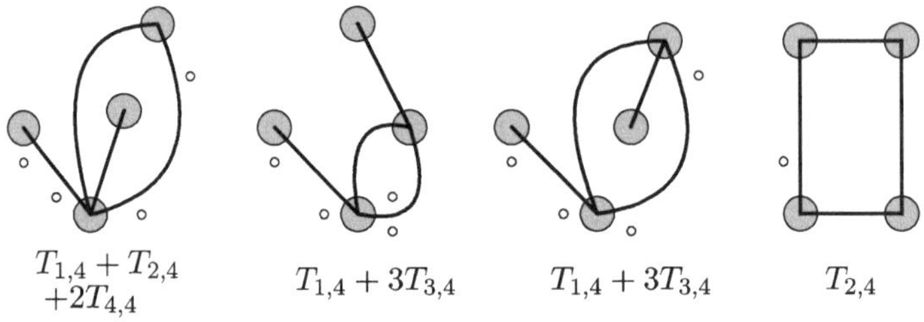

Figure 5.1.5: Classes of loopless planar rooted maps with size 4 III.

In Figures 5.1.4 and 5.1.5, the distinct non-isomorphic classes of loopless planar maps with size 4 are provided by

$$\begin{aligned}
(2T_{1,4} + 2T_{2,4} + 4T_{4,4}) &+ (T_{1,4} + 3T_{3,4} + 4T_{4,4}) \\
&+ (T_{2,4} + 3T_{3,4}) + (T_{4,4}) + (2T_{2,4} + 2T_{4,4}) \\
&= (2+1)T_{1,4} + (2+1+2)T_{2,4} + (3+3)T_{3,4} \\
&= +(4+4+1+2)T_{4,4}3T_{1,4} + 5T_{2,4} + 6T_{3,4} + 11T_{4,4},
\end{aligned}$$

$$\begin{aligned}
(T_{1,4} + T_{2,4} + 2T_{4,4}) &+ (T_{1,4} + 3T_{3,4}) + (T_{1,4} + 3T_{3,4}) + (T_{2,4}) \\
&= (1+1+1)T_{1,4} + (1+1)T_{2,4} + (3+3)T_{3,4} + (2)T_{4,4} \\
&= 3T_{1,4} + 2T_{2,4} + 6T_{3,4} + 2T_{4,4}.
\end{aligned}$$

In Figures 5.1.3–5.1.5, the distinct root-isomorphic classes of loopless planar maps with size 4 are provided by

$$13T_{1,4} + 20T_{2,4} + 21T_{3,4} + 14T_{4,4}.$$

This is $F_{*,n}$ for $n = 4$ in (5.1.13), *i. e.*,

$$F_4 = 13x + 20x^2 + 21x^3 + 14x^4.$$

5.2 Form of several straight differences

Consider the equation for $f \in \mathcal{R}\{y,z,t\}$

$$\begin{cases} f(1+yzt(f_{t=1}+f_{z=1})) = 1+yzt(\delta_{1,t}(tf)+\delta_{1,z}(zf)); \\ f|_{y=0\Rightarrow z=t=0} = 1. \end{cases} \tag{5.2.1}$$

This is equation (13) in Introduction when $a = b = c = d = 1$ because it is meaningful in the *dichrosum* (*i. e.*, dichromate sum) for ordinary planar maps.

In the equation, both a straight difference for z and a straight difference for t occur. So, this equation is said to be in the *form of several straight differences*.

Because of $f \in \mathcal{R}\{y,z,t\}$, whenever all $F_{*,n} = \partial_y^n f \in \mathcal{R}\{z,t\}$ for $n \geq 0$ are determined by equation (5.2.1), a solution of the equation is easily found.

For convenience, let us write

$$\begin{cases} [f|_{z=1}]_n = [f]_n|_{z=1} = F_{*,n}|_{z=1}; \\ [f|_{t=1}]_n = [f]_n|_{t=1} = F_{*,n}|_{t=1}, \end{cases} \tag{5.2.2}$$

and

$$\begin{cases} [\delta_{1,z}(zf)]_n = \dfrac{1}{1-z}([f|_{z=1}]_n - z[f]_n) \\ \qquad = \dfrac{1}{1-z}(F_{*,n}|_{z=1} - zF_{*,n}); \\ [\delta_{1,t}(tf)]_n = \dfrac{1}{1-t}([f|_{t=1}]_n - z[f]_n) \\ \qquad = \dfrac{1}{1-t}(F_{*,n}|_{t=1} - zF_{*,n}). \end{cases} \tag{5.2.3}$$

On the basis of (5.2.2) and (5.2.3), a procedure can be established to determine all $F_{*,n}$ for $n \geq 1$ from F_0 which is known by the initial condition as follows.

When $n = 0$,

$$y^0: \quad [f(1+yzt(f|_{z=1}+f|_{t=1}))]_0 = [1+yzt(\delta_{1,z}(zf)+\delta_{1,t}(tf))]_0.$$

Because of

$$[f(1+yzt(f|_{z=1}+f|_{t=1}))]_0 = [f]_0[1+yzt(f|_{z=1}+f|_{t=1})]_0 = F_0$$

and

$$[1+yzt(\delta_{1,z}(zf)+\delta_{1,t}(tf))]_0 = 1,$$

we have

$$F_{*.0} = 1 \Longrightarrow F_{*.0}|_{z=1} = 1 \text{ and } F_{*.0}|_{t=1} = 1. \tag{5.2.4}$$

When $n = 1$,

$$y^1: \quad [f(1 + yzt(f|_{z=1} + f|_{t=1}))]_1 = [1 + yzt(\delta_{1,z}(zf) + \delta_{1,t}(tf))]_1.$$

Because of

$$\begin{aligned}[f(1 + yzt(f|_{z=1} + f|_{t=1}))]_1 &= [f]_0[1 + yzt(f|_{z=1} + f|_{t=1})]_1 \\ &\quad + [f]_1[1 + yzt(f|_{z=1} + f|_{t=1})]_0, \quad \text{by (5.2.4)}, \\ &= zt[f|_{z=1} + f|_{t=1}]_0 + F_1, \quad \text{by (5.2.4)}, \\ &= F_1 + 2zt\end{aligned}$$

and

$$\begin{aligned}[1 + yzt(\delta_{1,z}(zf) + \delta_{1,t}(tf))]_1 &= zt[\delta_{1,z}(zf) + \delta_{1,t}(tf)]_0, \quad \text{by (5.2.3)}, \\ &= 2zt\end{aligned}$$

we have $F_1 + 2zt = 2zt$, *i. e.*,

$$F_{*,1} = 0 \implies F_{*,1}|_{z=1} = 0 \text{ and } F_{*,1}|_{t=1} = 0. \tag{5.2.5}$$

For $n \geq 2$, by equation (5.2.1), we have

$$y^n: \quad [f(1 + yzt(f|_{z=1} + f|_{t=1}))]_n = [1 + yzt(\delta_{1,z}(zf) + \delta_{1,t}(tf))]_n.$$

On the left hand side,

$$\begin{aligned}&[f(1 + yzt(f|_{z=1} + f|_{t=1}))]_n \\ &\quad = \sum_{i=0}^{n} [f]_i[1 - yzt(f|_{z=1} + f|_{t=1})]_{n-i} \\ &\quad = F_{*,n} + zt\sum_{i=0}^{n-1} F_{*,i}[f|_{z=1} + f|_{t=1}]_{n-i-1}, \quad \text{by (5.2.2)}, \\ &\quad = F_{*,n} + zt\sum_{i=0}^{n-1} F_{*,i}(F_{*,n-i-1}|_{z=1} + F_{*,n-i-1}|_{t=1}).\end{aligned}$$

On the right hand side, from $n \neq 0$,

$$\begin{aligned}&[1+yzt(\delta_{1,z}(zf) + \delta_{1,t}(tf))]_n \\ &\quad = zt[\delta_{1,z}(zf) + \delta_{1,t}(tf)]_{n-1}, \quad \text{by (5.2.3)}, \\ &\quad = zt\left(\frac{F_{*,n-1}|_{z=1} - zF_{*,n-1}}{1 - z} + \frac{F_{*,n-1}|_{t=1} - tF_{*,n-1}}{1 - t}\right).\end{aligned}$$

Therefore, we have

$$\begin{aligned}F_{*,n} = zt\Bigg(&\frac{F_{*,n-1}|_{z=1} - zF_{*,n-1}}{1 - z} + \frac{F_{*,n-1}|_{t=1} - tF_{*,n-1}}{1 - t} \\ &- \sum_{i=0}^{n-1} F_{*,i}(F_{*,n-i-1}|_{z=1} + F_{*,n-i-1}|_{t=1})\Bigg).\end{aligned} \tag{5.2.6}$$

Also, $F_{*,n} \in \mathcal{R}\{z, t\}$.

Theorem 5.2.1. *Equation* (5.2.1) *on* $\mathcal{R}\{y,z,t\}$ *is well-defined.*

Proof. It is easily seen that (5.2.4) leads to the initial condition of equation (5.2.1). From (5.2.6), for any $n \geq 1$, $F_{*,n}$ is determined by only F_i, $0 \leq i \leq n-1$. Hence, equation (5.2.1) has a solution on $\mathcal{R}\{y,z,t\}$.

From uniqueness of the procedure of finding $F_{*,n}$ for the given initial condition, the solution is the only one. □

As a matter of fact, (5.2.6) has already provided us a recursion for a finite sum with all terms positive.

Theorem 5.2.2. *The solution of equation* (5.2.1) $\mathcal{R}\{y,z,t\}$ *is* $f = 1$, *i. e.*,

$$\partial_y^n f = \begin{cases} 1, & \text{when } n = 0; \\ 0, & \text{when } n \geq 1. \end{cases} \tag{5.2.7}$$

Proof. When $n = 0$, (5.2.4) shows $F_{*,0} =$, the initial condition of equation (5.2.1). When $n = 1$, (5.2.5) shows $F_{*,1} = 0$.

For $n \geq 2$, we proceed by induction on n. Assume for any integer i, $n-1 \geq i \geq 1$, $F_{*,i} = 0$. We prove $F_{*,n} = 0$ by (5.2.6).

On account of the assumption,

$$\frac{F_{*,n-1}|_{z=1} - zF_{*,n-1}}{1-z} + \frac{F_{*,n-1}|_{t=1} - tF_{*,n-1}}{1-t} = 0$$

and

$$\begin{aligned} &\sum_{i=0}^{n-1} F_{*,i}(F_{*,n-i-1}|_{z=1} + F_{*,n-i-1}|_{t=1}) \\ &\quad = F_{*,0}(F_{*,n-1}|_{z=1} + F_{*,n-1}|_{t=1}) \\ &\quad = 0. \end{aligned}$$

Then, from (5.2.6), we have $F_{*,n} = 0$.

This is the conclusion. □

Example 1. Dichrosum equation of ordinary planar rooted maps. In Tutte WT [87] (1971), Tutte proposed the equation

$$\begin{aligned} \phi = 1 + \mu xz^2 t - xzt\left(\phi\phi_{t=1} - \frac{\phi_{t=1} - t\phi}{1-t}\right) \\ + \nu yzt^2 - yzt\left(\phi\phi_{z=1} - \frac{\phi_{z=1} - z\phi}{1-z}\right). \end{aligned} \tag{5.2.8}$$

By substituting $x = y$ into equation (5.2.8), we have

$$\begin{aligned} \phi = 1 + \mu yz^2 t - yzt\left(\phi\phi_{t=1} - \frac{\phi_{t=1} - t\phi}{1-t}\right) \\ + \nu yzt^2 - yzt\left(\phi\phi_{z=1} - \frac{\phi_{z=1} - z\phi}{1-z}\right). \end{aligned} \tag{5.2.9}$$

In fact, equation (5.2.1) is equation (5.2.9) for $\mu = \nu = 0$. Both equation (5.2.8) and equation (5.2.9) are in the form of several straight differences. Although their being well-defined can be addressed in a similar manner to that of equation (5.2.1), their solutions are much more complicated than the solution of equation (5.2.1) is.

Example 2. Dichrosum for $\mu = \nu = 0$. Because, for the dichromat (or Tutte polynomial) $\chi_{\mu,\nu}$, we have

$$\chi_{0,0}(M) = \begin{cases} 1, & \text{when } M = \vartheta, \text{ the vertex map;} \\ 0, & \text{otherwise,} \end{cases}$$

the dichrosum function of ordinary planar maps is the constant 1, the solution of equation (5.2.1). The fact in this section is proved only by Theorem 5.2.2.

5.3 With slope difference of one variable

Consider the equation for $f \in \mathcal{R}\{x, y\}$

$$\begin{cases} f = 1 + x^2yf^2 + y\partial_{1,x}(x^2f) - xyhf - (h-1)(f-1); \\ f|_{x=0\Rightarrow y=0} = 1 \quad \text{(the initial condition!),} \end{cases} \tag{5.3.1}$$

where $h = f|_{x=1} \in \mathcal{R}\{y\}$. This is equation (14) in Introduction when $a = b = c = d = 1$ because it is meaningful in the classification of simple planar maps.

On account of the slope difference for only one variable x, the equation is called a *slope difference of one variable.*

For $f \in \mathcal{R}\{x, y\}$, f is determined by

$$F_{*,n} = [f]_n = \partial_y^n f \in \mathcal{R}\{x\}, \quad n \geq 0. \tag{5.3.2}$$

Thus, the problem of evaluating f for equation (5.3.1) becomes that of extracting all $F_{*,n}$ for $n \geq$ from the initial condition of equation (5.3.1) on $\mathcal{R}\{x, y\}$.

If F_i, $0 \leq i \leq n$, are to be found, then

$$F_{*,n}^{[2]} = [f^2]_{*,n} = \partial_y^n f^2 = \sum_{i=0}^{n} F_{*,i}F_{*,n-i}. \tag{5.3.3}$$

Because of

$$[hf]_n = \partial_y^n(hf) = \sum_{i=0}^{n} [h]_i[f]_{n-i} \tag{5.3.4}$$

and

$$[h]_n = \begin{cases} F_{*,0}(= 1 = F_{*,0}|_{x=1}), & \text{when } n = 0 \text{ (the initial value!);} \\ F_{*,n}|_{x=1}, & \text{when } n \geq 1, \end{cases} \tag{5.3.5}$$

we have

$$[hf]_n = \sum_{i=0}^{n} F_i|_{x=1} F_{n-i}. \tag{5.3.6}$$

Furthermore,

$$[\partial_{1,x}(x^2 f)]_n = \frac{x}{1-x}(F_{*,n}|_{x=1} - xF_{*,n}) \tag{5.3.7}$$

and

$$\begin{aligned}[(h-1)(f-1)]_n &= \sum_{i=0}^{n} [h-1]_i [f-1]_{n-i}, \quad \text{the initial value,} \\ &= \begin{cases} 0, & \text{when } n = 0; \\ \sum_{i=1}^{n-1} [h]_i f_{n-i}, & \text{when } n \geq 1. \end{cases}\end{aligned} \tag{5.3.8}$$

On the basis of these, by equation (5.3.1),

$$\begin{aligned} y^0: \quad [f]_0 &= 1 - [(h-1)(f-1)]_0, \quad \text{by (5.3.8)}, \\ &= 1([h]_0 - 1)([f]_0 - 1), \quad \text{by (5.3.5)}, \\ &= 1 - (F_0 - 1)^2 \\ &\Longrightarrow F_0 = 1 - (F_0 - 1)^2 = 2F_0 - F_0^2 \\ &\Longrightarrow 0 = F_0(1 - F_0). \end{aligned}$$

By the initiation of equation (5.3.1), the only possibility is

$$F_0 = 1. \tag{5.3.9}$$

From equation (5.3.1), we also have

$$y^1: \quad [f]_1 = x^2[f^2]_0 + [\partial_{1,x}(x^2 f)]_0 - x[hf]_0 - [(h-1)(f-1)]_1. \tag{5.3.10}$$

By (5.3.9), $[h-1]_0 = [f-1]_0 = F_0 - 1 = 0$. For any integer $n \geq 2$,

$$[(h-1)(f-1)]_n = \sum_{i=1}^{n-1} [h]_i [f]_{n-i}. \tag{5.3.11}$$

When $n = 1$, $[(h-1)(f-1)]_1 = 0$. Thus, (5.3.10) becomes

$$[f]_1 = x^2[f^2]_0 + [\partial_{1,x}(x^2 f)]_0 - x[hf]_0.$$

By employing (5.3.3) and (5.3.7),

$$\begin{aligned} F_1 &= x^2 F_0^2 + x - xF_0^2, \quad \text{by (5.3.9)}, \\ &= x^2. \end{aligned} \tag{5.3.12}$$

If, for integer $n \geq 1$, all $F_{*,i}$, $0 \leq i \leq n-1$, are known, then, by equations (5.3.1) and (5.3.3)–(5.3.8),

$$\begin{aligned} y^n: \quad [f]_n &= x^2[f^2]_{n-1} + [\partial_{1,x}(x^2 f)]_{n-1} - x[hf]_{n-1} - [(h-1)(f-1)]_n \\ &= x^2 \sum_{i=0}^{n-1} F_i F_{n-1-i} + \frac{x}{1-x}(F_{n-1}|_{x=1} - xF_{n-1}) \\ &\quad - x \sum_{i=0}^{n-1} F_i|_{x=1} F_{n-1-i} - \sum_{i=1}^{n-1} F_i|_{x=1} F_{n-i}. \end{aligned} \tag{5.3.13}$$

Lemma 5.3.1. *For any integer $n \geq 1$, $F_{*,n}$ is a polynomial of x with degree $2n$ and minimum of degrees not less than 2.*

Proof. When $n = 1$, from (5.3.12), the conclusion is true.

For $n \geq 2$, we proceed by induction on n; assume that all $F_{*,i}$, $0 \leq i \leq n-1$, are polynomials of x with the minimum not less than 2 and maximum not greater than $2i$ of degrees. Denote by $d(P)$ the degree of the polynomial P. By the assumption,

$$\begin{aligned} d\left(\sum_{i=0}^{n-1} F_i F_{n-1-i}\right) &= 2i + 2(n-1-i) = 2(n-1), \\ d(F_{n-1}|_{x=1} - F_{n-1}) &= 1 + 2(n-1) = 2n-1, \\ d\left(\sum_{i=0}^{n-1} F_i|_{x=1} F_{n-1-i}\right) &\leq 2(n-1), \end{aligned}$$

and

$$d\left(\sum_{i=1}^{n-1} [h]_i [f]_{n-i}\right) \leq 2(n-1).$$

From (5.3.13),

$$d(F_{*,n}) = 2 + 2(n-1) = 2n. \tag{5.3.14}$$

By considering that $F_{*,n}$ has neither a term with degree 0 of x nor a term with degree 1 of x, the conclusion is drawn. □

This lemma enables us to express $F_{*,n}$, $n \geq 1$, in the form

$$F_{*,n} = \sum_{m=2}^{2n} F_{m,n} x^m, \quad F_{m,n} \in \mathcal{R}. \tag{5.3.15}$$

Hence, we have

$$\begin{aligned}\frac{F_{n-1}|_{x=1} - xF_{n-1}}{1-x} &= \frac{1}{1-x}\left(\sum_{m=2}^{2(n-1)} F_{m,n-1}(1-x^{m+1})\right)\\ &= \sum_{m=2}^{2(n-1)} F_{m,n-1}\sum_{i=0}^{m} x^i \\ &= \sum_{i=0}^{2(n-1)} \Lambda_{i,n-1}x^i\end{aligned} \tag{5.3.16}$$

where

$$\Lambda_{i,n-1} = \sum_{m=\max\{2,i\}}^{2(n-1)} F_{m,n-1}. \tag{5.3.17}$$

Theorem 5.3.2. *Equation* (5.3.1) *on* $\mathcal{R}\{x,y\}$ *is well-defined.*

Proof. Via (5.3.2)–(5.3.13), $F_{*,n}$, $n \geq 0$, as obtained provide a solution f of equation (5.3.1). Because of $F_{*,n} \in \mathcal{R}\{x\}, f \in \mathcal{R}\{x,y\}$. By considering the uniqueness of $F_{*,n}$ under the initial condition, f is the only solution. □

In order to clarify the solution and make it as simple as possible, its useful structures have to be further investigated.

Lemma 5.3.3. *For any integer* $n \geq 3$*, polynomial* $F_{*,n}$ *has its minimum of degrees not less than* n*.*

Proof. Although we checked from (5.3.13) that, for $n = 1$ and 2, the minimum degree of $F_{*,n}$ is greater than n, the minimum degree of F_3 is just 3.

For $n \geq 4$, we proceed by induction on n. Assume for any integer i, $3 \leq i \leq n-1$, that polynomial F_i has its minimum degree i. From Lemma 5.3.1,

$$F_i = \sum_{m=i}^{2i} F_{m,i}x^i. \tag{5.3.18}$$

Denote by $l(F_i)$ the minimum degree of F_i, then $l(F_i) = i$. Because of

$$\frac{F_{n-1}|_{x=1} - xF_{n-1}}{1-x} = \sum_{\substack{\langle n-1,i\rangle \leq m \leq 2(n-1)\\ 0 \leq i \leq 2(n-1)}} F_{m,n-1}x^i, \tag{5.3.19}$$

we have

$$\begin{aligned}[\partial_{1,x}(x^2 f)]_{n-1} &= x\sum_{i=0}^{2(n-1)}\left(\sum_{m=\langle n-1,i\rangle}^{2(n-1)} F_{m,n-1}\right)x^i \\ &= \sum_{\substack{\langle n-1,i\rangle \leq m \leq 2(n-1)\\ 0 \leq i \leq 2(n-1)}} F_{m,n-1}x^{i+1},\end{aligned} \tag{5.3.20}$$

where $\langle n-1, i\rangle = \max\{n-1, i\}$,

$$x[hf]_{n-1} = \sum_{m=0}^{2(n-1)} \Psi_{m,n-1}x^{m+1} \tag{5.3.21}$$

in which

$$\Psi_{m,n-1} = \begin{cases} \sum_{i=n-1-m}^{n-1-\lfloor m/2\rfloor} F_i|_{x=1}F_{m,n-1-i}, & 0 \le m \le n-1\,; \\ \sum_{i=0}^{n-1-\lfloor m/2\rfloor} F_i|_{x=1}F_{m,n-1-i}, & n \le m \le 2(n-1)\,, \end{cases}$$

and

$$[(h-1)(f-1)]_n = \sum_{m=1}^{2(n-1)} \Phi_{m,n-1}x^m \tag{5.3.22}$$

where

$$\Phi_{m,n-1} = \begin{cases} \sum_{i=n-m}^{n-1} F_i|_{x=1}F_{m,n-1-i}, & \text{when } 1 \le m \le 2; \\ \sum_{i=n-m}^{n-\lfloor m/2\rfloor} F_i|_{x=1}F_{m,n-1-i}, & \text{when } 3 \le m \le n-1, \\ \sum_{i=1}^{n-\lfloor m/2\rfloor} F_i|_{x=1}F_{m,n-1-i}, & \text{when } n \le m \le 2(n-1). \end{cases}$$

Observation 5.3.4. *For any integer $n-1 \ge m \ge 0$, $n \ge 3$,*

$$\Lambda_{m-1,n-1} - \Psi_{m-1,n-1} - \Phi_{m,n-1} = 0.$$

Proof. The result can be found in Example 1 of this section. □

On this basis, by (5.3.3), (5.3.19) and (5.3.13), the assumption leads to

$$\begin{aligned} l(F_{*,n}) &= \min\{2 + l([f^2]_{n-1}), l([\partial_{1,x}(x^2f)]_{n-1} - x[hf]_{n-1} - [(h-1)(f-1)]_n)\} \\ &\ge \min\{2 + (n-1), n\} \ge n. \end{aligned}$$

Therefore, the lemma is proved. □

From Lemma 5.3.1 and Lemma 5.3.3, we write

$$[f^2]_{n-1} = \sum_{m=n-1}^{2(n-1)} \Omega_{m,n-1}x^m, \tag{5.3.23}$$

$$\Omega_{m,n-1} = \sum_{\substack{l+t=m \\ (l,t)\in\mathcal{S}_{m,n-1}}} F_{l,i}F_{t,n-1-i}, \tag{5.3.24}$$

where $\mathcal{S}_{m,n-1} = \{(l,t) \mid i \le l \le 2i, n-1-i \le t \le 2(n-1-i), 0 \le i \le n-1\}$.

Observation 5.3.5. *For integer $2(n-1) \ge m \ge n-1$, $n \ge 3$,*

$$\Lambda_{m-1,n-1} \ge \Psi_{m-1,n-1} + \Phi_{m,n-1}.$$

Proof. The result can be found in Example 1 of this section. □

Lemma 5.3.6. *The polynomial $F_{*,n}$ has all its coefficients non-negative integers.*

Proof. From Observation 5.3.5, the conclusion is derived. □

Now, we are allowed to present the solution f of equation (5.3.1) determined by $F_{*,n}(n \geq 0)$ in the form of a finite sum with all terms non-negative integers.

Theorem 5.3.7. *The solutions f of equation* (5.3.1) *are determined by $F_{*,n} = \partial_y^n f$ for $n \geq 0$ as*

$$F_{*,n} = \begin{cases} 1, & \text{when } n = 0; \\ x^2, & \text{when } n = 1; \\ 2x^4, & \text{when } n = 2; \\ x^2 \sum_{m=n-1}^{2(n-1)} \Omega_{m,n-1} x^m + \sum_{m=n}^{2n} \Delta_{m,n-1} x^m, & \text{when } n \geq 3, \end{cases} \tag{5.3.25}$$

where

$$\Delta_{m,n-1} = \Lambda_{m-1,n-1} - \Psi_{m-1,n-1} - \Phi_{m,n-1} \tag{5.3.26}$$

and $\Omega_{m,n-1}$, $\Lambda_{m-1,n-1}$, $\Psi_{m-1,n-1}$ and $\Phi_{m,n-1}$ are given by, respectively, (5.3.24), (5.3.17), (5.3.21) *and* (5.3.22).

Proof. This is a result of what was described above. □

In what follows, the example shows equation (5.3.1) to be meaningful in combinatorics.

Example 1. Isomorphic classes of planar simple rooted maps by size and root-face valency. A map is said to be *simple* if neither a loop nor a multi-edge occurs.

Equation (5.3.1) is as a specific case of an equation from Liu YP [35] for determining the enufunction with face partition vector.

In Figures 5.3.1–5.3.3, $kS_{m,n}$ stands for a figure S whose maps are of size n, root-face valency m with k hollows (distinct classes). We have

$$\begin{aligned} &\text{when } n = 0, \quad 1S_{0,0} \Leftrightarrow F_0 = 1; \\ &\text{when } n = 1, \quad 1S_{2,1} \Leftrightarrow F_1 = x^2; \\ &\text{when } n = 2, \quad 2S_{4,2} \Leftrightarrow F_2 = 2x^4; \\ &\text{when } n = 1, \quad 1S_{3,3} + 5S_{6,3} \Leftrightarrow F_3 = x^3 + 5x^6; \\ &\text{when } n = 1, \quad 1S_{4,4} + 8S_{5,4} + 14S_{8,4} \Leftrightarrow F_4 = x^4 + 8x^5 + 14x^8. \end{aligned}$$

$S_{0,0}$ $S_{2,1}$ $2S_{4,2}$

Figure 5.3.1: Classes of planar simple rooted maps with sizes 0–2.

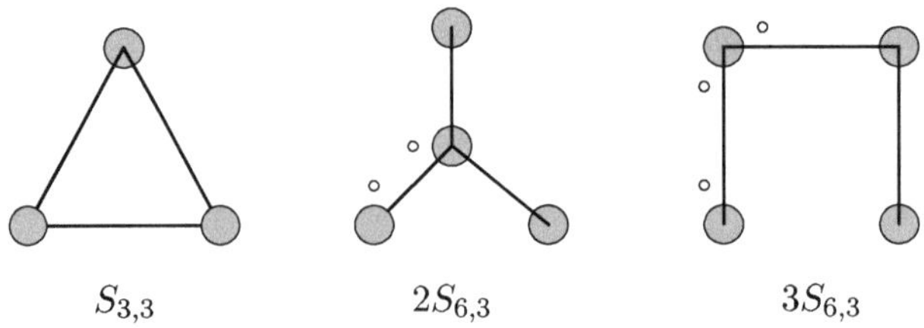

Figure 5.3.2: Classes of planar simple rooted maps with size 3.

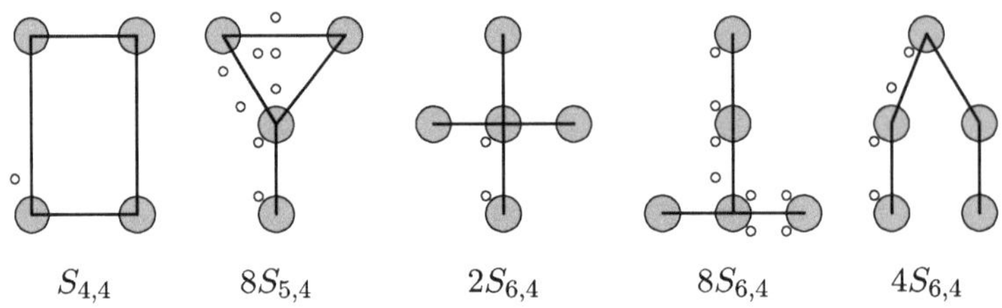

Figure 5.3.3: Classes of planar simple rooted maps with size 4.

As is seen, the corrections of Observation 5.3.4 and Observation 5.3.5 are natural.

Example 2. Equation

$$\begin{cases} f = 1 + zyf^2 + y\partial_{1,z}(zf); \\ f|_{z=y=0} = 1, \end{cases} \tag{5.3.27}$$

is also a type of straight difference with one variable.

In Liu YP [41], it is seen that the enufunction f of non-separable Euler planar rooted maps with size (power of y) and root-vertex valency (power of x) satisfies the equation

$$x^2y(1-x^2)f^2 - (1-x^2+x^2y)f + (1-x^2) + x^2yh = 0, \tag{5.3.28}$$

where $h = f|_{x=1}$.

By substitution $z = x^2$ to equation (5.3.28) and the cancelation law for $(1-x^2)$, equation (5.3.28) can be transformed into equation (5.3.27).

Thus, the enumeration of root-isomorphic classes of non-separable Euler planar maps can be done by solving a slope difference equation with one variable.

5.4 Form of several slope differences

In an equation of several variables, if the unknown comes to us with slope differences of two or more variables, it is said to be in the *form of several slope differences*.

Consider the equation for $f \in \mathcal{R}\{y, z, t\}$

$$\begin{cases} f = 2yz^2t + \dfrac{yzt\partial_{1,z}f}{1 - \frac{\partial_{1,z}f_{t=1}}{2}} - \dfrac{yzt\partial_{1,t}f}{1 - \frac{\partial_{1,t}f_{z=1}}{2}}, \\ f|_{y=0\Rightarrow z=t=0} = 0. \end{cases} \tag{5.4.1}$$

This is equation (15) in Introduction when $a = b = c = 1$ and $d = 0$ because it is meaningful in a classification of non-separable planar bipartite maps.

For $f \in \mathcal{R}\{z, t, y\}$, it is only necessary to determine $F_{*,n} = [f]_n = \partial_y^n f$, $n \geq 0$.

By the initial condition of equation (5.4.1),

$$F_0 = [f]_0 = f|_{y=0\Rightarrow z=t=0} = 0. \tag{5.4.2}$$

For any integer $n \geq 0$,

$$\begin{cases} [f_{z=1}]_n = F_{*,n}|_{z=1}, & n \geq 1 \text{ and } [f_{z=1}]_0 = 0; \\ [f_{t=1}]_n = F_{*,n}|_{t=1}, & n \geq 1 \text{ and } [f_{t=1}]_0 = 0. \end{cases} \tag{5.4.3}$$

Furthermore,

$$\begin{aligned} [\partial_{1,z}f]_n &= \frac{zF_{*,n}|_{z=1} - F_{*,n}}{1-z}, \quad \text{by (5.4.2) and (5.4.3)}, \\ &= \begin{cases} 0, & \text{when } n = 0; \\ \frac{tF_{*,n}|_{t=1} - F_{*,n}}{1-t}, \end{cases} \end{aligned} \tag{5.4.4}$$

$$\begin{aligned} [\partial_{1,t}f]_n &= \frac{tF_{*,n}|_{t=1} - F_{*,n}}{1-t}, \quad \text{by (5.4.2) and (5.4.3)}, \\ &= \begin{cases} 0, & \text{when } n = 0; \\ \frac{zF_{*,n}|_{z=1} - F_{*,n}}{1-z}, \end{cases} \end{aligned} \tag{5.4.5}$$

$$\begin{aligned} [\partial_{1,z}f_{t=1}]_n &= \frac{zF_{*,n}|_{t=1,z=1} - F_{*,n}|_{t=1}}{1-z}, \quad \text{by (5.4.2) and (5.4.3)}, \\ &= \begin{cases} 0, & \text{when } n = 0; \\ \frac{zF_{*,n}|_{t=1,z=1} - F_{*,n}|_{t=1}}{1-z}, \end{cases} \end{aligned} \tag{5.4.6}$$

and

$$\begin{aligned} [\partial_{1,t}f_{z=1}]_n &= \frac{tF_{*,n}|_{z=1,t=1} - F_{*,n}|_{z=1}}{1-t}, \quad \text{by (5.4.2) and (5.4.3)}, \\ &= \begin{cases} 0, & \text{when } n = 0; \\ \frac{tF_{*,n}|_{z=1,t=1} - F_{*,n}|_{z=1}}{1-t}. \end{cases} \end{aligned} \tag{5.4.7}$$

We notice that

$$\left(1-\frac{\partial_{1,z}f_{t=1}}{2}\right)^{-1}=\sum_{i\geq 0}\left(\frac{\partial_{1,z}f_{t=1}}{2}\right)^i.$$

From (5.4.6), for any integer $i \geq 0$,

$$y^i\left|\left(\frac{\partial_{1,z}f_{t=1}}{2}\right)^i,\right.$$

and hence

$$\left[\left(1-\frac{\partial_{1,z}f_{t=1}}{2}\right)^{-1}\right]_i=\left[\sum_{j=0}^{i}\left(\frac{\partial_{1,z}f_{t=1}}{2}\right)^i\right]_i. \tag{5.4.8}$$

Similarly, by (5.4.6), for any integer $i \geq 0$,

$$\left[\left(1-\frac{\partial_{1,t}f_{z=1}}{2}\right)^{-1}\right]_i=\left[\sum_{j=0}^{i}\left(\frac{\partial_{1,t}f_{z=1}}{2}\right)^j\right]_i. \tag{5.4.9}$$

Now, we are allowed to determine $F_{*,n}$ in the order of $n = 0, 1, 2, 3, \ldots$.
When $n = 0$, from equations (5.4.1) and (5.4.3)–(5.4.9),

$$\begin{aligned} y^0:\ F_0=0 \Rightarrow\ & [f_{z=1}]_0=[f_{t=1}]_0=0,\ [\partial_{1,z}f]_0=[\partial_{1,t}f]_0=0,\\ & [\partial_{1,z}f_{t=1}]_0=[\partial_{1,t}f_{z=1}]_0=0,\\ & \left[\left(1-\frac{\partial_{1,z}f_{t=1}}{2}\right)^{-1}\right]_0=1,\\ & \left[\left(1-\frac{\partial_{1,t}f_{z=1}}{2}\right)^{-1}\right]_0=1. \end{aligned} \tag{5.4.10}$$

This is the initial condition of equation (5.4.1).
When $n = 1$ from equations (5.4.1) and (5.4.10),

$$\begin{aligned} y^1:\ F_1 &= 2z^2t+zt([\partial_{1,z}f]_0-[\partial_{1,t}f]_0)=2z^2t\\ &\Rightarrow\ [f_{z=1}]_1=2t,\ [f_{t=1}]_1=2z^2,\ [\partial_{1,z}f]_1=2zt,\\ &\quad [\partial_{1,t}f]_1=0,\ [\partial_{1,z}f_{t=1}]_1=2,\ [\partial_{1,t}f_{z=1}]_1=0,\\ &\quad \left[\left(1-\frac{\partial_{1,z}f_{t=1}}{2}\right)^{-1}\right]_1=2,\\ &\quad \left[\left(1-\frac{\partial_{1,t}f_{z=1}}{2}\right)^{-1}\right]_0=0. \end{aligned} \tag{5.4.11}$$

When $n = 2$, from equations (5.4.1) and (5.4.11),

$$\begin{aligned}
y^2 : F_2 &= zt\left(\left[\frac{\partial_{1,z}f}{1-\frac{\partial 1, zf_{t=1}}{2}}\right]_1 - \left[\frac{\partial_{1,t}f}{1-\frac{\partial 1, tf_{z=1}}{2}}\right]_1\right) \\
&= zt[\partial_{1,z}f]_1 = 2z^2t^2 \\
&\Rightarrow [f_{z=1}]_2 = 2t^2,\ [f_{t=1}]_2 = 2z^2, \\
&\quad [\partial_{1,z}f]_2 = 2zt^2,\ [\partial_{1,t}f]_2 = 2z^2t, \\
&\quad [\partial_{1,z}f_{t=1}]_2 = 2,\ [\partial_{1,t}f_{z=1}]_2 = 2, \\
&\quad \left[\left(1-\frac{\partial_{1,z}f_{t=1}}{2}\right)^{-1}\right]_2 = 2zt + t^2, \\
&\quad \left[\left(1-\frac{\partial_{1,t}f_{z=1}}{2}\right)^{-1}\right]_2 = 2.
\end{aligned} \tag{5.4.12}$$

We now consider the general case of $n \geq 3$. First, we discuss how

$$\left(1-\frac{\partial_{1,z}f_{t=1}}{2}\right)^{-1} \quad \text{and} \quad \left(1-\frac{\partial_{1,t}f_{z=1}}{2}\right)^{-1}$$

can be expressed by F_i, $0 \leq i \leq n-1$.

From (5.4.8) and (5.4.9),

$$\left[\left(1-\frac{\partial_{1,z}f_{t=1}}{2}\right)^{-1}\right]_i = \sum_{j=0}^{i}\left[\left(\frac{\partial_{1,z}f_{t=1}}{2}\right)^j\right]_i \quad (i.e.,\ \ \Sigma_i^{(z)}) \tag{5.4.13}$$

and

$$\left[\left(1-\frac{\partial_{1,t}f_{z=1}}{2}\right)^{-1}\right]_i = \sum_{j=0}^{i}\left[\left(\frac{\partial_{1,t}f_{z=1}}{2}\right)^j\right]_i \quad (i.e.,\ \ \Sigma_i^{(t)}) \tag{5.4.14}$$

are, respectively, deduced.

Because of $\partial_{1,z}f_{t=1}, \partial_{1,t}f_{z=1} \in \mathcal{R}\{z,t\}$, by the multiplication principle,

$$\left[\left(\frac{\partial_{1,z}f_{t=1}}{2}\right)^j\right]_i, \left[\left(\frac{\partial_{1,t}f_{z=1}}{2}\right)^j\right]_i \in \mathcal{R}\{z,t\}.$$

By considering that they are only determined by $F_0, F_1, \ldots, F_i$, it is seen that (5.4.13) and (5.4.14) are also determined only by $F_0, F_1, \ldots, F_i$.

Theorem 5.4.1. *Equation* (5.4.1) *is well-defined on* $\mathcal{R}\{z,t,y\}$.

Proof. It is only necessary to determine $F_{*,n} = \partial_y^n f \in \mathcal{R}\{z,t\}$, $n \geq 0$. From (5.4.10)–(5.4.12), we see that $F_0 = 0$ (the initial condition!), $F_1 = 2z^2t$ and $F_2 = 2z^2t^2$. It is easily seen that they are in $\mathcal{R}\{z,t\}$.

For $n \geq 3$ we proceed by induction on n; assume that $F_i \in \mathcal{R}\{z,t\}$, $0 \leq i \leq n-1$, have been obtained. We evaluate $F_{*,n} \in \mathcal{R}\{z,t\}$.

By employing equation (5.4.1), from $n \geq 3$, we have

$$\begin{aligned} y^n : F_{*,n} &= zt \sum_{i=0}^{n-1} \Bigg([\partial_{1,z} f]_i \left[1 - \frac{\partial_{1,z} f_{t=1}}{2} \right]_{n-1-i} \\ &\quad - [\partial_{1,t} f]_i \left[1 - \frac{\partial_{1,t} f_{z=1}}{2} \right]_{n-1-i} \Bigg) \\ &= zt \sum_{i=0}^{n-1} ([\partial_{1,z} f]_i \Sigma_{n-1-i}^{(z)} - [\partial_{1,t} f]_i \Sigma_{n-1-i}^{(t)}) \end{aligned} \tag{5.4.15}$$

where $\Sigma_{n-1-i}^{(z)}$ and $\Sigma_{n-1-i}^{(t)}$ are, respectively, given by (5.4.13) and (5.4.14).

From (5.4.4)–(5.4.9), the right hand side in (5.4.15) is only dependent on F_i, $0 \leq i \leq n-1$. By the assumption, $F_{*,n} \in \mathcal{R}\{z,t\}$. Therefore, $f \in \mathcal{R}\{z,t,y\}$ is a solution of equation (5.4.1).

By considering the uniqueness of the procedure mentioned above for the initial value, equation (5.4.1) only has a single solution. □

In what follows, some useful structures of $F_{*,n}$ for $n \geq 0$, are investigated.

Lemma 5.4.2. *For any integer $n \geq 1$, $2|F_{*,n}$, i. e., 2 is a factor of $F_{*,n}$.*

Proof. We proceed by induction on n. From (5.4.11) and (5.4.11), when $n = 1$ and 2, $2|F_{*,n}$.

When $n \geq 3$, assume for any integer i, $2 \leq i \leq n-1$, $2|F_{*,n}$ are known, we prove the case of $i = n$.

From (5.4.4) and (5.4.5), the assumption leads to $2|\ [\partial_{1,z} f]_i$ and $2|\ [\partial_{1,t} f]_i$ for $0 \leq i \leq n-1$. From (5.4.15), $2|F_{*,n}$. This is the conclusion. □

Lemma 5.4.3. *For any integer $n \geq 2$, $F_{*,n}$ is a polynomial of not only z but also t with degree not greater than n and minimum degree not less than 2.*

Proof. We proceed by induction on $n \geq 1$. From (5.4.11) and (5.4.11), it is easily seen that the conclusion is true when $n = 1$ and 2.

For convenience, denote by $d_z(P)$ and $d_t(P)$ the degree of a polynomial $P \in \mathcal{R}\{z,t\}$ for, respectively, z and t.

When $n \geq 3$, assume that $d_z(F_{*,k}) = d_t(F_{*,k}) = k$ with minimum degree 2 for $2 \leq k \leq n-1$. We prove that $d_z(F_{*,n}) = d_t(F_{*,n}) = n$ with minimum degree 2. □

Observation 5.4.4. *For any integer $n \geq 3$,*

$$\sum_{i=0}^{n-1} [\partial_{1,z} f]_i \Sigma_{n-1-i}^{(z)} - \sum_{i=0}^{n-1} [\partial_{1,t} f]_i \Sigma_{n-1-i}^{(t)} \geq 0.$$

Proof. See Example 2 of this section. □

Observation 5.4.5. *For any integer s, $n-1 \geq s \geq 1$, $d_z(\Sigma_s^{(z)}) \leq s$.*

Proof. It is seen by induction on the basis of (5.4.13).

From (5.4.15),

$$\begin{aligned}
d_z(F_{*,n}) &= 1 + d_z\left(\sum_{i=0}^{n-1}([\partial_{1,z}f]_i\Sigma_{n-1-i}^{(z)} - [\partial_{1,t}f]_i\Sigma_{n-1-i}^{(t)})\right), \\
&\quad \text{by Observation 5.4.4,} \\
&= 1 + d_z\left(\sum_{i=0}^{n-1}[\partial_{1,z}f]_i\Sigma_{n-1-i}^{(z)}\right) \\
&= 1 + \max_{0\leq i\leq n-1} d_z([\partial_{1,z}f]_i) + d_z(\Sigma_{n-1-i}^{(z)}) \\
&= 1 + d_z([\partial_{1,z}f]_1) + d_z(\Sigma_{n-2}^{(z)}), \\
&\quad \text{by Observation 5.4.5,} \\
&= 1 + (1 + n - 2) = n,
\end{aligned} \tag{5.4.16}$$

and

$$\begin{aligned}
d_t(F_{*,n}) &= 1 + \max_{0\leq i\leq n-1} d_t([\partial_{1,z}f]_i) + d_t(\Sigma_{n-1-i}^{(z)}) \\
&= 1 + d_t([\partial_{1,z}f]_{n-1}) + d_t(\Sigma_0^{(z)}) \\
&= 1 + (n-1) + 0 = n.
\end{aligned} \tag{5.4.17}$$

Therefore, $F_{*,n}$ is a polynomial of degree n of not only z but also t.

Further from what was mentioned in the proof of Theorem 5.4.1, it is seen that, for $n \geq 2$, $zt|F_{*,n}$, *i. e.*, $F_{*,n}$ has a minimum degree not less than 2 of not only z but also t. □

Lemma 5.4.6. *For any integer $n \geq 1$, $F_{*,n}$ comes to us with all coefficients non-negative integers.*

Proof. From (5.4.15), it is seen that Observation 5.4.4 leads to all coefficients of $F_{*,n}$ being non-negative. As to integrity, it is deduced from $F_{*,n} \in \mathcal{R}\{z,t\}$. □

On the basis of Theorem 5.4.1 and Lemmas 5.4.2–5.4.6, let

$$[f]_n = F_{*,n} = \sum_{\substack{2\leq m\leq n \\ 2\leq s\leq n}} F_{m,s;n}z^m t^s, \quad 0 \leq F_{m,s;n} \in \mathcal{R}. \tag{5.4.18}$$

Then we have

$$\begin{cases}
[\partial_{1,z}f]_n = \displaystyle\sum_{\substack{1\leq m\leq n-1 \\ 2\leq s\leq n}}\left(\sum_{k=m+1}^{n} F_{k,s;n}\right)z^m t^s; \\
[\partial_{1,z}f_{t=1}]_n = \displaystyle\sum_{\substack{1\leq m\leq n-1 \\ 2\leq s\leq n}}\left(\sum_{k=m+1}^{n} F_{k,s;n}\right)z^m,
\end{cases} \tag{5.4.19}$$

and

$$\begin{cases} [\partial_{1,t}f]_n = \sum_{\substack{2\le m\le n\\ 1\le s\le n-1}} \left(\sum_{k=s+1}^{n} F_{m,k;n} \right) z^m t^s, \\ [\partial_{1,t}f_{z=1}]_n = \sum_{\substack{2\le m\le n\\ 1\le s\le n-1}} \left(\sum_{k=s+1}^{n} F_{m,k;n} \right) z^m. \end{cases} \tag{5.4.20}$$

Based on (5.4.18), for any integer $i \geq 2$, we can write

$$[f^i]_n = F_{*,n}^{[i]} = \begin{cases} \sum_{l=0}^{n} F_l F_{n-l}, & \text{when } i = 2, \\ \sum_{l=0}^{n} F_l F_{n-l}^{[i-1]}, & \text{when } i \geq 3, \end{cases} \tag{5.4.21}$$

then from (5.4.13) and (5.4.13),

$$\begin{aligned} \Sigma_i^{(z)} &= \begin{cases} 1, & \text{when } i = 0; \\ \sum_{j=1}^{i} [\frac{\partial_{1,z}f_{t=1}}{2}]_i^{[j]}, & \text{when } i \geq 1, \end{cases} \\ \Sigma_i^{(t)} &= \begin{cases} 1, & \text{when } i = 0; \\ \sum_{j=1}^{i} [\frac{\partial_{1,t}f_{z=1}}{2}]_i^{[j]}, & \text{when } i \geq 1. \end{cases} \end{aligned} \tag{5.4.22}$$

Therefore, from (5.4.19), (5.4.20) and (5.4.22),

$$\begin{cases} \sum_{i=0}^{n-1} [\partial_{1,z}f]_i \Sigma_{n-1-i}^{(z)} = \sum_{1\le m,s\le n-1} A_{m,s;n-1} z^m t^s; \\ \sum_{i=0}^{n-1} [\partial_{1,t}f]_i \Sigma_{n-1-i}^{(t)} = \sum_{1\le m,s\le n-1} B_{m,s;n-1} z^m t^s, \end{cases} \tag{5.4.23}$$

where $A_{m,s;n-1}, B_{m,s;n-1} \in \mathcal{R}\{z,t\}$, $A_{m,s;n-1} - B_{m,s;n-1} \geq 0$ (Observation 5.4.4!), being dependent only on F_i, $0 \leq i \leq n-1$, and given in (5.4.19)–(5.4.22).

Theorem 5.4.7. *For any integer* $n \geq 3$,

$$F_{*,n} = \sum_{2\le m,k\le n} F_{m,k;n} z^m t^k \tag{5.4.24}$$

where $F_{m,k;n} = A_{m-1,s-1;n-1} - B_{m-1,s-1;n-1} \geq 0$ *are given in* (5.4.23).

Proof. This is a direct result of the above. □

This theorem provides the solution of equation (5.4.1) with all coefficients in the form of a finite sum with all terms positive.

Example 1. Chromatic equation of non-separable planar rooted maps. In Liu YP [15], the equation

$$\begin{aligned} &(g - \lambda(\lambda-1)xz^2t)\left(1 - \frac{\partial_z g_{t=1}}{\lambda}\right)\left(1 - \frac{\partial_t g_{z=1}}{\lambda}\right) \\ &= yzt\partial_z g\left(1 - \frac{\partial_t g_{z=1}}{\lambda}\right) - xzt\partial_t g\left(1 - \frac{\partial_z g_{t=1}}{\lambda}\right) \end{aligned} \tag{5.4.25}$$

occurs where

$$g = \sum_{M \in \mathcal{M}} P(M : \lambda) x^{p(M)} y^{q(M)} r^{p(M)} t^{s(M)}$$

is a function of x, y, z and t such that $\mathcal{M}$ is the set of all non-separable planar rooted maps with $p(M)$ non-rooted vertices, $q(M)$ non-rooted faces, $r(M)$ root-vertex valency and $s(M)$ root-face valency of M. $P(M : \lambda)$ is the chromatic polynomial of M.

In equation (5.4.25), ∂_z and ∂_t are, respectively, the slope differences: $\partial_{1,z}$ and $\partial_{1,t}$.

If equation (5.4.25) has $x = y$ (the size $n(M) = p(M) + q(M)$!) and $\lambda = 2$, then it is equivalent to equation (5.5.1).

Example 2. Root-isomorphic classes of non-separable planar bipartite maps with given size root-vertex valency and root-face valency. Because a map has chromatic number 2 if, and only if, its underlying graph is bipartite, it is seen that, for any integer $n \geq 1$, $F_{*,n}/2$ in the solution f of equation (5.4.1) provides the number of root-isomorphic classes of non-separable planar bipartite maps with size n.

In Figure 5.4.1, a, b and c give $F_1/2 = z^2t$, $F_2/2 = z^2t^2$ and $F_3/2 = z^2t^3$. Here, the powers of z and t are, respectively, the root-face and root-vertex valencies. From d in Figure 5.4.1 to e in Figure 5.4.2, $F_4/2 = z^2t^4 + z^4t^2$. f–g in Figure 5.4.2 as

$$\begin{aligned} F_5/2 &= z^2t^5 + (z^2t^3 + 2z^4t^2 + 2z^4t^3) \\ &= z^2(t^3 + t^5) + z^4(2t^2 + 2t^3). \end{aligned}$$

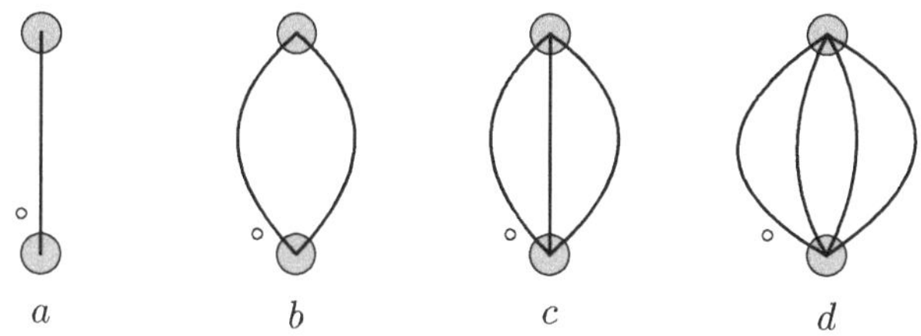

Figure 5.4.1: Classes of non-separable planar bipartite maps with sizes: 1–4.

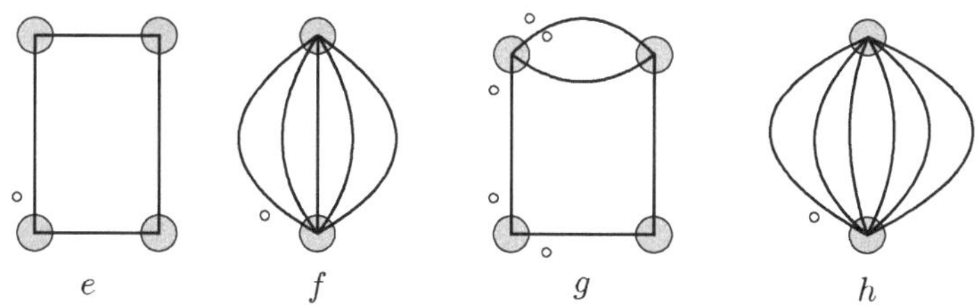

Figure 5.4.2: Classes of non-separable planar bipartite maps with sizes: 4–5.

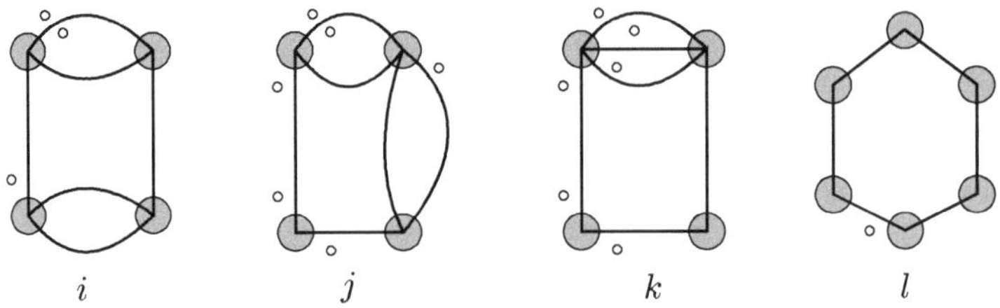

Figure 5.4.3: Classes of non-separable planar bipartite maps with size 6.

From h in Figure 5.4.2 to i–l in Figure 5.4.3,

$$\begin{aligned}F_6/2 &= z^2t^6 + (z^2t^3 + 2z^4t^3) + (z^2t^3 + 2z^4t^2 + 2z^4t^3 + z^4t^4)\\ &\quad + (2z^2t^4 + 2z^4t^2 + 2z^4t^4) + z^6t^2\\ &= z^2(4t^3 + 2t^4 + t^6) + z^4(4t^2 + 2t^3 + 3t^4) + z^6t^2.\end{aligned}$$

5.5 Mixed form of straight and slope differences

Consider the equation for $f \in \mathcal{R}\{x,y\}$

$$\begin{cases} f = 1 + xy\delta_{1,x}(xf) + \dfrac{x^2y(\delta_{1,x}f)^2}{1-(1+\partial_{1,x}f)}; \\ f|_{y=0\Rightarrow x=0} = 1. \end{cases} \tag{5.5.1}$$

This is equation (16) in Introduction when $a = b = c = d = 1$ because it is meaningful in a classification of simple planar maps.

In this equation, on account of both straight and slope differences being involved, it is called a *mixed form of straight and slope differences.*

Although $F_{*,n} \in \mathcal{R}\{x\}$ for $n \geq 1$ can be directly deduced in the form of a sum with all terms positive from the equation, a complication occurs for an infinite sum

$$\sum_{k\geq 0}\left[1 + \frac{\partial_{1,x}f}{1-x}\right]_i$$

where $i \geq 1$.

The aim of this section is to avoid the evaluation of an infinite sum. It is only necessary to expand $\delta_{1,x}f$, $\delta_{1,x}(xf)$ and $\partial_{1,x}f$ in equation (5.5.1) on $\mathcal{R}\{x,y\}$, for getting one of its equivalent expressions,

$$\begin{cases} f^2 - (xh + xy(1-x)h + 1)f + xh = 0; \\ f|_{y=0,x=0} = 1, \end{cases} \tag{5.5.2}$$

where $h = f|_{x=1}$.

Then the first line of equation (5.5.1) is transformed into

$$\begin{aligned} & f^2 - f - (xh + xy(1-x)h)f + xh = 0 \\ \Longrightarrow\ & f(f-1) - xh(f-1) = xy(1-x)hf \\ \Longrightarrow\ & (f - xh)(f-1) = xy(1-x)hf \\ \Longrightarrow\ & (1-f)\partial_{1,x}f = xyhf. \end{aligned} \tag{5.5.3}$$

For any integer $i \geq 1$, let $F_i = \partial_y^i f$ be a polynomial of x with degree m_i, from the initial condition of equation (5.5.1), we have

$$F_i = \sum_{m=1}^{m_i} F_{m,i} x^m \tag{5.5.4}$$

and

$$\left[\frac{xh - f}{1-x}\right]_i = \begin{cases} -1, & \text{when } i = 0; \\ \sum_{k=0}^{m_i - 1} (\sum_{l=\max\{k+1,2\}}^{m_l} F_{l,i}) x^k, & \text{when } i \geq 1. \end{cases} \tag{5.5.5}$$

By employing the third line of (5.5.3),

$$\begin{aligned} y^0 :\ & [1-f]_0[\partial_{1,x}f]_0 = 0, \quad \text{by (5.5.5)}, \\ \Longrightarrow\ & (1 - F_0)(-1) = 0 \\ \Longrightarrow\ & F_0 = 1 \quad \text{(the initial value of equation (5.5.1))}, \end{aligned} \tag{5.5.6}$$

and

$$\begin{aligned} y^1 :\ & [1-f]_0[\partial_{1,x}f]_1 + [1-f]_1[\partial_{1,x}f]_0 \\ & = x[hf]_0, \quad \text{by (5.5.5)}, \\ \Longrightarrow\ & (-F_1)(-1) = xF_0|_{x=1}F_0 \\ \Longrightarrow\ & F_1 = x. \end{aligned} \tag{5.5.7}$$

In general, for any integer $n \geq 2$,

$$\begin{aligned} y^n :\ & \sum_{i=0}^{n} [1-f]_i[\partial_{1,x}f]_{n-i} = x[hf]_{n-1}, \\ & \quad \text{by } [1-f]_0 = 0 \text{ and } [\partial_{1,x}f]_0 = -1, \\ \Longrightarrow\ & F_{*,n} = \sum_{i=1}^{n-1} F_i[\partial_{1,x}f]_{n-i} \\ & + x\sum_{i=0}^{n-1} F_i|_{x=1} F_{n-1-i}. \end{aligned} \tag{5.5.8}$$

Theorem 5.5.1. *Equation* (5.5.1) *is well-defined on* $\mathcal{R}\{x,y\}$.

Proof. By the principle of induction, based on (5.5.3) and (5.5.6)–(5.5.8), to determine $F_{*,n}$ for $n \geq 0$, a solution of equation (5.5.1) is found on $\mathcal{R}\{x, y\}$.

Because of the uniqueness of the procedure used under the given initial condition of equation (5.5.1), this solution is the only one. □

In order to seek a relatively simple expression of the solution, it is absolutely necessary to investigate some concrete structures of polynomials $F_{*,n}$ for $n \geq 1$.

Lemma 5.5.2. *For any integer $n \geq 1$, $F_{*,n}$ is polynomial of degree n with minimum degree 1 in $\mathcal{R}\{x\}$.*

Proof. For convenience, denote by $d_z(P)$ the degree of a polynomial P of z. If P has only one variable, substitute for $d(P)$. From (5.5.7), $d(F_{*,1}) = m_1 = 1$. The conclusion is true.

By the principle of induction, assume that, for any integer i, $1 \leq i \leq n-1$, $d(F_1) = m_i = i$ are known. We prove $d(F_{*,n}) = m_n = n$.

By (5.5.8),

$$\begin{aligned} d(F_{*,n}) &= \max\{\max\{d(F_i) + d(F_{n-i}) | 1 \leq i \leq n-1\}, \\ &\quad 1 + \max\{d(F_{n-1-i}) | 0 \leq i \leq n-1\}\}, \\ &\quad \text{by the induction assumption,} \\ &= \max\{\max\{i + (n-i) | 1 \leq i \leq n-1\}, \\ &\quad 1 + \max\{n - 1 - i | 0 \leq i \leq n-1\}\} \\ &= 1 + (n-1) = n. \end{aligned}$$

This is the conclusion for the degree. As for the minimum degree, the conclusion is drawn from (5.5.4) and (5.5.7). □

Therefore, (5.5.4) can be precisely written as

$$F_{*,i} = \sum_{m=1}^{i} F_{m,i} x^m. \tag{5.5.9}$$

Lemma 5.5.3. *For any integer $n \geq 1$, $F_{*,n}$ has all coefficients in $\mathbb{Z}_+$.*

Proof. Similarly, by induction, the conclusion can be done. □

For any polynomials p and q of x, write $X_i^{[pq]} = \partial_x^i(pq)$, then

$$X_i^{[pq]} = \sum_{\substack{i_1+i_2=i \\ 0 \leq i_1, i_2 \leq i}} X_{i_1}^{[p]} X_{i_2}^{[q]}. \tag{5.5.10}$$

Let $[\partial f] = [\partial_{1,x} f]$ and let

$$\Pi_{k,i} = \begin{cases} -1, & \text{when } i = 0; \\ \sum_{l=\max\{k+1,2\}}^{m_l} F_{l,i}, & \text{when } i \geq 1, \end{cases} \tag{5.5.11}$$

we have

$$[\partial_{1,x}f]_{n-i} = \sum_{j=1}^{n-i} \Pi_{j,n-i}x^j, \tag{5.5.12}$$

and by following the procedure in the proof of Lemma 5.5.2:

$$F_i[\partial_{1,x}f]_{n-i} = \sum_{j=1}^{n}\Big(\sum_{\substack{j_1+j_2=j\\ 0\le j_1,j_2\le n-1}} F_{j_1,i}\Pi_{j_2,n-i}\Big)x^j. \tag{5.5.13}$$

From (5.5.12),

$$X_{j,n-1}^{[F_i\partial F_{n-i}]} = \sum_{\substack{j_1+j_2=j\\ 0\le j_1,j_2\le n-1}} F_{j_1,i}\Pi_{j_2,n-i} \tag{5.5.14}$$

is determined by coefficients of polynomials F_i for $0 \le i \le n-1$.

Theorem 5.5.4. *In the solution f of equation* (5.5.1), $F_{*,n}$ *for* $n \ge 0$ *obeys an expression determined only by the coefficients of the polynomials* F_i *for* $0 \le i \le n-1$,

$$F_{*,n} = \begin{cases} 1, & \text{when } n = 0; \\ \sum_{m=1}^{n}\big(\sum_{i=1}^{n-1} X_{m,n-1}^{[F_i\partial F_{n-i}]} + \sum_{i=0}^{n-1} H_iF_{m-1,n-1-i}\big)x^m, \\ \qquad \text{when } n \ge 1, \end{cases} \tag{5.5.15}$$

where $X_m^{[F_i\partial F_{n-i}]}$ *is given by* (5.5.14) *and*

$$H_i = [h]_i = F_i|_{x=1} = \sum_{m=0}^{i} F_{m,i}. \tag{5.5.16}$$

Proof. On the basis of Theorem 5.5.4 and (5.5.8), by (5.5.11), (5.5.14) and (5.5.16), $F_{*,n}$ for $n \ge 0$, is determined by all coefficients of the polynomials F_i, $0 \le i \le n-1$. □

Example 1. Root-classification of dual simple planar maps by size and root-face valency. A map is said to be *comple* when there is neither a cut edge nor a cut pair of two edges. Because of that the dual of a comple map is a simple map, the name is for so.

From the uniqueness of duality, the equation for comple planar maps with given size and root-face valency is the same as for simple planar maps with given size and root-vertex valency as shown in equation (5.5.1).

Example 2. Root-classification of simple planar maps by size and root-vertex valency. By *simple map* is meant there to be neither a loop edge nor a multi-edge. It is easily seen that a cut edge or 2-circuit is allowed in a simple map.

Figures 5.5.1–5.5.3 show the root-isomorphic classes of simple planar maps with sizes: 0–4.

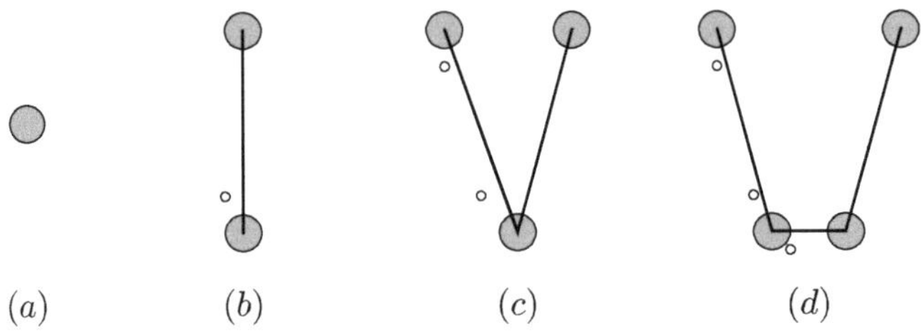

Figure 5.5.1: Root-isomorphic classes of simple planar maps with sizes: 0–3.

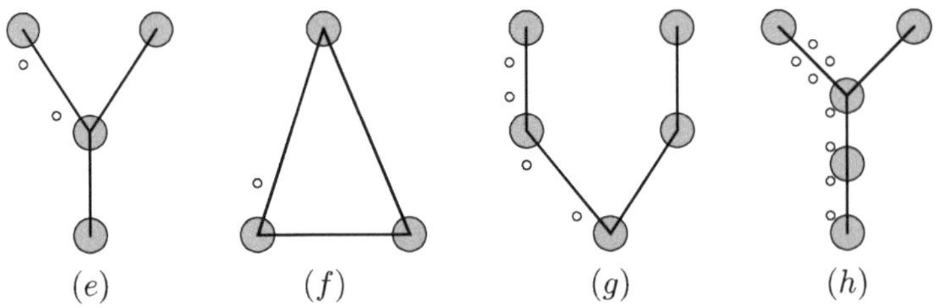

Figure 5.5.2: Root-isomorphic classes of simple planar maps with sizes: 3–4.

In Figure 5.5.1, (a) represents $F_{*,0} = 1$, *i. e.*, for a vertex map itself without edge. (b) and (c) provide $F_1 = x$ and $F_2 = x + x^2$, *i. e.*, the classes with, respectively, size 1 and size 2.

In Figure 5.5.1 (d), Figure 5.5.2 (e) and (f), we present $F_3 = (x+2x^2)+(x+x^3)+(x^2) = 2x + 3x^2 + x^3$, *i. e.*, a classification for size 3.

Cases (g) and (h) in Figure 5.5.2, (i), (j) and (k) in Figure 5.5.3 present

$$\begin{aligned} F_4 &= (x + 3x^2) + (3x + 2x^2 + 3x^3) + (x + x^4) + (x + 4x^2 + 3x^3) + (x^2) \\ &= 6x + 10x^2 + 6x^3 + x^4, \end{aligned}$$

i. e., the classification for size 4.

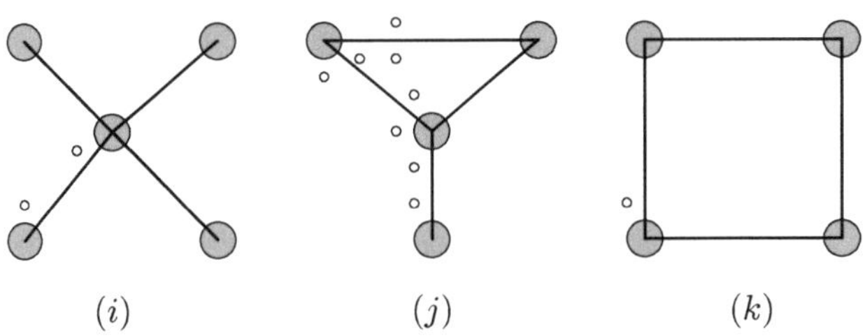

Figure 5.5.3: Root-isomorphic classes of simple planar maps with size 4.

5.6 Notes

5.6.1. In Section 5.1, the solution $f \in \mathcal{R}\{x, y\}$ of equation (5.5.1) is provided in the form of a finite sun with all terms positive. As a specific case, the number of non-isomorphic classes of loopless planar maps with given size and root-vertex valency.

The enumeration of loopless planar maps is from Liu YP [16] (1983). In Bender EA-Wormald NC [1] (1985), the topic is further investigated as well.

In Li ZX-Liu YP [10] (2002), by the enufunction of loopless planar maps with size (y) and root-face valency (x) as parameters, the equation

$$(1-x)x^2yf^2 - (1-x+xy(1-x)h+x^2y)f + 1 - x + xyh = 0 \tag{5.6.1}$$

is found where $h = f|_{x=1}$. Via transformation and the constant term of f considered, the equation

$$\begin{cases} f = 1 + x^2yf^2 - xyhf + xy\delta_{1,x}(xf); \\ f|_{y=0\Rightarrow x=0} = 1, \end{cases} \tag{5.6.2}$$

is obtained. Furthermore, for such maps on a projective plane, a straight difference equation of one variable is also established,

$$\begin{cases} f = x^2ySf - xy(Sh + S_1f) + xy\delta_{1,x}(xf) + \mathcal{L}(S); \\ f|_{y=0\Rightarrow x=0} = 0, \end{cases} \tag{5.6.3}$$

where $h = f|_{x=1}$, S is the solution of equation (5.6.2), $S_1 = S|_{x=1}$ and

$$\mathcal{L}(S) = x^2y\left(\frac{\partial(xS)}{\partial x} - S^2\right).$$

5.6.2. In Cai JL-Liu YP [7], from enumerating ordinary planar rooted maps with size (y), root-vertex valency (z) and root-face valency (x), an equation for g is extracted as

$$((1-x)(1-z)(f^* - f) + x - z)g = (1-x)f^* - (1-z)f \tag{5.6.4}$$

where $f(x, y) = g|_{z=1}$ $f^* = g|_{x=1}$.

By transformation and the constant term of g considered, a straight difference equation of several variables

$$\begin{cases} (f^* - f)g = (f^* - f)\dfrac{z-x}{(1-z)(1-x)} + \delta_{1,z}g - \delta_{1,x}g; \\ g|_{y=0\Rightarrow x=0,z=0} = 0, \end{cases} \tag{5.6.5}$$

is found. Because of the occurrence of y, it looks as if there is no way as in Section 5.2 to solve it yet.

However, via decomposition of maps, another equation for g arises:

$$g = 1 + x^2yzfg + xyz\delta_{1,x}(xg) - (1-z)xyzf^*g \tag{5.6.6}$$

is induced. Because of the symmetry between x and z in g with constant term considered, it becomes

$$\begin{cases} g = 1 + xyz\left(xf + zf^* - \dfrac{f^*+f}{2}\right)g + xyz(\delta_{1,x}(xg) + \delta_{1,z}(zg)); \\ g|_{y=0\Rightarrow x=0,z=0} = 1. \end{cases} \tag{5.6.7}$$

This is a straight difference equation of several variables which can be solved in the way of Section 5.2.

5.6.3. In investigating the enufunction of planar simple bipartite maps with size (y) and root-face valency (x, $z = x^2$), an equation for f arises:

$$zyf^2 - \left(\frac{zy}{1-z} + f^*\right) + \left(\frac{zy}{1-z} + 1\right)f^* = 0 \tag{5.6.8}$$

is extracted where $f^* = |_{z=1}$ refer to Liu YP [44] (equation (7.4.21)).

Via a transformation and the initial value considered, the equation for f

$$\begin{cases} f = 1 + zyf^2 + y\partial_{1,x}(zf) - (f^*-1)(f-1); \\ f_{y=0\Rightarrow z=0} = 1, \end{cases} \tag{5.6.9}$$

is attained. This is a slope difference equation of one variable. By the method used in 5.3, a recursion as a sum of finite positive terms, and an explicision can be done.

5.6.4. The slope difference equation of several variables

$$\begin{cases} f = 6yz^2t + \dfrac{yzt\partial_{1,z}f}{1 - \frac{\partial_{1,z}f_{t=1}}{3}} - \dfrac{yzt\partial_{1,t}f}{1 - \frac{\partial_{1,t}f_{z=1}}{3}}, \\ f|_{y=0\Rightarrow z=t=0} = 0, \end{cases} \tag{5.6.10}$$

can by the method used in Section 5.4 be addressed to evaluate its solution by a recursion in the form of a sum of finite positive terms, even an explicision.

In fact, equation (5.6.10) is (5.4.25) when $\lambda = 3$. So, equation (5.6.10) is meaningful in combinatorics.

5.6.5. The enufunction of non-separable Euler planar rooted maps with the number of non-root-vertices (z), the number of non-root-faces (y) and root-vertex valency ($x, t = x^2$) is shown to satisfy the equation

$$f = tz + \frac{ty(f - f^*)}{t(1+f^*)^2 - (1+f)^2} \tag{5.6.11}$$

where $f^* = f|_{t=1}$.

The following mixed straight and slope difference equation for $f \in \mathcal{R}\{t,y,z\}$ is shown to be well-defined and its solution satisfies equation (5.6.11) as well:

$$\begin{cases} f = tz + \dfrac{ty\delta_{1,t}f}{(1-\partial_{1,t}f)^2 - t\delta_{1,t}^2 f}; \\ f|_{z=0,y=0\Rightarrow t=0} = 0. \end{cases} \tag{5.6.12}$$

6 Ordinary differential equations

6.1 Parametric equations

On the basis of equivalent equations, or equation systems transformed by characteristic curves, or surfaces from a function, or functional equation, to establish a differential equation for evaluating a solution (or rather *the* solution, if it is known to be well-defined!); that is, a local (restricted) solution via a differential equation deduced from the original equation.

In Liu YP [13] (*i. e.*, (3.8) with interchange between x and y!), one finds the equation for f

$$(x-1)x^2yf^2 + (x^2y - x + 1)f + x - 1 - xyh = 0 \tag{6.1.1}$$

where $h = f|_{x=1}$.

From this equation, a parameter θ is introduced to express both h and y such that h is expressed as a function of only y, *i. e.*,

$$\begin{cases} h = \dfrac{4\theta - 3}{(3\theta - 2)^2}; \\ y = (1-\theta)(3\theta - 2). \end{cases} \tag{6.1.2}$$

In order to determine h, observe what is the equation satisfied by h. Because of

$$\begin{cases} \dfrac{dh}{d\theta} = \dfrac{2(5-6\theta)}{(3\theta-2)^3}; \\ \dfrac{dy}{d\theta} = 5 - 6\theta, \end{cases} \tag{6.1.3}$$

we have

$$\frac{dh}{dy} = \frac{2}{(3\theta-2)^3}. \tag{6.1.4}$$

From (6.1.2) and (6.1.4),

$$\begin{cases} y\dfrac{dh}{dy} = 2\tau\left(y\dfrac{dh}{dy} + h\right); \\ \tau = \dfrac{y}{1-3\tau}, \end{cases} \tag{6.1.5}$$

where $\tau = 1 - \theta$.

Lemma 6.1.1. *In equation* (6.1.5), $\tau \in \mathcal{R}_+\{y\}$.

Proof. Let $T_n = \partial_y^n \tau$, where integer $n \geq 0$. From the second case of (6.1.5) and $\tau = y + 3\tau^2$,

$$\begin{aligned} y^0 : T_0 = 3T_0^2 &\Longrightarrow T_0(1-3T_0) = 0, \;\; T_0 \in \mathcal{R}_+, \\ &\Longrightarrow T_0 = 0; \\ y^1 : T_1 = 1 + 6T_0T_1 &\Longrightarrow T_1 = 1. \end{aligned}$$

https://doi.org/10.1515/9783110625837-006

For $n \geq 2$,

$$y^n: \; T_n = 3\sum_{i=1}^{n-1} T_i T_{n-i}.$$

Hence, for $n \geq 0$,

$$T_n = \begin{cases} 0, & \text{when } n = 0; \\ 1, & \text{when } n = 1; \\ 3\sum_{i=1}^{n-1} T_i T_{n-i}, & \text{otherwise.} \end{cases} \tag{6.1.6}$$

Because of $T_0, T_1 \in \mathcal{R}_+$ and (6.1.6), $T_i \in \mathcal{R}_+ (i \leq n-1)$ leads, for any integer $n \geq 2$, to $T_n \in \mathcal{R}_+$, $\tau \in \mathcal{R}_+\{y\}$ as can be seen by induction. □

From (6.1.6),

$$\tau = y + 3y^2 + 18y^3 + 135y^4 + \cdots. \tag{6.1.7}$$

Furthermore, by induction on n, it is shown that, for integer $n \geq 1$,

$$T_n = \frac{3^{n-1}}{n}\binom{2n-2}{n-1}. \tag{6.1.8}$$

By comparing with (6.1.6) and (6.1.8), the combinatorial identity is done: for integer $n \geq 2$,

$$\binom{2n-2}{n-1} = \sum_{i=1}^{n-1} \frac{3n}{i(n-i)}\binom{2i-2}{i-1}\binom{2(n-i)-1}{n-i-1}. \tag{6.1.9}$$

Theorem 6.1.2. *Equation*

$$\begin{cases} y\dfrac{\mathrm{d}h}{\mathrm{d}y} = 2\tau\left(2y\dfrac{\mathrm{d}h}{\mathrm{d}y} + h\right); \\ h|_{y=0} = 1, \end{cases} \tag{6.1.10}$$

is well-defined on $\mathcal{R}_+\{y\}$.

This is equation (17) in Introduction when $a = b = d = 1$ because it is meaningful in a classification for ordinary planar maps.

Proof. Let $H_n = \partial_y^n h$, $n \geq 0$. Write $d = 2y\frac{\mathrm{d}h}{\mathrm{d}y} + h$ and $D_n = \partial_y^n d$ for $n \geq 0$. Because of

$$\left[y\frac{\mathrm{d}h}{\mathrm{d}y}\right]_n = \partial_y^n\left(y\frac{\mathrm{d}h}{\mathrm{d}y}\right) = nH_n, \quad n \geq 1, \tag{6.1.11}$$

and, for $n \geq 0$,

$$\left[2y\frac{\mathrm{d}h}{\mathrm{d}y} + h\right]_n = \partial_y^n d = D_n$$

where

$$D_n = \begin{cases} H_0, & n = 0; \\ (2n+1)H_n, & n \geq 1, \end{cases} \tag{6.1.12}$$

for $n \geq 0$,

$$[\tau d]_n = \begin{cases} 0, & \text{when } n = 0; \\ H_1, & \text{when } n = 1; \\ \sum_{i=1}^{n} T_i D_{n-i}, & \text{when } n \geq 2. \end{cases} \tag{6.1.13}$$

On the basis of equation (6.1.10), by (6.1.6), (6.1.11) and (6.1.13), we have

$$\begin{aligned} y^1 : \left[y\frac{\mathrm{d}h}{\mathrm{d}y} \right]_1 = 2[\tau d]_1 &\Longrightarrow H_1 = 2T_1 H_0 = 0, \\ &\quad \text{by initial condition: } H_0 = 1, \\ &\Longrightarrow H_1 = 2T_1 = 2, \\ y^2 : \left[y\frac{\mathrm{d}h}{\mathrm{d}y} \right]_2 = 2[\tau d]_2 &\Longrightarrow 2H_2 = 2(3H_1 + 3) \\ &\Longrightarrow H_2 = 3H_1 + 3 = 9, \end{aligned} \tag{6.1.14}$$

and, for $n \geq 2$,

$$y^n : \left[y\frac{\mathrm{d}h}{\mathrm{d}y} \right]_n = 2[\tau d]_n \Longrightarrow nH_n = 2\sum_{i=1}^{n} T_i D_{n-i}.$$

Therefore

$$H_n = \frac{2}{n}\sum_{i=1}^{n} T_i D_{n-i}. \tag{6.1.15}$$

In consequence, for integer $n \geq 0$,

$$H_n = \begin{cases} 1, & \text{when } n = 0; \\ 2, & \text{when } n = 1; \\ \frac{2}{n}\sum_{i=1}^{n} T_i D_{n-i}, & \text{when } n \geq 2. \end{cases} \tag{6.1.16}$$

On the basis of (6.1.16), from (6.1.8) and (6.1.12), it is seen that H_n is determined by H_i, $i \leq n-1$. Thus, $H_n \in \mathcal{R}_+$. This leads to the fact that $h \in \mathcal{R}_+\{y\}$ is a solution of equation (6.1.10). Further by considering the uniqueness of the procedure starting from the initial condition, $h \in \mathcal{R}_+\{y\}$ is the only solution of equation (6.1.10). □

In fact, from (6.1.8) and (6.1.12), (6.1.16) becomes

$$\partial_y^n h = \begin{cases} 1, & \text{when } n = 0; \\ 2, & \text{when } n = 1; \\ \frac{2}{n}\sum_{i=1}^{n} \frac{3^{i-1}(2n-2i+1)(2i-2)!}{i!(i-1)!} H_{n-i}, & \text{when } n \geq 2, \end{cases} \tag{6.1.17}$$

for $n \geq 0$.

On the other hand, for integer $n \geq 1$, write $h' = \frac{dh}{dy}$ and $H'_n = \partial_y^n h'$. From

$$\begin{cases} h' = \dfrac{2}{(1-3\tau)^3}; \\ \tau = \left(\dfrac{1}{1-3\tau}\right) y, \end{cases} \tag{6.1.18}$$

and by induction on n, it is shown that

$$H'_n = \frac{2 \cdot 3^{n+1}(2n+2)!}{n!(n+3)!}. \tag{6.1.19}$$

Because of $nH_n = H'_n$ for $n \geq 1$ and by (6.1.19), it is seen that

$$\partial_y^n h = \begin{cases} 1, & \text{when } n = 0 \\ & \text{(initial condition of equation (6.1.10));} \\ \frac{2 \cdot 3^n (2n)!}{n!(n+2)!}, & \text{when } n \geq 1. \end{cases} \tag{6.1.20}$$

Via (6.1.17) and (6.1.20), the identity

$$\frac{3(2n)!}{2(n+2)!n!} = \sum_{i=1}^{n} \frac{(2n-2i+1)!(2i-2)!}{(n-i+2)!(n-i)!} \tag{6.1.21}$$

is found for integer $n \geq 1$.

One might see that by τ, h' is determined. Then it is much simpler than the case that h is determined directly by θ as in Liu YP [13]. This suggests us carefully to choose a parameter, or adaptively to substitute a parameter so that sophistication occurs as little as possible.

The choice of an equation for a function considered is also essential for us to reduce the complexity in the procedure of solving it. For example, in Liu YP [12], the function $Y(x)$ is treated so as to satisfy a differential equation of first order instead of the equation of second order in Tutte WT [80] to make the procedure of determining Y much simpler.

In addition, for proving that an equation is well-defined, usually it is necessary to transform the equation into a suitable equivalence so that the procedure of seeking a solution from the initial condition is as simple as possible.

In what follows, several examples are chosen to address the universality for the methods mentioned in this section.

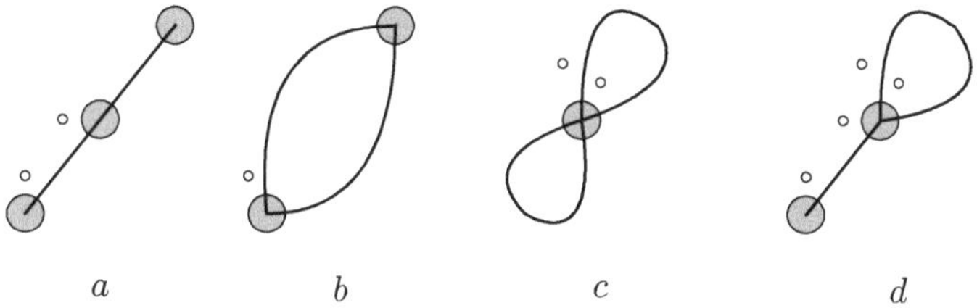

Figure 6.1.1: Root-classes of ordinary planar maps of size 2.

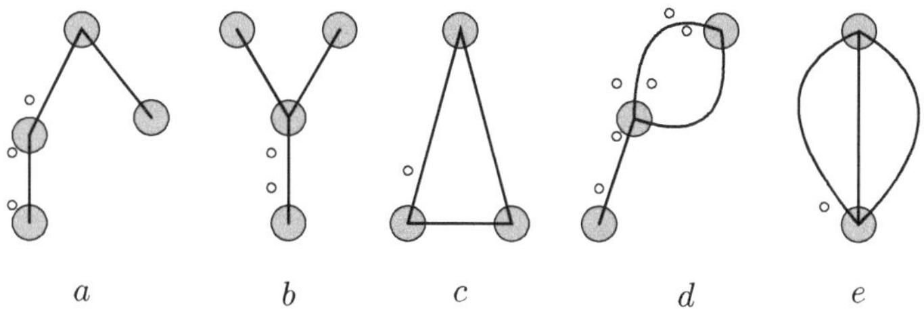

Figure 6.1.2: Root-classes of ordinary planar maps of size 3 I.

Example 1. Classification of ordinary planar rooted maps by size. By (6.1.20), the number of root-isomorphic classes of ordinary planar maps with given size is known. For example, when $n = 0$, 1 represents that ordinary planar maps without edge has 1 root-isomorphic class. This is the vertex map itself. When $n = 1$, 2 shows that such a map of size 1 has 2 classes. They are the link map and the loop map seen, respectively, in the left and the right in Figure 4.2.1. Such maps of size 2 have 9 classes as $2a + b + 2c + 4d$, as shown in Figure 6.1.1.

Ordinary planar maps of size 3 have 54 root-classes. They have $3a + 2b + c + 6d + e$ (*i. e.*, $3 + 2 + 1 + 6 + 1 = 13$) root-classes shown in Figure 6.1.2, $6f + 6g + 3h + 6i + 6j$ ($6 + 6 + 3 + 6 + 6 = 27$) in Figure 6.1.3, $3k + 6l + 3m + 2n$ ($3 + 6 + 3 + 2 = 14$) shown in Figure 6.1.4. The total sum is $13 + 27 + 14 = 54$.

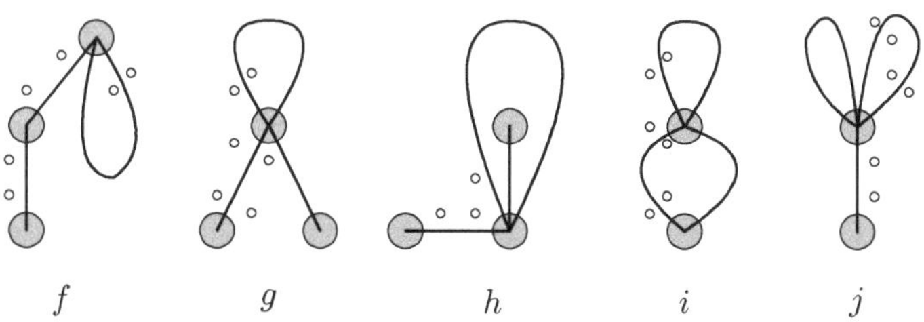

Figure 6.1.3: Root-classes of ordinary planar maps of size 3 II.

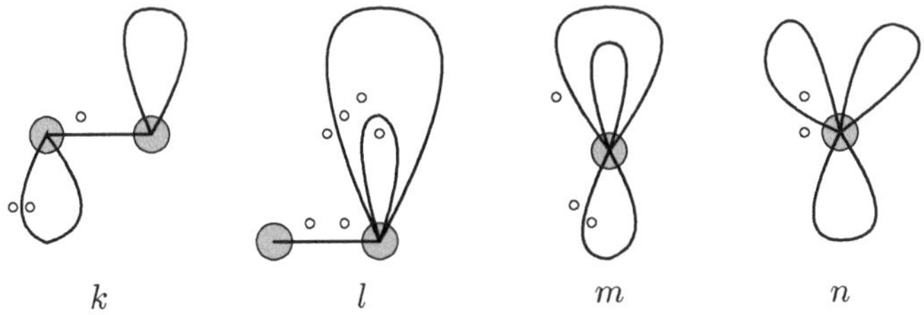

Figure 6.1.4: Root-classes of ordinary planar maps of size 3 III.

Example 2. In Liu YP [29] and Liu YP [31], one might see the equation for $f \in \mathcal{R}\{x, y\}$

$$f^2 - (xh + xy(1 - x)h + 1) + xh = 0 \tag{6.1.22}$$

where $h = f|_{x=1}$.

It is well-known that h and y obey expressions of t as

$$\begin{cases} h = t^2(2 - t); \\ y = \dfrac{(t - 1)(2 - t)}{t^2}. \end{cases} \tag{6.1.23}$$

Because of

$$\begin{cases} \dfrac{\mathrm{d}h}{\mathrm{d}t} = t(4 - 3t); \\ \dfrac{\mathrm{d}y}{\mathrm{d}t} = \dfrac{4 - 3t}{t^3}, \end{cases}$$

we have

$$\frac{\mathrm{d}h}{\mathrm{d}y} = t^4. \tag{6.1.24}$$

Furthermore, the ordinary differential equation

$$\begin{cases} y\dfrac{\mathrm{d}h}{\mathrm{d}y} = (t - 1)h; \\ t = 1 + \dfrac{t^2}{2 - t}y, \end{cases} \tag{6.1.25}$$

is obtained.

Example 3. In Liu YP [36], one might see the equation for $f \in \mathcal{R}\{x, y\}$ with $h = f|_{x=1} \in \mathcal{R}\{x\}$

$$x^2yf^2 + (1 - x + xyh)f + x - 1 = 0. \tag{6.1.26}$$

It is well-known that h and y obey expressions of t (refer to Liu YP [16] and Liu YP [14]), the equation

$$\begin{cases} h = t^2(2 - t); \\ y = \dfrac{t - 1}{t^4}, \end{cases} \tag{6.1.27}$$

is obtained. Because of

$$\begin{cases} \dfrac{dh}{dt} = t(4-3t); \\ \dfrac{dy}{dt} = \dfrac{4-3t}{t^5}, \end{cases}$$

we have

$$\frac{dh}{dy} = t^6. \tag{6.1.28}$$

On the basis of (6.1.27) and (6.1.28), the ordinary differential equation

$$\begin{cases} y\dfrac{dh}{dy} = (t-1)\left(y\dfrac{dh}{dy} + h\right); \\ t = 1 + t^4 y, \end{cases} \tag{6.1.29}$$

is obtained.

Example 4. In Liu YP [19], one finds the equation for $f \in \mathcal{R}\{x,y\}$ with $h = f|_{x=1} \in \mathcal{R}\{x\}$

$$f^2 + ((1+xy)(1-x) - xh)f + x^2(1-x)y(1+h) = 0. \tag{6.1.30}$$

It is well-known that h and y obey expressions of η as a parameter as

$$\begin{cases} h = \dfrac{1}{3}\eta(4-\eta); \\ y = \dfrac{1}{27}(\eta-1)(4-\eta)^2. \end{cases} \tag{6.1.31}$$

Because of

$$\begin{cases} \dfrac{dh}{d\eta} = \dfrac{2}{3}(2-\eta); \\ \dfrac{dy}{d\eta} = \dfrac{1}{9}(\eta-2)(\eta-4), \end{cases}$$

we have

$$\frac{dh}{dy} = \frac{6}{4-\eta}. \tag{6.1.32}$$

On the basis of (6.1.31) and (6.1.32), the ordinary differential equation as

$$\begin{cases} (4-\eta)\dfrac{dh}{dy} = 6; \\ \eta = 1 + \dfrac{27}{(4-\eta)^2}y, \end{cases} \tag{6.1.33}$$

is obtained.

Example 5. In Tutte WT [80], one might see the enufunction $h = h(y)$ of simple planar rooted triangulations and y expressed by s as

$$\begin{cases} h = \dfrac{s(1+s)}{(1-s)^2}; \\ y = -s(1+s)^2. \end{cases} \tag{6.1.34}$$

Because of

$$\begin{cases} \dfrac{dh}{ds} = \dfrac{3s+1}{(1-s)^3}; \\ \dfrac{dy}{ds} = -(1+3s)(1+s), \end{cases}$$

we have

$$\frac{dh}{dy} = -\frac{1}{(1-s)^3(1+s)}. \tag{6.1.35}$$

On the basis of (6.1.34) and (6.1.35), the ordinary differential equation

$$\begin{cases} (1-s)y\dfrac{dh}{dy} = h; \\ s = -\dfrac{y}{(1+s)^2}, \end{cases} \tag{6.1.36}$$

is obtained.

Example 6. In Tutte WT [80], the enufunction $h = h(y)$ of strict planar rooted triangulations and y are expressed by λ as a parameter as

$$\begin{cases} h = \lambda(3-2\lambda); \\ y = \lambda^3(1-\lambda). \end{cases} \tag{6.1.37}$$

Because of

$$\begin{cases} \dfrac{dh}{ds} = \lambda^2(3-4\lambda); \\ \dfrac{dy}{ds} = \lambda^3(1-\lambda), \end{cases}$$

we have

$$\frac{dh}{dy} = \frac{1}{\lambda^2}. \tag{6.1.38}$$

On the basis of (6.1.37) and (6.1.38), the ordinary differential equation

$$\begin{cases} (3-2\lambda)y\dfrac{dh}{dy} = (1-\lambda)h; \\ \lambda = 1-\dfrac{y}{\lambda^3}, \end{cases} \tag{6.1.39}$$

is obtained.

6.2 Sum of petal bundles

For determining the number of root-isomorphic classes of petal bundles on orientable surfaces, the following first order differential equation occurs (see Liu YP [47]):

$$\begin{cases} 2x^2\dfrac{\mathrm{d}h}{\mathrm{d}x} = -1 + (1-x)h; \\ h_0 = h|_{x=0} = 1. \end{cases} \tag{6.2.1}$$

This is equation (18) in Introduction when $a = c = d = 1$ because it is meaningful in a classification for petal bundles on all orientable surfaces.

Theorem 6.2.1. *Equation* (6.2.1) *is well-defined on* $\mathcal{R}_+\{x\}$.

Proof. The first case of equation (6.2.1) is transformed into one of its equivalences as

$$h = 1 + xh + 2x^2\frac{\mathrm{d}h}{\mathrm{d}x}. \tag{6.2.2}$$

Let $H_n = \partial_x^n h$, $n \geq 0$. From the initial condition,

$$H_0 = h_0 = 1 \in \mathbb{Z}_+. \tag{6.2.3}$$

Because of

$$\begin{cases} \partial_x^n xh = \partial_x^{n-1} h = H_{n-1}; \\ \partial_x^n x^2\dfrac{\mathrm{d}h}{\mathrm{d}x} = \partial_x^{n-2}\dfrac{\mathrm{d}h}{\mathrm{d}x} = (n-1)H_{n-1} \end{cases}$$

for $n \geq 1$, by (6.2.2),

$$H_n = H_{n-1} + 2(n-1)H_{n-1} = (2n-1)H_{n-1} \in \mathbb{Z}_+. \tag{6.2.4}$$

Thus, $h \in \mathcal{R}_+\{x\}$ obtained by (6.2.3) and (6.2.4) satisfies equation (6.2.2), and hence is a solution of equation (6.2.1).

From the uniqueness of H_n ($n \geq 1$) for H_0, this solution is the only one. □

In fact, from the recursion deduced from (6.2.3) and (6.2.4), it is seen that, for any integer $n \geq 1$,

$$H_n = (2n-1)!! = \frac{(2n-1)!}{2^{n-1}(n-1)!}. \tag{6.2.5}$$

In Liu YP [47], one might see the following ordinary equation for g:

$$\begin{cases} 4x^2\dfrac{\mathrm{d}g}{\mathrm{d}x} = (1-2x)g - x\left(h + 2x\dfrac{\mathrm{d}h}{\mathrm{d}x}\right); \\ \left.\dfrac{\mathrm{d}g}{\mathrm{d}x}\right|_{x=0} = 1, \end{cases} \tag{6.2.6}$$

where h determined by (6.2.5) is the solution of equation (6.2.1).

Note carefully that h satisfies (6.2.1), which is equivalent to

$$\begin{cases} 4x^2\dfrac{\mathrm{d}g}{\mathrm{d}x} = (1-2x)g - h + 1; \\ g_0 = g|_{x=0} = 0. \end{cases} \tag{6.2.7}$$

Theorem 6.2.2. *Equation* (6.2.7) *is well-defined on* $\mathcal{R}_+\{x\}$.

Proof. The first case of equation (6.2.7) is equivalent to

$$g = 4x^2\frac{\mathrm{d}g}{\mathrm{d}x} + 2xg + h - 1. \tag{6.2.8}$$

Let $G_n = \partial_x^n g$, $n \geq 0$. Because of

$$\begin{aligned} \partial_x^n xg &= \partial_x^{n-1} g = G_{n-1}, \ n \geq 1; \\ \partial_x^n \frac{\mathrm{d}g}{\mathrm{d}x} &= G_{n+1}\partial_x^n \frac{\mathrm{d}x^{n+1}}{\mathrm{d}x} = (n+1)G_{n+1}, \quad n \geq 0; \\ \partial_x^n x^2\frac{\mathrm{d}g}{\mathrm{d}x} &= \partial_x^{n-2}\frac{\mathrm{d}g}{\mathrm{d}x} = (n-1)G_{n-1}, \end{aligned}$$

for (6.2.5),

$$y^0 : G_0 = H_0 - 1 \implies G_0 = 0(\text{initial condition!}) \tag{6.2.9}$$

and, for $n \geq 1$,

$$\begin{aligned} y^n : G_n &= 4(n-1)G_{n-1} + 2G_{n-1} + H_n \\ &= (4n-2)G_{n-1} + H_n. \end{aligned} \tag{6.2.10}$$

From (6.2.9) and (6.2.10), $G_0 \in \mathbb{Z}_+$ and $G_n \in \mathbb{Z}_+$ are obtained, respectively. Thus, $g \in \mathcal{R}_+\{x\}$ satisfies (6.2.8) and hence is a solution of equation (6.2.7).

From the uniqueness of G_n $(n \geq 1)$ for G_0, this solution is the only one. □

On the basis of (6.2.9) and (6.2.10), we find that, for $n \geq 2$,

$$G_n = H_n + \prod_{i=2}^{n}(4i-2) + \sum_{i=2}^{n-1} H_i \prod_{j=i+1}^{n}(4j-2). \tag{6.2.11}$$

On the other hand, we look for an equation satisfied by $f = g + h$ where h and g are, respectively, the solutions of equation (6.2.1) and equation (6.2.7).

Because h and g are, respectively, the solutions of equations (6.2.1) and (6.2.7), h and g satisfy, respectively, (6.2.2) and (6.2.8), and hence

$$g + h = \left(-1 + h + 2xg + 4x^2\frac{\mathrm{d}g}{\mathrm{d}x}\right) + \left(1 + xh + 2x^2\frac{\mathrm{d}h}{\mathrm{d}x}\right).$$

From (6.2.2), by substituting $xh + 2x^2\frac{dh}{dx}$ in the first parenthesis, of $-1 + h$,

$$f = 1 + 2xf + 4x^2\frac{df}{dx}. \tag{6.2.12}$$

By considering $f_0 = f|_{x=0} = g_0 + h_0 = 1$ and then using (6.2.12), f satisfies the equation

$$\begin{cases} 4x^2\dfrac{df}{dx} = -1 + (1-2x)f; \\ f_0 = f|_{x=0} = 1. \end{cases} \tag{6.2.13}$$

Theorem 6.2.3. *Equation* (6.2.13) *is well-defined on* $\mathcal{R}_+\{x\}$.

Proof. Take $f = g+h$ such that g and h are, respectively, the solutions of equation (6.2.1) and equation (6.2.7), then from Theorems 6.2.1 and 6.2.2, $g \in \mathcal{R}_+\{x\}$ and $h \in \mathcal{R}_+\{x\}$. Thus, $f \in \mathcal{R}_+\{x\}$. As mentioned above, f is a solution of equation (6.2.13) in $\mathcal{R}_+\{x\}$.

On the other hand, if f determined by $F_n = \partial_x^n f$, $n \geq 0$, there is another solution of equation (6.2.13) in $\mathcal{R}_+\{x\}$, of the form

$$\begin{aligned} \partial_x^n xf &= \partial_x^{n-1} f = F_{n-1}, \ n \geq 1; \\ \partial_x^n \frac{df}{dx} &= F_{n+1}\partial_x^n \frac{dx^{n+1}}{dx} = (n+1)F_{n+1}, \quad n \geq 0; \\ \partial_x^n x^2\frac{df}{dx} &= \partial_x^{n-2}\frac{df}{dx} = (n-1)F_{n-1}, \end{aligned}$$

and considering (6.2.12), it is seen that

$$y^0 : F_0 = 1 \text{ (the initial condition of equation (6.2.13))} \tag{6.2.14}$$

and that, for $n \geq 1$,

$$\begin{aligned} y^n : F_n &= 2F_{n-1} + 4(n-1)F_{n-1} \\ &= (4n-2)F_{n-1}. \end{aligned} \tag{6.2.15}$$

Because of the uniqueness of $F_n \in \mathbb{Z}_+$ ($n \geq 1$) in the procedure shown as (6.2.15) on $F_0 = 1 \in \mathbb{Z}_+$, it is the only solution of equation (6.2.13). □

On the basis of Theorem 6.2.3, and from $F_0 = 1$ and (6.2.15), the only possibility is that $f = g + h$.

Because of $(4n-2)F_{n-1} = 2(2n-1)F_{n-1}$, on comparing with (6.2.4) and $F_0 = H_0 = 1$, we have, for $n \geq 1$,

$$F_n = 2^n H_n = \frac{2^n(2n-1)!}{2^{n-1}(n-1)!}. \tag{6.2.16}$$

By considering $f = g + h$ (*i. e.*, $g = f - h$) with (6.2.5) and (6.2.16), we have, for $n \geq 1$,

$$G_n = (2^n - 1)H_n = \frac{(2^n-1)(2n-1)!}{2^{n-1}(n-1)!}. \tag{6.2.17}$$

On account of (6.2.11), we obtain, for $n \geq 2$,

$$2(2^{n-1} - 1)H_n = \prod_{i=2}^{n}(4i - 2) + \sum_{i=2}^{n-1} H_i \prod_{j=i+1}^{n}(4j - 2) \tag{6.2.18}$$

where H_i, $2 \leq i \leq n$, is given by (6.2.5). Thus, the identity

$$2(2^{n-1} - 1)(2n - 1)!! = \prod_{i=2}^{n}(4i - 2) + \sum_{i=2}^{n-1} (2i - 1)!! \prod_{j=i+1}^{n}(4j - 2) \tag{6.2.19}$$

is concluded to.

Example 1. Root-isomorphic classification of orientable petal bundles on surfaces by size. The case of genus 0 surface (*i. e.*, S_0) is seen in Figure 3.1.4. Orientable petal bundles of sizes 0, 1, 2, 3 and 4 on S_0, have, respectively, 1, 1, 2, 5 and 14 classes, in total $1 + 1 + 2 + 5 + 14 = 23$ classes.

In Figure 6.2.1, we show the cases of genus 1 (*i. e.*, S_1) and genus 2 (*i. e.*, S_2) surfaces. In the figure, i and $\bar{i}$ (or i) represent the edge i on which two sides are with different directions (or the same direction) for $i = 1, 2, 3$ and 4. On S_1, because of no map (the petal bundle is a specific case) with 0 1 edges, only observe those of sizes 2, 3 and 4. For size 2, only $1a$ represents 1 class. For size 3, $4b + 3c + 3d$, we have a total of 10 classes. For size 4, $8e + 8f + 8g + 8h + 16i + 8j + 4k + 4l + 4m + 2n$, we have a total of 70 classes. On S_2, only size 4 is a possibility. In this case, $4o + 8p + 8q + 1r$, and we have in total 21 classes. See Figure 6.2.2.

Example 2. Root-isomorphic classification of non-orientable petal bundles on surfaces by size. In Figures 6.2.3 and 6.2.4, it is seen that petal bundles of size 1 only occur on a non-orientable surface of genus 1, *i. e.*, $S_{\tilde{1}}$; Size 2 is only possible on $S_{\tilde{1}}$ and $S_{\tilde{2}}$; and size 3 is only possible on $S_{\tilde{1}}$, $S_{\tilde{2}}$ and $S_{\tilde{3}}$.

In Figure 6.2.3, it is seen that we have petal bundles of size 1 on $S_{\tilde{1}}$, only $1a$, *i. e.*, 1 class. There are petal bundles of size 2 on $S_{\tilde{1}}$ and $S_{\tilde{2}}$, respectively, $1b + 3c$ and $2d + 2e$, *i. e.*, 4 and 4 classes.

In Figure 6.2.4, we have root-isomorphic classes of petal bundles on non-orientable surfaces with size 3. In the figure, e, like a, b, c and d, shows such classes of petal bundles on surface $S_{\tilde{1}}$.

Thus, the number of such classes for petal bundles on non-orientable surfaces of genus $\tilde{1}$ is $6a + 6b + 6c + 3d + 1e$, in total $6 + 6 + 6 + 3 + 1 = 22$; f–m show the classes on surface $S_{\tilde{2}}$, and n–u on the surface $S_{\tilde{3}}$.

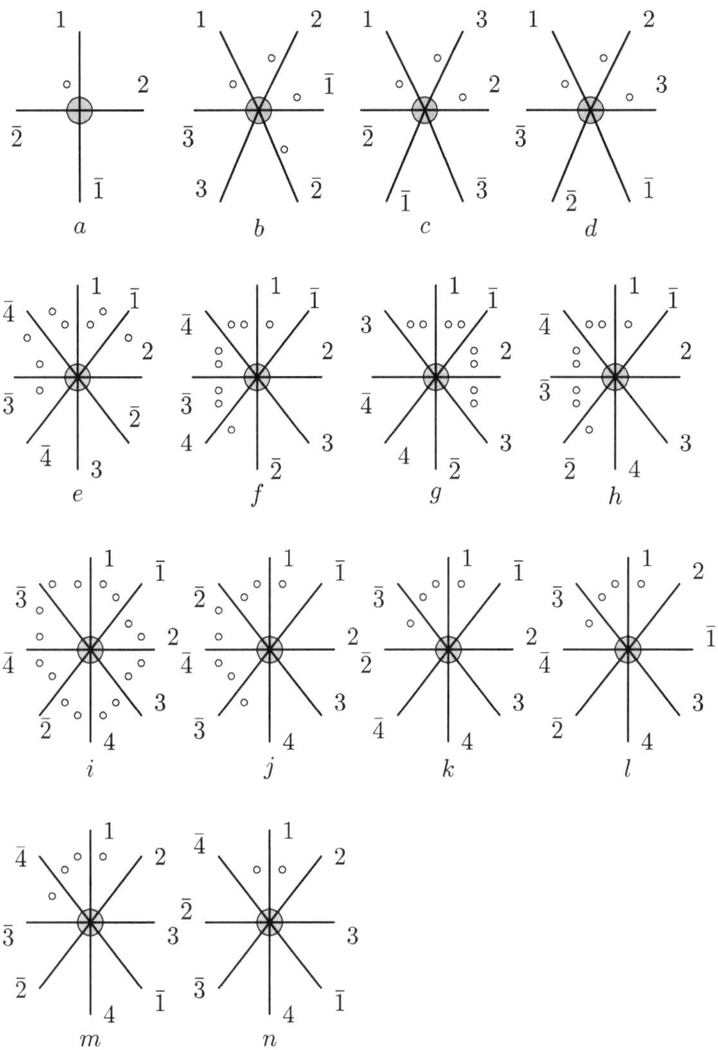

Figure 6.2.1: Petal bundles on orientable surface of genus 1 with sizes: 2–4.

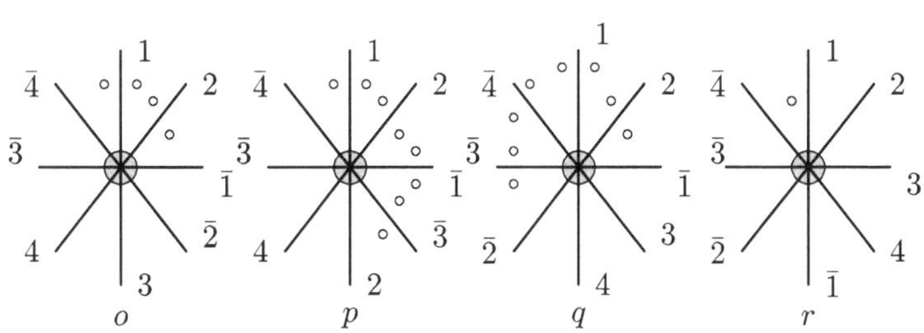

Figure 6.2.2: Petal bundles on orientable surface of genus 2 with size 4.

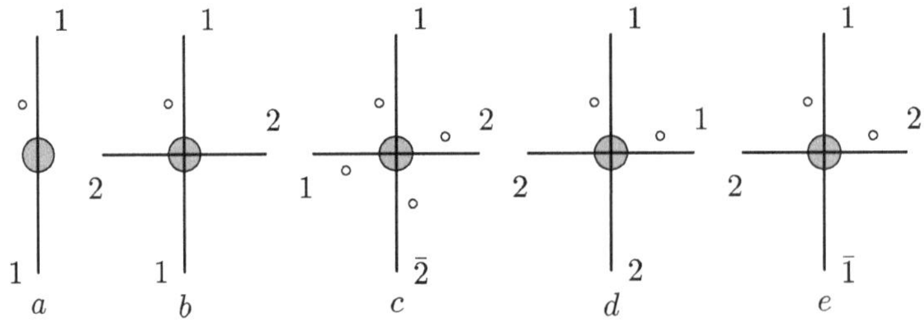

Figure 6.2.3: Petal bundles on non-orientable surfaces with sizes 1–2.

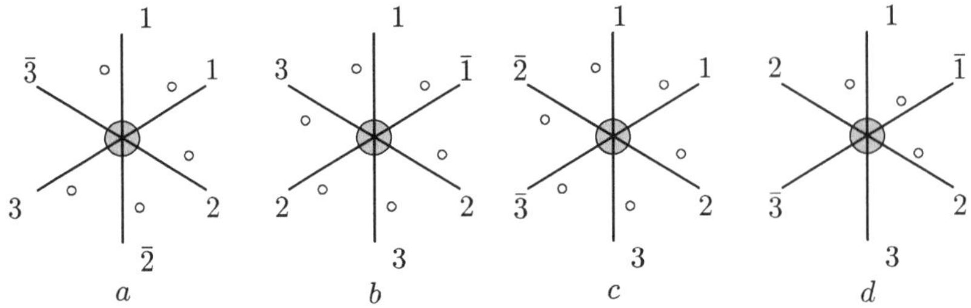

Figure 6.2.4: Petal bundles on non-orientable surfaces with size 3.

Hence, the number of such classes of petal bundles with size 3 on the non-orientable surface of genus $\tilde{2}$ is $6f+12g+6h+6i+3j+3k+3l+3m$, in total $6+12+6+6+3+3+3+3 = 42$ and on the non-orientable surface of genus $\tilde{3}$, $2n + 6o + 12p + 3q + 6r + 6s + 3t + 3u$, a total of $2 + 6 + 12 + 3 + 6 + 6 + 3 + 3 = 41$. See Figure 6.2.4.

6.3 Orientable sum

In Liu YP [44] (1999, Theorem 8.5.1, p. 268), the equation for $f \in \mathcal{R}\{x\}$

$$\begin{cases} 2x^2 \dfrac{\mathrm{d}f}{\mathrm{d}x} = -1 + (1-x)f - xf^2; \\ f_0 = f|_{x=0} = 1, \end{cases} \tag{6.3.1}$$

occurs. The aim is a solution in $\mathcal{R}_+\{x\}$ if there is any.

This is equation (19) in Introduction when $a = b = c = d = 1$ because it is meaningful in a classification for ordinary maps on all orientable surfaces.

Equation (6.3.1) is transformed into its equivalence

$$\begin{cases} f = 1 + xf + 2x^2 \dfrac{\mathrm{d}f}{\mathrm{d}x} + xf^2; \\ f_0 = f|_{x=0} = 1. \end{cases} \tag{6.3.2}$$

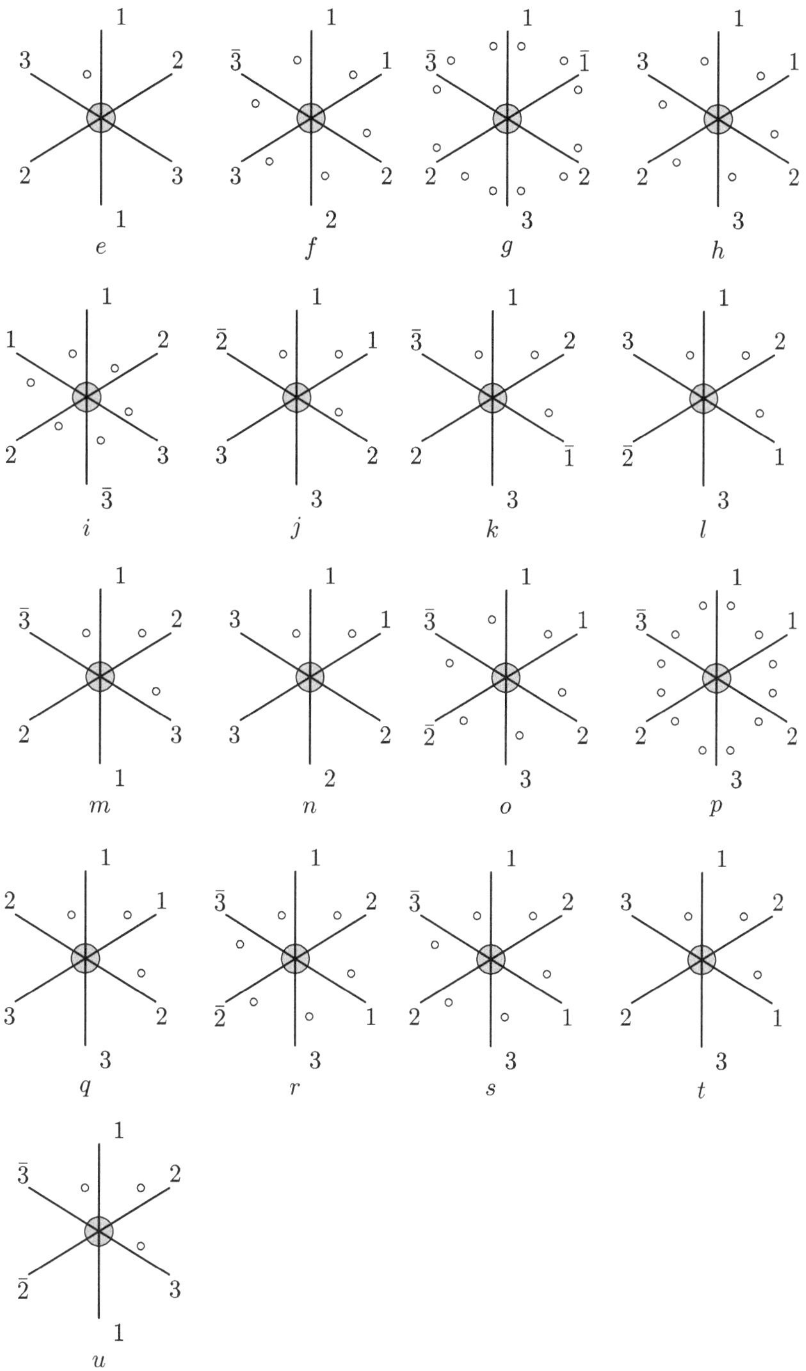

Figure 6.2.4: Continued.

Because of $f \in \mathcal{R}_+\{x\}$, f is determined by $[f]_n = \partial_x^n f$ for $n \geq 0$. Let $F_n = [f]_n$ for $n \geq 0$, then

$$[xf]_n = [f]_{n-1} = \begin{cases} 0, & \text{when } n = 0; \\ F_{n-1}, & \text{when } n \geq 1, \end{cases} \tag{6.3.3}$$

and

$$[xf^2]_n = [f^2]_{n-1} = \sum_{i=0}^{n} F_n F_{n-i}, \quad n \geq 0. \tag{6.3.4}$$

Since, for any integer $n \geq 0$,

$$\left[\frac{\mathrm{d}f}{\mathrm{d}x}\right]_n = \begin{cases} 0, & \text{when } n = 0; \\ (n+1)F_{n+1}, & \text{when } n \geq 1, \end{cases}$$

we have

$$\left[x^2\frac{\mathrm{d}f}{\mathrm{d}x}\right]_n = \left[\frac{\mathrm{d}f}{\mathrm{d}x}\right]_{n-2} = \begin{cases} 0, & \text{when } 2 \geq n \geq 0; \\ (n-1)F_{n-1}, & \text{when } n \geq 3. \end{cases} \tag{6.3.5}$$

On the basis of (6.3.3)–(6.3.5), by equation (6.3.2), we have

$$x^0 : F_0 = 1 + [xf]_0 + 2\left[x^2\frac{\mathrm{d}f}{\mathrm{d}x}\right]_0 + [xf^2]_0 = 0 \Longrightarrow F_0 = 1, \tag{6.3.6}$$

$$x^1 : F_1 = [xf]_1 + 2\left[x^2\frac{\mathrm{d}f}{\mathrm{d}x}\right]_1 + [xf^2]_1 = F_0 + F_0^2 \Longrightarrow F_1 = 1 + 1 = 2, \tag{6.3.7}$$

$$x^2 : \left\{\begin{aligned} F_2 &= [xf]_2 + 2\left[x^2\frac{\mathrm{d}f}{\mathrm{d}x}\right]_2 + [xf^2]_2 \\ &= F_1 + 2F_1 + 2F_0F_1 \\ &= 2 + 4 + 4 \end{aligned}\right\} \Longrightarrow F_2 = 10, \tag{6.3.8}$$

and, for $n \geq 2$,

$$x^n : \left\{\begin{aligned} F_n &= [xf]_n + 2\left[x^2\frac{\mathrm{d}f}{\mathrm{d}x}\right]_n + [xf^2]_n \\ &= F_{n-1} + 2(n-1)F_{n-1} + \sum_{i=0}^{n-1} F_iF_{n-1-i} \end{aligned}\right\} \tag{6.3.9}$$

$$\Longrightarrow F_n = (2n-1)F_{n-1} + \sum_{i=0}^{n-1} F_iF_{n-1-i}.$$

Lemma 6.3.1. *For any integer $n \geq 0$,*

$$\left.\begin{array}{rl} 1 & (\text{when } n = 0) \\ 2 & (\text{when } n = 1) \\ (2n-1)F_{n-1} + \sum_{i=0}^{n-1} F_iF_{n-1-i} & (\text{when } n \geq 2) \end{array}\right\} = F_n \in \mathbb{Z}_+. \tag{6.3.10}$$

Proof. Because of $F_0 = 1$, $F_1 = 2 \in \mathbb{Z}_+$, the conclusion is true for $n = 0$ and 1. For $n \geq 2$, by employing induction assumption we find that, for any i, $0 \leq i \leq n-1$, $F_i \in \mathbb{Z}_+$. Because of all $F_{n-1}, F_i, F_{n-1-i} \in \mathbb{Z}_+$, (6.3.10) leads to $F_n \in \mathbb{Z}_+$. This is the conclusion. □

This tells us that the function f determined by (6.3.10) in Lemma 6.3.1 obeys $f \in \mathcal{R}_+\{x\}$.

Theorem 6.3.2. *Equation* (6.3.1) *is well-defined on* $\mathcal{R}\{x\}$.

Proof. It is easily understood that the function f obtained by (6.3.10) is a solution of equation (6.3.1) because of $F_0 = 1$, the initial condition of the equation. Then we consider the uniqueness of the procedure for the initial condition. This solution is the only one of equation (6.3.1). □

Example 1. On all orientable surfaces, root-isomorphic classes of ordinal maps with given size. Because of the results in, e. g., Liu YP [44] (1999, Theorem 8.5.1, p. 268), or in dedeal, in Liu, YP [48] (2001, Theorem 9.5.1, p. 314), the enufunction of ordinary maps with size as a parameter on all orientable surfaces satisfies equation (6.3.1), and hence, Theorem 6.3.2 leads to the solution f determined by (6.3.10) as the enufunction. Here, only those for size at most 3 are shown.

From (6.3.10), we have $F_0 = 1$, $F_1 = 2$, $F_2 = 10$ and $F_3 = 74$. Among $F_0 + F_1 + F_2 = 1 + 2 + 10 = 13$ maps, only 1 is not on S_0 (sphere, or plane). This map has 2 edges with 1 vertex, shown in a of Figure 6.3.1 on S_1 (orientable surface of genus 1, or torus).

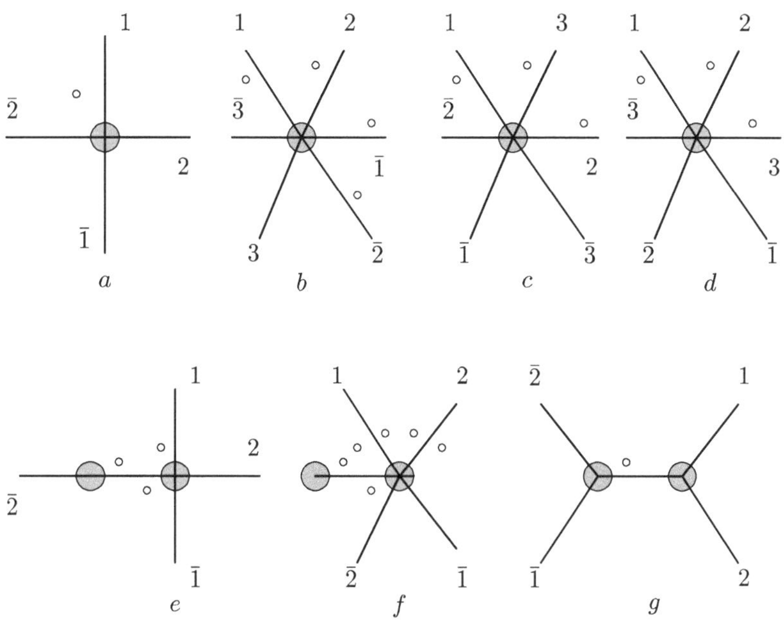

Figure 6.3.1: Classes of maps of size 3 on orientable surfaces of genus not 0.

Because all maps of size 3 only are allowed to be on orientable surfaces of genus at most 1, among all $F_3 = 74$ maps, only 20 are non-planar. They are all on S_1. As shown in Figure 6.3.1, see b–g, *i. e.*, $4b+3c+3d+3e+6f+1g$, we have a total of $4+3+3+3+6+1 = 20$.

Example 2. Consider the generalized equation

$$\begin{cases} ax^2\dfrac{\mathrm{d}f}{\mathrm{d}x} = -b + b(1-x)f - cxf^2; \\ f|_{x=0} = 1, \end{cases} \tag{6.3.11}$$

where $a, b, c \in \mathbb{Z}_+$, $b|a\ b|c$.

For convenience, the first case of equation (6.3.11) is replaced by its equivalence,

$$bf = b + bxf + ax^2\frac{\mathrm{d}f}{\mathrm{d}x} + cxf^2.$$

From $b|a$, $\alpha = a/b \in \mathbb{Z}_+$. From $b|c$, $\beta = c/b \in \mathbb{Z}_+$. The equation becomes

$$f = 1 + xf + \alpha x^2\frac{\mathrm{d}f}{\mathrm{d}x} + \beta xf^2. \tag{6.3.12}$$

On the basis of (6.3.3)–(6.3.5), (6.3.12) leads to

$$\begin{aligned} x^0 : F_0 &= 1 + [xf]_0 + \alpha\left[x^2\frac{\mathrm{d}f}{\mathrm{d}x}\right]_0 + \beta[xf^2]_0 = 0 \\ &\Longrightarrow F_0 = 1, \end{aligned} \tag{6.3.13}$$

$$\begin{aligned} x^1 : F_1 &= [xf]_1 + \alpha\left[x^2\frac{\mathrm{d}f}{\mathrm{d}x}\right]_1 + \beta[xf^2]_1 = F_0 + \beta F_0^2 \\ &\Longrightarrow F_1 = 1 + \beta, \end{aligned} \tag{6.3.14}$$

and, for $n \geq 2$,

$$\begin{aligned} x^n : &\begin{Bmatrix} F_n = [xf]_n + \alpha\left[x^2\dfrac{\mathrm{d}f}{\mathrm{d}x}\right]_n + \beta[xf^2]_n \\ = F_{n-1} + \alpha(n-1)F_{n-1} + \beta\displaystyle\sum_{i=0}^{n-1} F_iF_{n-1-i} \end{Bmatrix} \\ &\Longrightarrow F_n = (\alpha n - \alpha + 1)F_{n-1} + \beta\sum_{i=0}^{n-1} F_iF_{n-1-i}. \end{aligned} \tag{6.3.15}$$

Lemma 6.3.3. *For any integer $n \geq 0$,*

$$\left.\begin{aligned} 1 \quad &(when\ n = 0) \\ 1+\beta \quad &(when\ n = 1) \\ (\alpha n - \alpha + 1)F_{n-1} + \beta\textstyle\sum_{i=0}^{n-1} F_iF_{n-1-i} \quad &(when\ n \geq 2) \end{aligned}\right\} = F_n \in \mathbb{Z}_+. \tag{6.3.16}$$

Proof. Because of $F_0 = 1$, $F_1 = 1 + \beta \in \mathbb{Z}_+$, and for any integer $n \geq 2$, by the induction assumption, $F_i \in \mathbb{Z}_+, 0 \leq i \leq n-1$, then (6.3.16) leads to $F_n \in \mathcal{R}_+$. This is the conclusion. □

This lemma shows that $f \in \mathcal{R}_+\{x\}$ is determined by (6.3.16).

Theorem 6.3.4. *Equation* (6.3.11) *is well-defined on* $\mathcal{R}\{x\}$.

Proof. In (6.3.16), since $F_0 = 1$ is just the initial condition of equation (6.3.11), from the function $f \in \mathcal{R}_+\{x\} \subseteq \mathcal{R}\{x\}$ determined by (6.3.16), it is seen that f is a solution of equation (6.3.11).

Furthermore, from the uniqueness of f as determined by (6.3.16) for the initial condition F_0, f is the only solution of equation (6.3.11). □

Because of F_n, $n \geq 2$, being of the form of a finite sum of positive terms as shown in (6.3.16), all coefficients of f are expressed in the form of a finite sum of positive integers.

6.4 Non-orientable sum

In Liu YP [46] (2003, Theorem 9.6, p. 209. Attention should be paid to the expression of $b(x)$, "–" being replaced by "+"!); one might find

$$\begin{cases} 4x^2\dfrac{\mathrm{d}f}{\mathrm{d}x} = \alpha(x)f - xf^2 - 2x\beta(x); \\ \left.\dfrac{\mathrm{d}f}{\mathrm{d}x}\right|_{x=0} = 1, \end{cases} \tag{6.4.1}$$

where

$$\begin{cases} \alpha(x) = 1 - 2x - 2xf_{\mathrm{Orien}}; \\ \beta(x) = f_{\mathrm{Orien}} + 2x\dfrac{\mathrm{d}f_{\mathrm{Orien}}}{\mathrm{d}x}. \end{cases} \tag{6.4.2}$$

Here, f_{Orien} as determined by (6.3.10) is the solution of equation (6.3.1).

This is equation (20) in Introduction when $a = c = d = 1$ because it is meaningful in a classification for petal bundles on all non-orientable surfaces.

For convenience, the first case of equation (6.3.1) is transformed into one of its equivalences by

$$\begin{aligned} f &= xb(x) + 2x(1 + f_{\mathrm{Orien}})f + 4x^2\frac{\mathrm{d}f}{\mathrm{d}x} + xf^2 \\ &= A(x) + 2B(x)f + 2xf_{\mathrm{Orien}}f + xf^2 \end{aligned} \tag{6.4.3}$$

where

$$\begin{cases} A(x) = x\left(f_{\mathrm{Orien}} + 2x\dfrac{\mathrm{d}f_{\mathrm{Orien}}}{\mathrm{d}x}\right); \\ B(x) = x\left(f + 2x\dfrac{\mathrm{d}f}{\mathrm{d}x}\right). \end{cases} \tag{6.4.4}$$

For integer $n \geq 0$, let $O_n = [f_{\text{Orien}}]_n$, *i. e.*, $\partial_x^n f_{\text{Orien}}$. From (6.3.10),

$$O_n = \begin{cases} 1, & \text{when } n = 0; \\ 2, & \text{when } n = 1; \\ (2n-1)O_{n-1} + \sum_{i=0}^{n-1} O_i O_{n-1-i}, & \text{when } n \geq 2. \end{cases} \tag{6.4.5}$$

For any integer $n \geq 1$,

$$\begin{aligned} [A(x)]_n &= \left[f_{\text{Orien}} + 2x \frac{\mathrm{d}f_{\text{Orien}}}{\mathrm{d}x} \right]_{n-1} \\ &= O_{n-1} + 2\left[\frac{\mathrm{d}f_{\text{Orien}}}{\mathrm{d}x} \right]_{n-2} \\ &= O_{n-1} + 2(n-1)O_{n-1} \\ &= (2n-1)O_{n-1} \end{aligned} \tag{6.4.6}$$

and

$$\begin{aligned} [B(x)]_n &= \left[f + 2x \frac{\mathrm{d}f}{\mathrm{d}x} \right]_{n-1} \\ &= F_{n-1} + 2\left[\frac{\mathrm{d}f}{\mathrm{d}x} \right]_{n-2} \\ &= F_{n-1} + 2(n-1)F_{n-1} \\ &= (2n-1)F_{n-1} \end{aligned} \tag{6.4.7}$$

where $F_n = [f]_n$, $n \geq 0$.

On the basis of equation (6.4.4), for integer $n \geq 0$,

$$[f]_n = [A(x)]_n + 2[B(x)f]_n + 2[f_{\text{Orien}} f]_{n-1} + [f^2]_{n-1}. \tag{6.4.8}$$

By employing (6.4.6) and (6.4.7),

$$x^0 : F_0 = 0 + 0 + 0 + 0 = 0,$$

by noticing that $O_0 = 1$ and $F_0 = 0$,

$$x^1 : F_1 = 1 + 0 + 0 + 0 = 1$$

and, for any integer $n \geq 2$, by noticing that $O_0 = 1$ and $F_0 = 0$ as well,

$$\begin{aligned} F_n &= (2n-1)O_{n-1} + 2(2n-1)F_{n-1} + 2\sum_{i=1}^{n-1} F_i O_{n-1-i} + \sum_{i=1}^{n-2} F_i F_{n-1-i} \\ &= (2n-1)O_{n-1} + 4nF_{n-1} + \sum_{i=1}^{n-2} F_i (2O_{n-1-i} + F_{n-1-i}). \end{aligned}$$

In consequence, for integer $n \geq 0$,

$$F_n = \begin{cases} 0, & \text{when } n = 0; \\ 1, & \text{when } n = 1; \\ (2n-1)O_{n-1} + 4nF_{n-1} & \\ \quad + \sum_{i=1}^{n-2} F_i(2O_{n-1-i} + F_{n-1-i}), & \text{when } n \geq 2. \end{cases} \tag{6.4.9}$$

Lemma 6.4.1. *For any integer $n \geq 0$, $F_n \in \mathbb{Z}_+$ is determined by* (6.4.9).

Proof. First, $F_0 = 0 \in \mathbb{Z}_+$ and $F_1 = 1 \in \mathbb{Z}_+$. Then, for integer $n \geq 2$, by induction, assume $F_i \in \mathcal{R}_+$, $0 \leq i \leq n-1$. By Lemma 6.3.1, $O_n \in \mathbb{Z}_+$, $n \geq 0$. By the assumption, $F_{n-1} \in \mathbb{Z}_+$ and $F_i, F_{n-1-i} \in \mathbb{Z}_+$, $1 \leq i \leq n-2$. From (6.4.9), $F_n \in \mathbb{Z}_+$ is deduced. The conclusion is drawn. □

This lemma tells us that the function f determined by (6.4.9) is in $\mathcal{R}_+\{x\} \subseteq \mathcal{R}\{x\}$.

Theorem 6.4.2. *Equation* (6.4.1) *is well-defined on* $\mathcal{R}\{x\}$.

Proof. Because f is determined by (6.4.9) satisfying (6.4.3) the first case of equation (6.4.1) with $F_1 = 1 = \frac{\mathrm{d}f}{\mathrm{d}x}|_{x=0}$, Lemma 6.4.1 leads us to see that f is a solution of equation (6.4.1).

Furthermore, by considering the uniqueness of $\{F_n \mid n \geq 0\}$ determined from (6.4.9) for the initial condition $F_1 = 1$, f determined by $\{F_n \mid n \geq 0\}$ is the only solution of equation (6.4.1). □

It is shown from the theorem that $f = f_{\mathrm{Non}}$ determined by (6.4.9) is just the solution of equation (6.4.1).

Example 1. Root-isomorphic classes of non-orientable maps. Since a map whose underline graph without circuit is never non-orientable (a loop is a circuit in its own right!), the maps of their underline graphs as trees do not occur in our case. Denote by G_{x-y-z} the zth item in all graphs with size x and order y. The solution f_{Non} of equation (6.4.1) is just the enufunction of all non-orientable rooted maps with size as parameter. From (6.4.9), $F_n = \partial_x^n f_{\mathrm{Non}}$, denoted by N_n, $n \geq 0$. Then we have

$$N_0 = 0, \quad N_1 = 1, \quad N_2 = 14, \quad \text{and} \quad N_3 = 223.$$

In Figures 6.4.1–6.4.3, all root-isomorphic classes of non-orientable maps with size not greater 3 are shown.

In Figure 6.4.1, a shows that maps of size 1 have only 1 class. In maps of size 2, maps of underline graph G_{2-1-1} have $4b + 1c$, in total $4 + 1 = 6 = 5$ classes on $S_{\tilde{1}}$ and $2d + 2e$, in total $2 + 2 = 4$ classes on $S_{\tilde{2}}$, a total of 9 classes. Maps of the underline graph G_{2-2-1} have $4f$, in total 4 classes and $1f$, in total 1 class, in summary 5 classes on $S_{\tilde{1}}$. Thus, maps of size 2 have $5 + 5 = 10$ classes on $S_{\tilde{1}}$ and 4 classes on $S_{\tilde{2}}$, in summary $10 + 4 = 14$ classes.

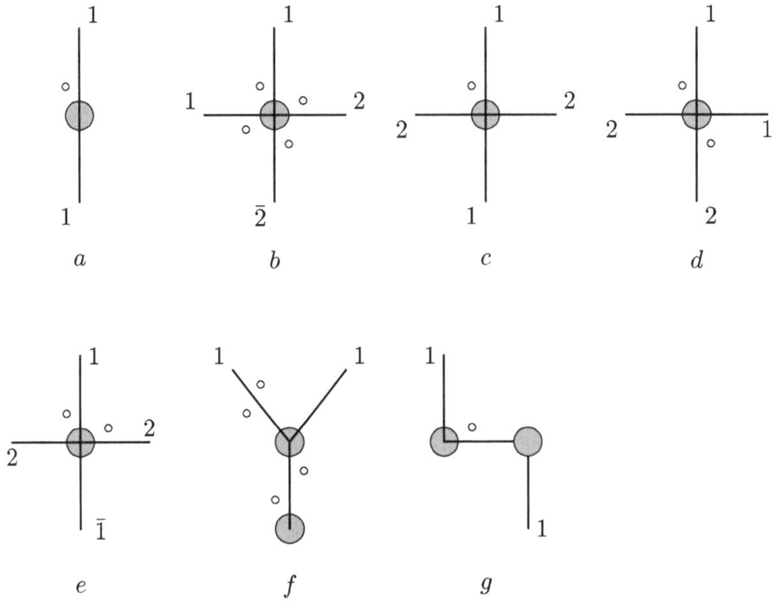

Figure 6.4.1: Classes of non-orientable maps with sizes 1–2.

Figures 6.4.2–6.4.3 provide classes of non-orientable maps with size 3. Among them, Figure 6.4.2 shows the classes of maps considered of underline graph G_{3-1-1} with size 3 and Figure 6.4.3, all other classes of maps with size 3. Attention: A figure without hollow on one side of each edge stands for 12 (*i. e.*, 4×3) hollows.

In Figure 6.4.2, $6a + 6b + 6c + 3d + 1e$, in total $6 + 6 + 6 + 3 + 1 = 22$ classes on $S_{\tilde{1}}$, $6f + 12g + 6h + 6i + 3j + 3k + 3l + 3m$, in total $6 + 12 + 6 + 6 + 3 + 3 + 3 + 3 = 42$ classes on $S_{\tilde{2}}$ and $2n + 6o + 12p + 3q + 6r + 6s + 3t + 3u$, a total of $2 + 6 + 12 + 3 + 6 + 6 + 3 + 3 = 41$ classes on $S_{\tilde{3}}$.

In Figure 6.4.3, of the underline graph G_{3-2-1}, we have $12a + 6b + 6c + 6d$, in total 30 classes on $S_{\tilde{1}}$ and $6e + 12f + 6g$, in total 24 classes on $S_{\tilde{2}}$.

Of the underline graph G_{3-2-2}, $6h$, *i. e.*, classes on $S_{\tilde{1}}$ and $3i$, *i. e.*, 3 classes on $S_{\tilde{2}}$.

Of the underline graph G_{3-2-3}, $6j + 6k + 3l$, *i. e.*, we have 15 classes on $S_{\tilde{1}}$ and $6m + 3n + 3o$, *i. e.*, 12 classes on $S_{\tilde{2}}$.

Of G_{3-2-4}, $3p$, *i. e.*, 3 classes on $S_{\tilde{1}}$ and $3q$, *i. e.*, 3 classes on $S_{\tilde{2}}$.

Of G_{3-3-1}, $6r$, *i. e.*, 6 classes on $S_{\tilde{1}}$.

Of G_{3-3-2}, $6s + 3t$, *i. e.*, 9 classes on $S_{\tilde{1}}$.

Of G_{3-3-3}, $6u$, *i. e.*, 6 classes on $S_{\tilde{1}}$.

On G_{3-3-4}, $1v$, *i. e.*, 1 classes on $S_{\tilde{1}}$.

In consequence, non-orientable maps of size 3 have $22+30+6+15+3+6+6+6+1 = 96$ classes on $S_{\tilde{1}}$, $42 + 24 + 3 + 12 + 3 = 86$ classes on $S_{\tilde{2}}$, 41 on $S_{\tilde{3}}$. On all non-orientable surfaces, maps of size 3 have $96 + 86 + 41 = 223$ classes in all.

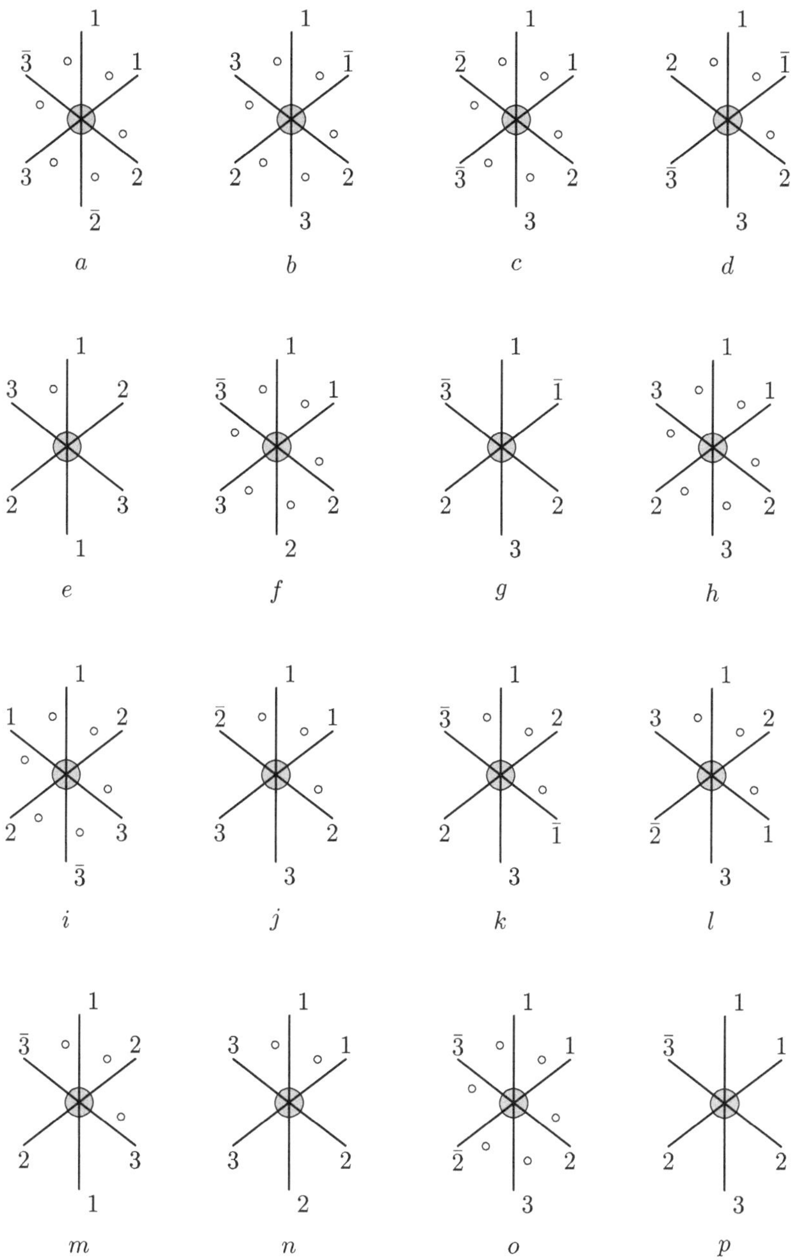

Figure 6.4.2: Classes of non-orientable maps with underline graph G_{3-1-1}.

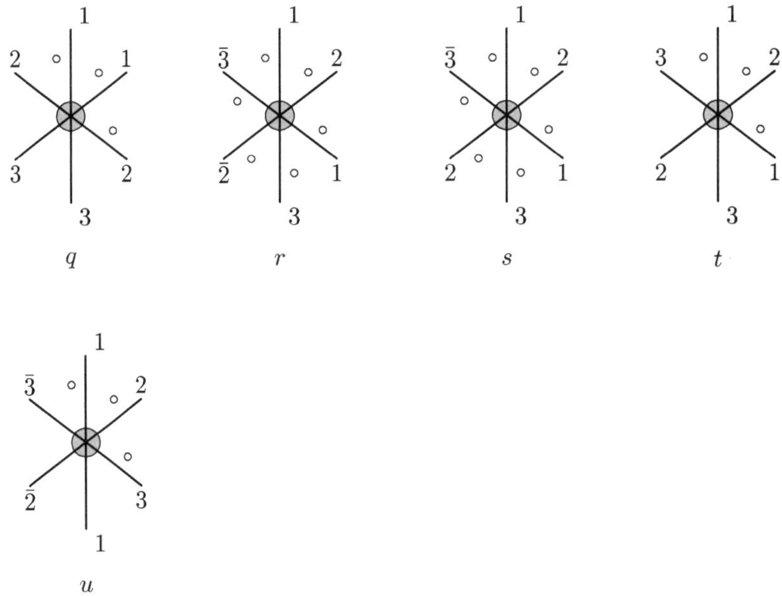

Figure 6.4.2: Continued.

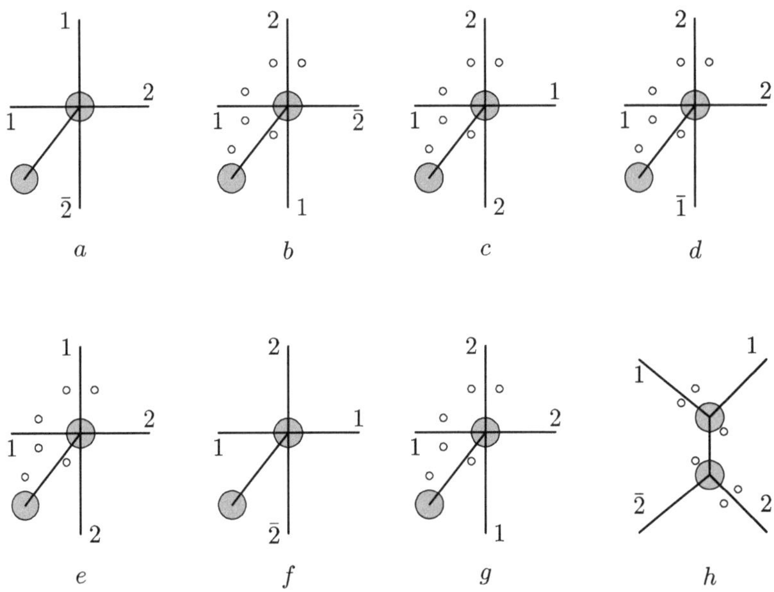

Figure 6.4.3: Other classes of non-orientable maps with size 3.

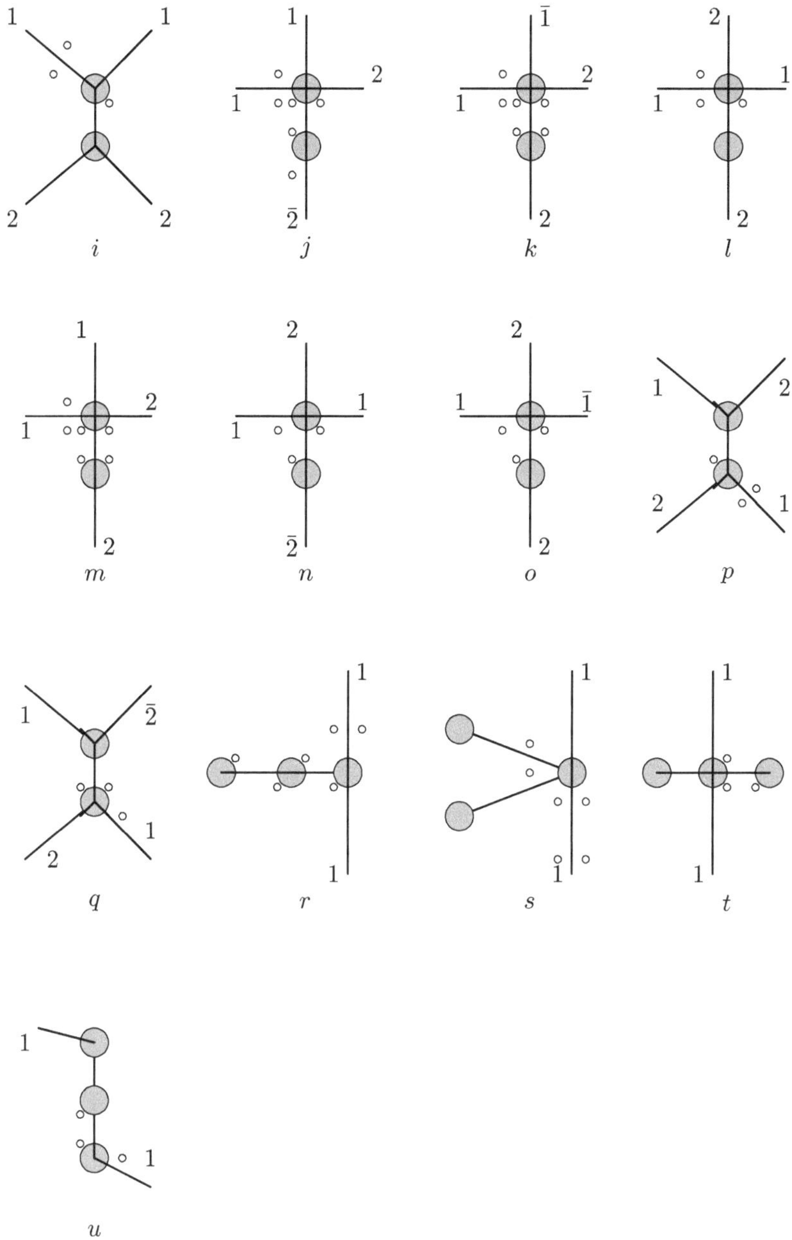

Figure 6.4.3: Continued.

6.5 Ordinary total sum

Consider the equation for $f \in \mathcal{R}\{x\}$

$$\begin{cases} ax^2\dfrac{\mathrm{d}f}{\mathrm{d}x} = -1 + (1 - bx)f - cxf^2; \\ f|_{x=0} = 1, \end{cases} \tag{6.5.1}$$

where $a, b, c \in \mathbb{Z}_+$.

This is equation (21) in Introduction when $d = 1$ and then $a = 4$, $b = 2$, $c = 1$ because it is meaningful in a classification for ordinary maps on all surfaces (orientable and non-orientable).

The following equivalence of the first case of equation (6.5.1) is adopted:

$$f = 1 + bxf + ax^2\frac{\mathrm{d}f}{\mathrm{d}x} + cxf^2. \tag{6.5.2}$$

From equation (6.5.2),

$$\begin{aligned} [f]_n &= [1]_n + b[xf]_n + a\left[x^2\frac{\mathrm{d}f}{\mathrm{d}x}\right]_n + c[xf^2]_n \\ &= \begin{cases} 1, & \text{when } n = 0; \\ b[f]_{n-1} + a[\frac{\mathrm{d}f}{\mathrm{d}x}]_{n-2} + c[f^2]_{n-1}, & \text{when } n = 0, \end{cases} \end{aligned} \tag{6.5.3}$$

for any integer $n \geq 0$.

Let $F_n = [f]_n = \partial_x^n f$ and

$$\begin{aligned} \left[\frac{\mathrm{d}f}{\mathrm{d}x}\right]_{n-2} &= (n-1)F_{n-1}; \\ [f^2]_{n-1} &= \sum_{i=0}^{n-1} F_i F_{n-1-i}. \end{aligned}$$

Based on (6.5.3),

$$x^0 : F_0 = 1, \tag{6.5.4}$$

$$x^1 : F_1 = bF_0 + 0 + cF_0^2 \Longrightarrow F_1 = b + c, \tag{6.5.5}$$

and, for $n \geq 2$,

$$\begin{aligned} x^n : &\left\{\begin{aligned} F_n &= b[f]_{n-1} + a\left[\frac{\mathrm{d}f}{\mathrm{d}x}\right]_{n-2} + c[f^2]_{n-1} \\ &= bF_{n-1} + a(n-1)F_{n-1} + c\sum_{i=0}^{n-1} F_i F_{n-1-i} \end{aligned}\right\} \\ &\Longrightarrow F_n = (an - a + b)F_{n-1} + c\sum_{i=0}^{n-1} F_i F_{n-1-i}. \end{aligned} \tag{6.5.6}$$

Lemma 6.5.1. *For any integer $n \geq 0$,*

$$\left.\begin{array}{r} 1 \quad (when\ n = 0) \\ b + c \quad (when\ n = 1) \\ (an - a + b)F_{n-1} + c\sum_{i=0}^{n-1} F_i F_{n-1-i} \\ (when\ n \geq 2) \end{array}\right\} = F_n \in \mathbb{Z}_+ \tag{6.5.7}$$

Proof. Because of $F_0 = 1$, $F_1 = b + c \in \mathbb{Z}_+$, the conclusion is true for $n = 0$ and 1. We proceed by induction. For integer $n \geq 2$, assume $F_i \in \mathbb{Z}_+$, $0 \leq i \leq n-1$. Because F_n is determined only by F_i, $0 \leq i \leq n-1$, from the third case of (6.5.7), the assumption leads to $F_n \in \mathbb{Z}_+$. This is the conclusion. □

This lemma tells us that the function f determined by (6.5.7) is in $\mathcal{R}_+\{x\} \subseteq \mathcal{R}\{x\}$.

Theorem 6.5.2. *Equation* (6.5.1) *is well-defined on* $\mathcal{R}\{x\}$.

Proof. On the basis of (6.5.4)—(6.5.6), from Lemma 6.5.1, known that $f \in \mathcal{R}_+\{x\} \subseteq \mathcal{R}\{x\}$ determined by (6.5.7) is a solution of equation (6.5.2) and hence equation (6.5.1).

Furthermore, by considering the uniqueness of the procedure for evaluating F_n, $n \geq 0$, by (6.5.7) from the initial condition F_0, f is the only solution of equation (6.5.1). □

Because of each F_n, $n \geq 2$, is a finite sum of positive integers, the function f is an expression in the form of a sum with all terms positive.

Example 1. When $a = 4$, $b = 2$ and $c = 1$, equation (6.5.1) becomes the equation in Liu YP [46] (2003, Theorem 9.7, p. 213)

$$\begin{cases} 4x^2 \dfrac{\mathrm{d}f}{\mathrm{d}x} = -1 + (1 - 2x)f - xf^2; \\ f_0 = f|_{x=0} = 1, \end{cases} \tag{6.5.8}$$

used for determining the number of root-isomorphic classes of ordinary maps with size as parameter on all surfaces (orientable and non-orientable). The number is called the *ordinary sum*.

Example 2. Because equation (6.5.8) is the sum of equation (6.3.1) and equation (6.4.1), the solution of equation (6.5.8) can be directly derived by (6.3.10) and (6.4.9).

6.6 Four color sum for triangulations on sphere

Consider the quadratic equation with ordinary differentiation of second order for $f \in \mathcal{R}\{x\}$

$$\begin{cases} \left(2z + 5f - 3z\frac{\mathrm{d}f}{\mathrm{d}z}\right)\frac{\mathrm{d}^2 f}{\mathrm{d}z^2} = 48z; \\ f|_{z=0} = 0, \quad \left.\frac{\mathrm{d}f}{\mathrm{d}z}\right|_{z=0} = 0. \end{cases} \tag{6.6.1}$$

This is equation (22) in Introduction when $a = c = 1$ and $d = 0$ because it is meaningful in four color sums for planar triangulations as addressed by Tutte.

On the extension of the integral domain $\mathcal{R}\{y\}$, for equivalently transforming equation (6.6.1) about f into a system of equations about $\{F_n = \partial_y^n \mid \text{integer } n \geq 0\}$, let us write

$$\begin{aligned} s &= 2z + 5f - 3z\frac{\mathrm{d}f}{\mathrm{d}z}, \quad S_n = \partial_y^n s; \\ f'' &= \frac{\mathrm{d}^2 f}{\mathrm{d}z^2}, \quad F_n'' = \partial_y^n f'', \end{aligned} \tag{6.6.2}$$

where the integer $n \geq 0$. Thus, for $n \geq 0$,

$$\begin{aligned} S_n &= \begin{cases} 5F_0, & \text{when } n = 0, \\ 2(1 + F_1), & \text{when } n = 1, \\ (5 - 3n)F_n, & \text{when } n \geq 2; \end{cases} \\ F_n'' &= (n + 2)(n + 1)F_{n+2}. \end{aligned} \tag{6.6.3}$$

Furthermore, for integer $n \geq 0$, let

$$\Delta_n = \sum_{i=0}^{n} S_i F_{n-i}''. \tag{6.6.4}$$

On the basis of (6.6.2)–(6.6.4), from equation (6.6.1),

$$\begin{aligned} z^0 : \quad & \Delta_0 = S_0 F_0'' = 0, \\ & \quad \text{by } S_0 = 5F_0 \text{ and } F_0'' = 2F_2, \\ \Longrightarrow\ & \Delta_0 = (5F_0)(2F_2'') \end{aligned} \tag{6.6.5}$$

for $n = 0$. From the initial condition of equation (6.6.1) $F_0 = 0$, the equality $\Delta_0 = 0$ holds.

For $n = 1$,

$$\begin{aligned} z^1 : \quad & \Delta_1 = S_0 F_1'' + S_1 F_0'', \\ & \quad \text{by } S_0 = 0 \text{ and } F_1'' = 6F_3, \\ \Longrightarrow\ & \Delta_1 = S_1 F_0'', \\ & \quad \text{by } S_1 = 2(1 + F_1) \text{ and } F_0'' = 2F_2, \\ \Longrightarrow\ & \Delta_1 = 2(1 + F_1)2F_2. \end{aligned} \tag{6.6.6}$$

From the initial condition $F_1 = 0$ of equation (6.6.1), the equality $\Delta_1 = 48$ holds when $F_2 = 12$.

For $n = 2$,

$$\begin{aligned} z^2: \quad & \Delta_2 = S_0 F_2'' + S_1 F_1'' + S_2 F_0'', \\ & \quad \text{by } S_0 = 0 \text{ and } F_1 = 0, \\ \Longrightarrow\ & \Delta_2 = 2(6F_3) + S_2 F_0'', \\ & \quad \text{by } S_2 = (5-6)F_2 \text{ and } F_0'' = 2F_2, \\ \Longrightarrow\ & \Delta_2 = 12F_3 - F_2(2F_2). \end{aligned} \tag{6.6.7}$$

Because of the condition $F_2 = 12$ for $\Delta_1 = 48$, $\Delta_2 = 0$ holds only when $F_3 = 2F_2 = 24$.

For $n = 3$,

$$\begin{aligned} z^3: \quad & \Delta_3 = S_0 F_3'' + S_1 F_2'' + S_2 F_1'' + s_3 F_0'', \\ & \quad \text{by } S_0 = 0 \text{ and } S_1 = 2, \\ \Longrightarrow\ & \Delta_3 = 2(12F_4) + S_2(6F_3) + S_3(2F_2), \\ & \quad \text{by } S_2 = -F_2 \text{ and } S_3 = -4F_3, \\ \Longrightarrow\ & \Delta_3 = 2(12F_4) - F_2(6F_3) - 4F_3(2F_2). \end{aligned} \tag{6.6.8}$$

Because of the condition $F_3 = 24$ for $\Delta_2 = 0$, the quality $\Delta_3 = 0$ holds only when $F_4 = 6F_2 + 8F_2 = 168$.

In general, for $n \geq 4$, because of $S_0 = 0$, $S_1 = 2$ and $F''_{n-1} = (n+1)nF_{n+1}$,

$$\begin{aligned} z^n: \quad \Delta_n &= 2(n+1)nF_{n+1} + \sum_{i=2}^{n} S_i F''_{n-i}, \\ & \quad \text{by the first case of equation (6.6.3)}, \\ &= 2(n+1)nF_{n+1} - \sum_{i=2}^{n} (3i-5) F_i F''_{n-i}, \\ & \quad \text{by the second case of equation (6.6.3)}, \\ &= 2(n+1)nF_{n+1} - \sum_{i=2}^{n} \lambda_{n,i} F_i F_{n-i+2}, \end{aligned} \tag{6.6.9}$$

where

$$\lambda_{n,i} = (3i-5)(n-i+2)(n-i+1).$$

Only when

$$F_{n+1} = \frac{1}{2(n+1)n} \sum_{i=2}^{n} \lambda_{n,i} F_i F_{n-i+2}, \tag{6.6.10}$$

the equality $\Delta_n = 0$ holds.

Theorem 6.6.1. *Equation* (6.6.1) *for* f *is equivalent to the system of equations for* $\{F_n \mid n \geq 0\}$

$$\Delta_n = \begin{cases} 0, & \text{when } n = 0; \\ 48, & \text{when } n = 1; \\ 0, & \text{when } n \geq 2, \end{cases} \tag{6.6.11}$$

with the conditions $F_0 = 0$ *and* $F_1 = 0$, *where* Δ_n *is given by* (6.6.9).

Proof. By the procedure from (6.6.5) through (6.6.10), it is seen that a solution of equation (6.6.1) can be transformed into a solution of the system of equations equation (6.6.11). Conversely, from (6.6.7)–(6.6.10), a solution of the system of equations (6.6.11), and hence a solution of equation (6.6.1), can be derived. □

The theorem enables us to extract a solution of equation (6.6.11) for getting a solution of equation (6.6.1).

Lemma 6.6.2. *In a solution* $F_n (n \geq 0)$ *of equation* (6.6.11), $F_n \geq 0$ *for any integer* $n \geq 0$.

Proof. We proceed by induction. With a view on the discussion of (6.6.5)–(6.6.8), from $F_0 = F_1 = 0$, $F_2 = 12$, $F_3 = 24$ and $F_4 = 168$, it is seen that the conclusion is true for $n \leq 4$. For $n \geq 5$, assume that $F_j \geq 0$ for $0 \leq j \leq n-1$. By (6.6.10), on account of $\lambda_{n,i} \geq 0$ for $2 \leq i \leq n-1$, the assumption leads to $F_n \geq 0$. □

Theorem 6.6.3. *Equation* (6.6.1) *is well-defined on* $\mathcal{R}_+\{z\}$.

Proof. From Theorem 6.6.1 and Lemma 6.6.2, the function f determined by the procedure of doing (6.6.5)–(6.6.10) is a solution in $\mathcal{R}_+\{z\}$. Then, by considering the uniqueness of the procedure by running from (6.6.5) through (6.6.10) based on the initial conditions $F_0 = F_1 = 0$, we find that the solution is the only one. □

Furthermore, the solution of equation (6.6.1) enables us to get its expression in the form of a finite sum with all terms positive.

Theorem 6.6.4. *The solution of equation* (6.6.1) *has its expression in the form of a finite sum with all terms positive,*

$$F_n = \begin{cases} 0, & \text{when } n = 0, 1; \\ 12, & \text{when } n = 2; \\ 24, & \text{when } n = 3; \\ 168, & \text{when } n = 4; \\ \sum_{i=2}^{n-1} \frac{(3i-5)(n-i+1)(n-i)}{2(n-1)n} F_i F_{n-i+1}, & \text{when } n \geq 5. \end{cases} \tag{6.6.12}$$

Proof. From the initial conditions of equation (6.6.1), by (6.6.6)–(6.6.10), (6.6.12) is then obtained. □

Next, as an example, we show an application of equation (6.6.1).

Example 1. Four color sum on root-isomorphic classes of planar triangulations. In Tutte WT [86] is shown a solution of equation (6.6.1): F_n, $n \geq 0$ such that

$$h = \sum_{n \geq 1} F_{n+2} z^n$$

is the four color sum function of planar rooted triangulations, *i. e.*, $H_n = \partial_z^n h = F_{n+2}$, $n \geq 1$, is the total sum of colorations by 4 colors over all non-separable planar rooted triangulations with $2n$ faces.

Lemma 6.6.5. *In the solution $F_n (n \geq 0)$ of the system of equations* (6.6.11), *for any integer $n \geq 4$,*

$$2(n+1)n \,\Big|\, \sum_{i=2}^{n} (3i-5)(n-i+2)(n-i+1) F_i F_{n-i+2}. \tag{6.6.13}$$

Proof. By (6.6.12),

$$2(n+1)nF_{n+1} = \sum_{i=2}^{n} (3i-5)(n-i+2)(n-i+1) F_i F_{n-i+2}.$$

From the combinatorial meaning of F_{n+1}, F_{n+1} is a positive integer, *i. e.*, $F_{n+1} \in \mathbb{Z}_+$. Therefore, the conclusion is drawn. □

The conclusion above can be directly proved by (6.6.12) itself, however, it looks that some complication might be involved.

Theorem 6.6.6. *On the solution $F_n (n \geq 0)$ of equation* (6.6.1) *as determined by* (6.6.12), *for any integer $n \geq 2$, $F_n \in \mathbb{Z}_+$.*

Proof. This is a direct result of Lemma 6.6.5. □

In Figure 6.6.1, it is seen that $H_1 = F_3$ and $H_2 = F_4$ are meaningful in combinatorics, particularly, in the four color sum of maps. For example, a shows that non-separable

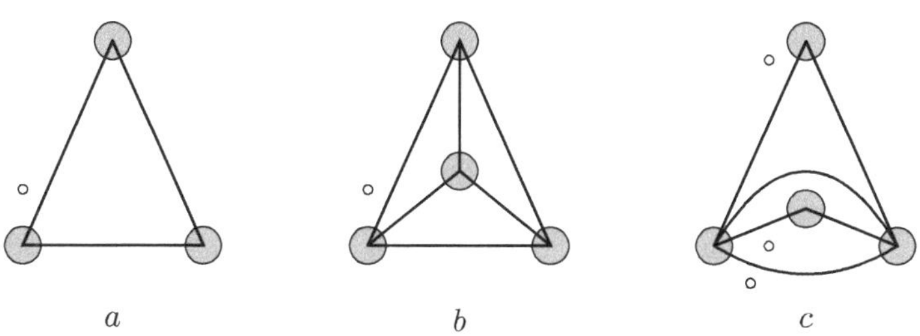

Figure 6.6.1: Four color sum of root-classes of triangulations on sphere.

planar triangulations of 2 faces have only 1 root-isomorphic class. Its chromatic polynomial is $P(a) = \lambda(\lambda-1)(\lambda-2)$. When $\lambda = 4$, $P(a)|_{\lambda=4} = 4\times3\times2 = 24$. This is $H_1 = F_3 = 24$. From b and c, it is seen that non-separable planar triangulations of 4 faces have 4 root-isomorphic classes. Thus,

$$\begin{aligned} H_2 = F_4 &= 1P(b)|_{\lambda=4} + 3P(c)|_{\lambda=4} \\ &= \lambda(\lambda-1)(\lambda-2)(\lambda-3)|_{\lambda=4} + 3\lambda(\lambda-1)(\lambda-2)^2|_{\lambda=4} \\ &= 24 + 3\times12\times2^2 = 7\times24 = 168. \end{aligned}$$

6.7 Notes

6.7.1. All differential functional equations mentioned in 6.1 can be directly solved by integration. However, certain complications are often encountered in trying to simplify.

How to find the simplest differential equation is a problem absolutely necessary to do further research on via suitable parametric expressions. For example, in Tutte WT [84], a function for planar c-nets (or 3-connected planar maps) is extracted from a second order ordinary differential equation. However, in Liu YP [12], one works by a first order ordinary differential equation to get a result that is very simple.

6.7.2. In Liu YP [44] (1999, Theorem 8.5.2, p. 271), the first order ordinary differential equation

$$\begin{cases} 4x^2\dfrac{dg}{dx} = (1-x)g - x(1+h); \\ g|_{x=0} = 0, \end{cases} \tag{6.7.1}$$

is provided for counting the root-isomorphic classes of non-orientable petal bundles with size as parameter. Here h is determined by equation (6.2.1). Notably, h is the solution of equation (6.2.1) if the constant term is omitted. It can be shown that the solution g of equation (6.6.1) is just determined by (6.2.17).

6.7.3. The expression for F_n in the form of a finite sum of all terms positive shown by (6.3.10) should be done further for getting an explicision of the summation free case.

6.7.4. On the basis of the explicision of F_n determined by (6.3.10), we evaluate an explicision in the summation free form of F_n determined by (6.4.9).

6.7.5. On the basis of (6.3.10) and (6.4.9), we evaluate an explicision in the form of a finite sum with all terms positive of F_n as determined by (6.5.7).

6.7.6. Equation (6.6.1) firstly occurs in Tutte WT [86] (1982). However, no suitable expression, particularly an explicision, has been found yet up to now.

7 Partial differential equations

7.1 Quadrangulations on sphere

Consider the equation (applied later!) for $f \in \mathcal{R}\{x,y\}$

$$\begin{cases} x^4yf^2 + (y - x^2)f - x^2yf^* + x^2 - y = 0; \\ f|_{x=y=0} = 1, \end{cases} \tag{7.1.1}$$

where $f^* = \partial_x^2 f$. This is equation (23) in Introduction when $a = b = c = d = 1$. It is meaningful for quadrangulations on the sphere.

This equation is derived from a theory of decomposition appearing in Liu YP [40] (1992). More clearly this is found in Liu YP [44] (Section 5.4, 1999). Some definitions relevant for y have to be attended to.

Let $F_n = [f]_n = \partial_y^n f$, $n \geq 0$, then, for any integer $n \geq 0$,

$$F_n^* = [f^*]_n = [\partial_x^2 f]_n = \partial_x^2 [f]_n = \partial_x^2 F_n. \tag{7.1.2}$$

In order to evaluate F_n, $n \geq 0$, the first case of equation (7.1.1) is transformed into one of its equivalences by

$$\begin{aligned} x^2 f &= x^4 y f^2 + yf - x^2 yf^* + x^2 - y \\ &= x^4 y f^2 + y(f - x^2 f^* - 1) + x^2. \end{aligned} \tag{7.1.3}$$

Because, for any integer $n \geq 0$,

$$[f^2]_n = \sum_{i=0}^{n} F_i F_{n-i} \tag{7.1.4}$$

and

$$[f - x^2 f^* - 1]_n = \begin{cases} F_0 - x^2 F_0^* - 1, & \text{when } n = 0; \\ F_n - x^4 F_n^*, & \text{when } n \geq 1, \end{cases} \tag{7.1.5}$$

we have, from (7.1.3),

$$\begin{aligned} y^0: \quad & x^2 [f]_0 = [x^2]_0 \Longrightarrow x^2 F_0 = x^2; \\ & \Longrightarrow F_0 = 1, \quad F_0^* = 0, \end{aligned} \tag{7.1.6}$$

$$\begin{aligned} y^1: \quad & x^2 [f]_1 = x^4 [f^2]_0 + [f - x^2 f^* - 1]_0, \\ & \quad \text{by (7.1.4)–(7.1.5)}, \\ & = x^4 F_0 F_0 + (F_0 - x^2 F_0^* - 1), \\ & \quad \text{by (7.1.6)}, \\ & = x^4 \Longrightarrow x^2 F_1 = x^4 \\ & \Longrightarrow F_1 = x^2, \quad F_1^* = 1, \end{aligned} \tag{7.1.7}$$

https://doi.org/10.1515/9783110625837-007

$$
\begin{aligned}
y^2: \quad x^2[f]_2 &= x^4[f^2]_1 + [f - x^2 f^* - 1]_1,\\
&\quad \text{by (7.1.4)–(7.1.5)},\\
&= x^4(2F_0F_1) + (F_1 - x^2F_1^*),\\
&\quad \text{by (7.1.6)–(7.1.7)},\\
&= 2x^6 \implies x^2F_2 = 2x^6\\
&\implies F_2 = 2x^4, \quad F_2^* = 0,\\
y^3: \quad x^2[f]_3 &= x^4[f^2]_2 + [f - x^2 f^* - 1]_2,\\
&\quad \text{by (7.1.4)–(7.1.5)},\\
&= x^4(F_1^2 + 2F_0F_2) + (F_2 - x^2F_2^*),\\
&\quad \text{by (7.1.6)–(7.1.8)},\\
&= x^4(x^4 + 4x^4) + 2x^4\\
&\implies x^2F_2 = 5x^8 + 3x^4\\
&\implies F_3 = 5x^6 + 2x^2, \quad F_3^* = 2,
\end{aligned}
\tag{7.1.8}
$$

$$
\tag{7.1.9}
$$

and

$$
\begin{aligned}
y^4: \quad x^2[f]_4 &= x^4[f^2]_3 + [f - x^2 f^* - 1]_3,\\
&\quad \text{by (7.1.4)–(7.1.5)},\\
&= x^4(2F_1F_2 + 2F_0F_3) + (F_3 - x^2F_3^*),\\
&\quad \text{by (7.1.6)–(7.1.9)},\\
&= x^4(14x^6 + 4x^2) + 5x^6\\
&\implies x^2F_2 = 14x^{10} + 9x^6\\
&\implies F_3 = 14x^8 + 9x^4, \quad F_4^* = 0.
\end{aligned}
\tag{7.1.10}
$$

As a matter of fact, for any integer $n \geq 2$,

$$
\begin{aligned}
x^2F_n &= x^4[f^2]_{n-1} + [f - x^2 f^*]_{n-1}\\
&= x^4 \sum_{i=0}^{n-1} F_iF_{n-1-i} + F_{n-1} - x^2F_{n-1}^*.
\end{aligned}
\tag{7.1.11}
$$

Lemma 7.1.1. *For any integer $n \geq 1$, F_n has a factor x^2.*

Proof. Because f is an even function for x, f has no term of x^{2i+1} where $i \geq 0$ is an integer.

First, from (7.1.7) and (7.1.8), both F_1 and F_2 have a factor x^2.

Then we proceed by induction for $n \geq 3$. For any i, $2 \leq i \leq n-1$, assume F_i has x^2 as a factor. By (7.1.11) and the assumption, F_n has x^2 as a factor (attention: $x^4 | (F_{n-1} - x^2F_{n-1}^*)$!). Therefore, the conclusion is drawn. □

This lemma tells us that on the basis of (7.1.11), for any integer $n \geq 2$,

$$x^2 | (x^4 [f^2]_{n-1} + [f - x^2 f^*]_{n-1}) \tag{7.1.12}$$

and, for $n \geq 0$, $F_n \in \mathcal{R}_+\{x\}$.

Lemma 7.1.2. *For any integer $n \geq 1$, F_n is a polynomial of x with degree at most $2n$.*

Proof. From (7.1.6) and (7.1.7), F_0 and F_1 obey the conclusion.

We proceed by induction for $n \geq 3$. Assume for $1 \leq i \leq n-1$, F_i is a polynomial of x with degree at most $2i$. Then from Lemma 7.1.1, the degree of $[f - x^2 f^*]_{n-1}$ is $d([f - x^2 f^*]_{n-1}) \leq 2(n-1) - 2 = 2n - 4$. Because the degree of $x^4 [f^2]_{n-1}$ is $d(x^4 [f^2]_{n-1}) \leq 4 + 2(n-1) = 2n + 2$, from (7.1.11), $d(F_n) \leq d(x^4 [f^2]_{n-1}) - 2 \leq (2n+2) - 2 = 2n$. Therefore, based on the induction principle, the conclusion is easily drawn. □

Lemma 7.1.3. *For any integer $n \geq 1$, F_n has no term of x with odd degree.*

Proof. This is a result of f being even for x. □

The three lemmas above with the procedures of the proofs enable us to express F_n for $n \geq 1$ in the form of

$$\begin{aligned} F_n &= \sum_{\substack{m=2 \\ 2|m}}^{2n} F_{m,n} x^m \\ &= \sum_{m=1}^{n} F_{2m,n} x^{2m} \end{aligned} \tag{7.1.13}$$

where $F_{2m,n} \in \mathcal{R}_+$. Thus, because of

$$[f - x^2 f^*]_n = \sum_{m=2}^{n} F_{2m,n} x^{2m}, \tag{7.1.14}$$

by (7.1.11), we have $F_n \in \mathcal{R}_+\{x\}$.

Theorem 7.1.4. *Equation* (7.1.1) *is well-defined on $\mathcal{R}\{x, y\}$.*

Proof. First, because of (7.1.3) equivalent to the first case of equation (7.1.1) and the initial condition $F_0 = 1$ of equation (7.1.1), on the basis of the lemmas above, f: F_n for $n \geq 0$ evaluated from (7.1.6), (7.1.7) and (7.1.11), provide a solution of equation (7.1.1) on $\mathcal{R}\{x, y\}$.

Then, by considering the uniqueness of the procedure in evaluating $F_n (n \geq 1)$ on $\mathcal{R}\{x, y\}$ for the initial condition of equation (7.1.1), f is known to be the only solution of equation (7.1.1). □

In what follows, we discuss expressions of F_n, $n \geq 2$, in the form of a finite sum with all terms positive.

Let us write $F_n^{[2]} = [f^2]_n$, $n \geq 0$. On the basis of (7.1.6) and (7.1.7), for $n \geq 2$,

$$\begin{aligned} F_n^{[2]} &= \sum_{i=0}^{n} F_i F_{n-i} \\ &= 2F_n + 2x^2 F_{n-1} + \sum_{i=2}^{n-2} F_i F_{n-i}. \end{aligned} \tag{7.1.15}$$

Denote by Σ_n the summation for i from 2 through $n-2$; then from (7.1.13), $n \geq 5$, we have

$$\begin{aligned} \Sigma_n &= \sum_{i=2}^{n-2} \sum_{t=2}^{n-i+1} \left(\sum_{l=1}^{t} F_{2l,i} F_{2(t-l),n-i} \right) x^{2t} \\ &= \sum_{t=2}^{n-1} \Phi_{2t,n} x^{2t}, \end{aligned} \tag{7.1.16}$$

where

$$\Phi_{2t,n} = \sum_{l=1}^{t} \left(\sum_{i=2}^{\min\{n-2,n-t+1\}} F_{2l,i} F_{2(t-l),n-i} \right). \tag{7.1.17}$$

On this basis, from (7.1.11), we have, for $n \geq 5$,

$$F_n = x^2 (2(F_{n-1} + x^2 F_{n-2}) + \Sigma_{n-1}) + \frac{F_{n-1} - x^2 F_{n-1}^*}{x^2}. \tag{7.1.18}$$

Meanwhile, from (7.1.14),

$$\begin{aligned} \frac{F_{n-1} - x^2 F_{n-1}^*}{x^2} &= \frac{1}{x^2} \sum_{l=2}^{n-1} F_{2l,n-1} x^{2l} \\ &= \sum_{l=1}^{n-2} F_{2(l+1),n-1} x^{2l}. \end{aligned} \tag{7.1.19}$$

From (7.1.13),

$$\begin{aligned} 2x^2 (F_{n-1} + x^2 F_{n-2}) &= 2 \left(F_{2,n-1} x^4 + \sum_{l=3}^{n} \Lambda_l x^{2l} \right) \\ &= 2F_{2,n-1} x^4 + \sum_{l=3}^{n} 2\Lambda_l x^{2l} \end{aligned} \tag{7.1.20}$$

where $\Lambda_l = F_{2(l-1),n-1} + F_{2(l-2),n-2}$ for $3 \leq l \leq n$ and from (7.1.16),

$$\begin{aligned} x^2 \Sigma_{n-1} &= x^2 \sum_{t=2}^{n-2} \Phi_{2t,n-1} x^{2t} \\ &= \sum_{t=3}^{n-1} \Phi_{2(t-1),n-1} x^{2t}. \end{aligned} \tag{7.1.21}$$

Theorem 7.1.5. *Let $f = f_{\text{0-quad}}$ be the solution of equation* (7.1.1), *determined by $Q_n = \partial_y^n f_{\text{0-quad}} \in \mathcal{R}_+\{x\}$, for $n \geq 0$, then*

$$Q_n = \begin{cases} 1, & \text{when } n = 0; \\ x^2, & \text{when } n = 1; \\ 2x^4, & \text{when } n = 2; \\ 5x^6 + 2x^2, & \text{when } n = 3; \\ 14x^8 + 9x^4, & \text{when } n = 4; \\ \sum_{m=1}^{n} Q_{2m,n}x^{2m}, & \text{when } n \geq 5, \end{cases} \tag{7.1.22}$$

where

$$Q_{2m,n} = \begin{cases} Q_{4,n-1}, & \text{when } m = 1; \\ Q_{6,n-1} + 2F_{2,n-1}, & \text{when } m = 2; \\ 2(Q_{2(m-1),n-1} + Q_{2(m-2),n-2}) & \\ \quad + Q_{2(m+1),n-1} + \Phi_{2(m-1),n-1}, & \text{when } 3 \leq m \leq n-2; \\ 2(Q_{2(n-2),n-1} + Q_{2(n-3),n-2}) & \\ \quad + \Phi_{2(n-1),n-1}, & \text{when } m = n-1; \\ 2(Q_{2(n-1),n-1} + Q_{2(n-2),n-2}), & \text{when } m = n, \end{cases} \tag{7.1.23}$$

and $\Phi_{2t,n}$ determined by (7.1.17).

Proof. By substituting (7.1.19)–(7.1.21) into (7.1.18), the conclusion is easily drawn. □

Lemma 7.1.6. *For any integer $n \geq 0$,*

$$\left.\begin{array}{ll} \text{when } n = 0 \pmod 2, & Q_{2,n} \\ \text{when } n = 1 \pmod 2, & Q_{4,n} \end{array}\right\} = 0.$$

Proof. For $n \leq 4$, from (7.1.22), it is seen that the conclusion is true. For $n \geq 5$, by induction on n. Assume for any integer $i \leq n-1$, the conclusion is true. When n is even. Because of $n-1$ odd. From the assumption, $Q_{4,n-1} = 0$. From (7.1.22), $Q_{2,n} = Q_{4,n-1} = 0$. n is odd, because $n-1$ even. By the assumption, $Q_{4,n-1} = 0$. Hence, according to the principle of induction, the conclusion is easily drawn. □

Lemma 7.1.7. *For any integer $n \geq 1$, $Q_{2(n-1),n} = 0$.*

Proof. When $n = 1, \ldots, 5$. From (7.1.19), the lemma is true. When $n \geq 6$, we proceed by induction on n. Assume for any integer $i \leq n-1$, $F_{2(i-1),i} = 0$. By (7.1.23),

$$Q_{2(n-1),n} = 2(Q_{2(n-2),n-1} + Q_{2(n-3),n-2}) + \Phi_{2(n-1),n-1};$$

by the assumption, $Q_{2(n-2),n-1} = 0$ and $Q_{2(n-3),n-2} = 0$. On the basis of (7.1.17), when $t = n - 1$,

$$\begin{aligned}\Phi_{2(n-1),n-1} &= \sum_{l=1}^{n-1}\left(\sum_{i=2}^{\min\{n-3,2\}} Q_{2l,i}Q_{2(n-2),n-i-1}\right)\\ &= \sum_{l=1}^{n-1}(Q_{2l,2}Q_{2(n-2),n-3}).\end{aligned}$$

Because of $Q_{2(n-2),n-3} = 0$, $\Phi_{2(n-1),n-1} = 0$. Therefore, by the principle of induction, the lemma is true. □

Corollary 7.1.8. *For integer $n \geq 5$, the sequence $\{Q_n \mid n \geq 1\}$ determined by* (7.1.22) *and* (7.1.23) *satisfies the identity*

$$Q_{2(n-2),n-1} + Q_{2(n-3),n-2} + \sum_{\substack{2\leq i\leq \min\{n-3,2\}\\ 1\leq l\leq n-1}} Q_{2l,i}Q_{2(n-2),n-i-1} = 0. \tag{7.1.24}$$

Proof. By Theorem 7.1.5 and Lemma 7.1.7, the conclusion is drawn. □

Example 1. Root-isomorphic classes of planar near-quadrangulations.

In Figure 7.1.1, root-isomorphic classes planar near-quadrangulations with size and root-face valency as parameters are provided. The cases for size 0, 1 and 2 show no difference with trees. They are already shown in Figure 3.1.3 as, respectively, $L_{0,1}$, $L_{1,1}$ and $L_{2,1}$. From $L_{3,1}$ and $L_{3,2}$, it is also seen that planar near-quadrangulations of size 3 and root-face valency 6 have $3L_{3,1} + 2L_{3,2}$, in total 5 classes. From $4L_{4,1} + 8L_{4,3} + 2L_{4,3}$, it is seen that those of size 4 and root-face valency 8 have 14 classes. Their contributions to Q_3 and Q_4 are, respectively, $5x^6$ and $14x^8$.

From a of Figure 7.1.1, it is found that planar near-quadrangulations of size 3 and root-face valency 2 have $2a$, a total of 2 classes. Their contribution to Q_3 is $2x^2$. From b, c and d of Figure 7.1.1, it is found that planar near-quadrangulations of size 4 and root-valency 4 have $4b + 4c + 2d$, a total of 9 classes. The contribution to Q_4 is $9x^4$.

In summary, we have the two parts $Q_3 = 5x^6 + 2x^2$ and $Q_4 = 14x^8 + 9x^4$. Here, the effect of Lemma 7.1.6 and Lemma 7.1.7 might be seen.

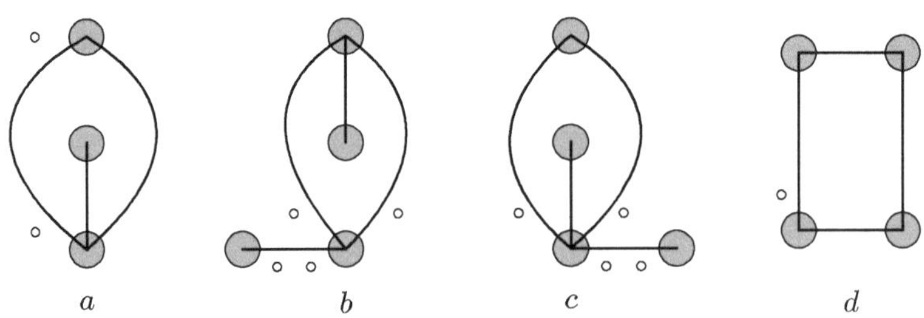

Figure 7.1.1: Classes of planar near-quadrangulations of sizes 3–4.

Example 2. In Liu YP [44] (Section 5.4, 1999), an equation for f is given:

$$\begin{cases} x^4yf^2 + (1-x^2)f + x^2 - x^2f^* - 1 - 0; \\ f|_{x=y=0} = 1, \end{cases} \tag{7.1.25}$$

where $f^* = \partial_x^2 f$.

For this equation, attention has to be paid to the fact that, because $[f]_n = \partial_y^n f$ $(n \geq 0)$ is not a polynomial of x, trouble occurs in determining f. However, it can be avoided by restricting the power of y.

Example 3. In Liu YP-Cai JL [6] (2000), an equation with three variables occurs:

$$\begin{cases} x^4zf^2 + (y-x^2)f + x^2 - x^2yf^* - y = 0; \\ f|_{x=y=z=0} = 1, \end{cases} \tag{7.1.26}$$

where $f^* = \partial_x^2 f$.

For this equation and the like, f can be determined by $[f]_{s,t} = \partial_{y,z}^{s,t} f$ for $s,t \geq 0$, more or less, without cumbersome dealings.

Example 4. Lack-1 face near-quadrangulations on the sphere. Consider the equation

$$\begin{cases} (x^2 - 2x^4yq - y)f = \dfrac{x^3zy}{z-x}(zq|_{x=z} - xq) + yf_{i_x \geq 3}; \\ f|_{x=z=y=0} = 0, \end{cases} \tag{7.1.27}$$

where $q = q(x,y)$ is the solution of equation (7.1.1), and $f_{i_x \geq 3}$ is the function obtained by truncating all terms with degree of x less than 3 from f.

For convenience, it is necessary to discuss the issue of being well-defined and the procedure of evaluating the solution of an equation considered.

First, equation (7.1.27) is transformed into one of its equivalences:

$$f = 2x^2yf_{\text{O-nq}}f + \frac{xzy}{z-x}(zf_{\text{O-nq}}|_{x=z} - xf_{\text{O-nq}}) + x^{-2}yf_{i_x \geq 3} \tag{7.1.28}$$

where $f_{\text{O-nq}}$ ($[f_{\text{O-nq}}]_n = \partial_y^n f_{\text{O-nq}} = Q_n$, $n \geq 0$) determines the solution of equation (7.1.1), given by (7.1.22) and (7.1.23).

On the basis of (7.1.28), for $F_n = [f]_n$, $n \geq 0$, the procedure of evaluating them follows:

$$\begin{aligned} y^0:\quad & [f]_0 = 0 (\text{by } y \text{ as a factor on the right}); \\ & \Longrightarrow F_0 = 0, \quad [f_{\text{O-nq}}f]_0 = 0, \quad [f_{i_x \geq 3}]_0 = 0, \end{aligned} \tag{7.1.29}$$

$$\begin{aligned} y^1:\quad [f]_1 &= \frac{xz}{z-x}[zf_{\text{O-nq}}|_{x=z} - xf_{\text{O-nq}}]_0 \\ &= \frac{xz}{z-x}(z[f_{\text{O-nq}}]_0|_{x=z} - x[f_{\text{O-nq}}]_0) \\ &= \frac{xz}{z-x}(zQ_0|_{x=z} - xQ_0), \\ & \quad \text{by (7.1.22) and (7.1.23)}, \\ &\Longrightarrow F_1 = xz, \quad [f_{\text{O-nq}}f]_1 = xz, \quad [f_{i_x \geq 3}]_1 = 0, \end{aligned} \tag{7.1.30}$$

$$
\begin{aligned}
y^2: \quad [f]_2 &= 2x^2[f_{\text{0-nq}}f]_1 + \frac{xz}{z-x}[zf_{\text{0-nq}}|_{x=z} - xf_{\text{0-nq}}]_1 \\
&= 2x^2(xz) + \frac{xz}{z-x}(zQ_1|_{x=z} - xQ_1), \\
&\quad \text{by (7.1.22) and (7.1.23),} \\
&= 2x^3z + xz(z^2 + xz + x^2) \\
&\Longrightarrow F_2 = 3x^3z + x^2z^2 + xz^3, \\
&\quad [f_{\text{0-nq}}f]_2 = 4x^3z + x^2z^2 + xz^3, \\
&\quad [f_{i_x\geq 3}]_2 = 0,
\end{aligned}
\tag{7.1.31}
$$

$$
\begin{aligned}
y^3: \quad [f]_3 &= 2x^2[f_{\text{0-nq}}f]_2 + \frac{xz}{z-x}[zf_{\text{0-nq}}|_{x=z} - xf_{\text{0-nq}}]_2 \\
&= 2x^2(4x^3z + x^2z^2 + xz^3) \\
&\quad + \frac{xz}{z-x}(zQ_2|_{x=z} - xQ_2), \\
&\quad \text{by (7.1.22) and (7.1.23)}, \\
&\Longrightarrow F_3 = 9x^5z + 4x^4z^2 + 4x^3z^3 + 2x^2z^4 + 2xz^5,
\end{aligned}
\tag{7.1.32}
$$

and, for any integer $n \geq 4$,

$$
\begin{aligned}
y^n: \quad [f]_n &= 2x^2[f_{\text{0-nq}}f]_{n-1} + \frac{xz}{z-x}[zf_{\text{0-nq}}|_{x=z} \\
&\quad - xf_{\text{0-nq}}]_{n-1} + x^{-2}[f_{i_x\geq 3}]_{n-1} \\
&\Longrightarrow F_n = x^2\sum_{i=0}^{n-1} Q_iF_{n-1-i} \\
&\quad + \sum_{j=1}^{2(n-1)} x^j\left(\sum_{m\langle 1,j-1\rangle}^{2n-3} Q_{m,n-1}z^{m-j+1}\right) \\
&\quad + \sum_{m=2}^{2n-5} x^m\partial_x^{m+2}F_{n-1}.
\end{aligned}
\tag{7.1.33}
$$

Lemma 7.1.9. *For any integer $n \geq 1$, F_n is a polynomial of x or y with both degrees neither less than 1 nor greater than $2n-1$ on $\mathcal{R}_+\{x,z\}$.*

Proof. Similarly to the proofs of Lemma 7.1.1–Lemma 7.1.3. □

On the basis of this lemma, for any integer $n \geq 1$ and $F_{s,t}^{(n)} = \partial_{x,z}^{s,t}f \in \mathcal{R}_+$, we have

$$
F_n = \sum_{1\leq s,t\leq 2n-1} F_{s,t}^{(n)}x^sz^t = \sum_{s=1}^{2n-1} F_{s,*}^{(n)}x^s = \sum_{t=1}^{2n-1} F_{*,t}^{(n)}z^t. \tag{7.1.34}
$$

Theorem 7.1.10. *Let $f_{\text{mis-1}}$ be the solution of equation (7.1.27), then, for any integer $n \geq 0$, $[f_{\text{mis-1}}]_n = F_n$ we have*

$$
F_n = x^2\sum_{i=0}^{n-1} Q_iF_{n-1-i} + \sum_{j=1}^{2(n-1)} A_{j,n-1}x^j + \sum_{m=2}^{2n-5} x^mF_{m+2,n-1}, \tag{7.1.35}
$$

where

$$A_{j,n-1} = \sum_{m=\langle 1,j-1\rangle}^{2n-3} Q_{m,n-1} z^{m-j+1}, \quad 1 \le j \le 2(n-1). \tag{7.1.36}$$

Proof. On the basis of (7.1.34), see Theorem 7.1.4 and Theorem 7.1.5. □

In Figure 7.1.2, a shows the root-classes of lack-1 planar near-quadrangulations with size 1: xz, *i. e.*, $F_1 = xz$ obtained by (7.1.30). b and c show the root-classes of lack-1 planar near-quadrangulations with size 2: $3x^3z+xz^3(b)+x^2z^2(c) = 3x^3z+x^2z^2+xz^3 = F_2$, as given by (7.1.31).

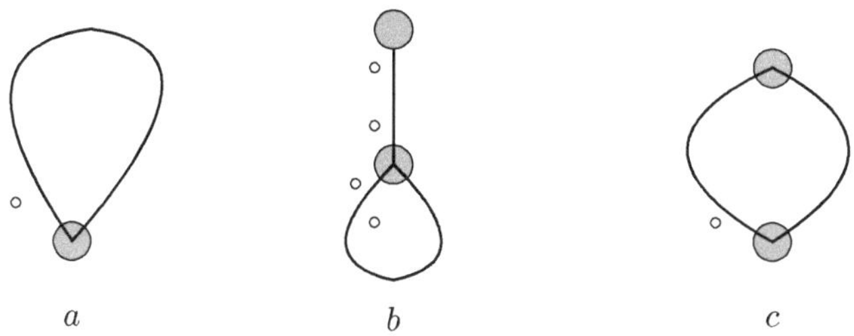

Figure 7.1.2: Classes of lack-1 planar near-quadrangulations with sizes 1–2.

Similarly, in Figure 7.1.3, one might see that the root-classes of lack-1 planar near-quadrangulations with size 3:

$$\begin{aligned} &2xz(a) + xz(b) + x^3z^3(c) + 5x^5z + xz^5(d) + 5x^5z + xz^5(e) \\ &\quad + 3x^3z^3(f) + 4x^4z^2 + 2x^2z^4(g) \\ &= 10x^5z + 4x^4z^2 + 4x^3z^3 + 2x^2z^4 + 2xz^5 + 3xz = F_3, \end{aligned}$$

as given by (7.1.32).

7.2 Quadrangulations on projective plane

In Liu YP [51] (Section 4.5, 2009), one finds the system of equations for g and f

$$\begin{cases} g = \dfrac{x^4y(f + x\frac{\partial f}{\partial x}) - yx^2g^*}{x^2 - y - 2x^4yf}; \\ f = \dfrac{x^4yf^2 - x^2yf^* + x^2 - y}{x^2 - y}; \\ f|_{x=y=0} = 1, \quad g|_{x=y=0} = 0, \end{cases} \tag{7.2.1}$$

where $f^* = \partial_x^2 f$ and $g^* = \partial_x^2 g$. This is equation (24) in Introduction which is meaningful in a classification for quadrangulations on the projective plane.

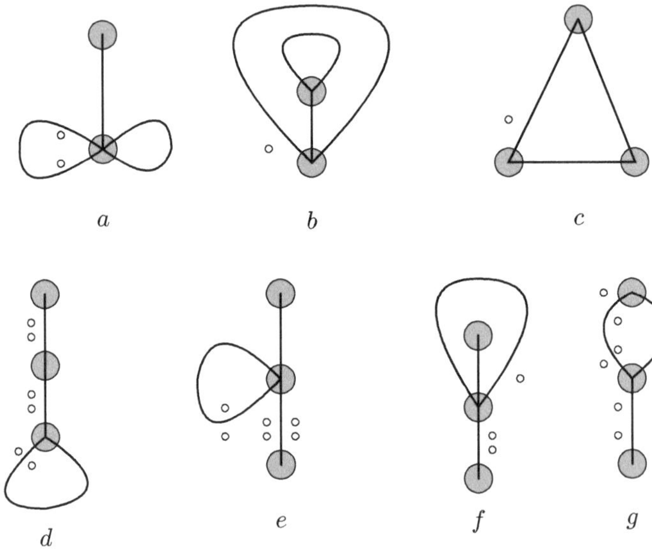

Figure 7.1.3: Classes of lack-1 planar near-quadrangulations with size 3.

In the system of equations, although the differential function of f occurs, because of the independence of the second equation and its solution given in Section 7.1, this equation is already known.

In fact, $f = f_{\text{0-quad}}$ is determined by $Q_n = \partial_y^n f_{\text{0-quad}}$, for $n \geq 0$, all of which are shown in (7.1.22).

From (7.1.22), we have, for any integer $n \geq 0$,

$$\begin{aligned}\left[x\frac{\partial f}{\partial x}\right]_n &= x\left[\frac{\partial f}{\partial x}\right]_n \\ &= x\frac{\partial Q_n}{\partial x}, \quad \text{by (7.1.23)}, \\ &= \sum_{m=1}^{2n} 2mQ_{2m,n}x^{2m}.\end{aligned} \tag{7.2.2}$$

Thus, equation (7.2.1) is equivalently transformed into

$$\begin{cases} g = \dfrac{x^4y(f + x\frac{\partial f}{\partial x}) - yx^2g^*}{x^2 - y - 2x^4yf}; \\ g|_{x=y=0} = 0, \end{cases} \tag{7.2.3}$$

where $f = f_{\text{0-quad}}$ is given in Section 7.1 and $g^* = \partial_x^2 g \in \mathcal{R}\{y\}$.

The first case of equation (7.2.3) is further transformed into a suitable equivalence:

$$\begin{aligned} x^2 g &= x^4 y\left(f + x\frac{\partial f}{\partial x}\right) - x^2 y g^* + yg + 2x^4 yfg \\ &= x^4 y\left(f - 1 + x\frac{\partial f}{\partial x}\right) + y(g - x^2 g^*) + 2x^4 yfg. \end{aligned} \tag{7.2.4}$$

Because of $[f]_0 = Q_0 = 1$ and $\frac{\mathrm{d}Q_0}{\mathrm{d}x} = 0$, for any integer $n \geq 0$,

$$\begin{aligned} \left[f + x\frac{\partial f}{\partial x}\right]_0 &= [f]_0 + x\frac{\mathrm{d}[f]_0}{\mathrm{d}x} \\ &= Q_0 + x\frac{\mathrm{d}Q_0}{\mathrm{d}x} \\ &= 1. \end{aligned}$$

Moreover,

$$\left[f + x\frac{\partial f}{\partial x}\right]_n = \begin{cases} 1, & \text{when } n = 0; \\ Q_n + x\frac{\mathrm{d}Q_n}{\mathrm{d}x}, & \text{when } n \geq 1. \end{cases} \tag{7.2.5}$$

Let $G_n = [g]_n = \partial_y^n g$, then from $[g]_0 = 0$, $[g^*]_0 = 0$. Hence for $n \geq 0$,

$$[g - x^2 g^*]_n = \begin{cases} 0, & \text{when } n = 0; \\ [g]_n - x^2 [g^*]_n, & \text{when } n \geq 1, \end{cases} \tag{7.2.6}$$

and

$$[fg]_n = \sum_{i=0}^{n} Q_i G_{n-i}. \tag{7.2.7}$$

On the basis of (7.2.4)–(7.2.7),

$$\begin{aligned} y^0 : \quad & x^2 [g]_0 = x^4 \left[y\left(f + x\frac{\partial f}{\partial x}\right)\right]_0 + [y(g - x^2 g^*)]_0 \\ & \quad + 2x^4 [yfg]_0 = 0 \\ & \Longrightarrow G_0 = 0 \text{ and } G_0^* = 0, \end{aligned} \tag{7.2.8}$$

$$\begin{aligned} y^1 : \quad & x^2 [g]_1 = x^4 \left[f + x\frac{\partial f}{\partial x}\right]_0 + [g - x^2 g^*]_0 + 2x^4 [fg]_0 \\ & = x^4 \left(Q_0 + x\frac{\mathrm{d}Q_0}{\mathrm{d}x}\right) + (G_0 - x^2 G_0^*) \\ & \quad + 2x^4 Q_0 G_0 \\ & = x^4 \\ & \Longrightarrow G_1 = x^2 \text{ and } G_1^* = 1, \end{aligned} \tag{7.2.9}$$

$$
\begin{aligned}
y^2: \quad x^2[g]_2 &= x^4\left[f + x\frac{\partial f}{\partial x}\right]_1 + [g - x^2g^*]_0 + 2x^4[fg]_1 \\
&= x^4\left(Q_1 + x\frac{\mathrm{d}Q_1}{\mathrm{d}x}\right) + (G_1 - x^2G_1^*) \\
&\quad + 2x^4(Q_0G_1 + Q_1G_0) \\
&= x^4(x^2 + 2x^2) + 2x^6 \\
&= 5x^6 \\
\Longrightarrow G_2 &= 5x^4 \text{ and } G_2^* = 0,
\end{aligned}
\tag{7.2.10}
$$

and

$$
\begin{aligned}
y^3: \quad x^2[g]_3 &= x^4\left[f + x\frac{\partial f}{\partial x}\right]_2 + [g - x^2g^*]_2 + 2x^4[fg]_2 \\
&= x^4\left(Q_2 + x\frac{\mathrm{d}Q_2}{\mathrm{d}x}\right) + (G_2 - x^2G_2^*) \\
&\quad + 2x^4(Q_0G_2 + Q_1G_1 + Q_2G_0) \\
&= x^4(2x^4 + 8x^4) + 5x^4 + 2x^4(5x^4 + x^4) \\
&= 5x^4 + 22x^8 \\
\Longrightarrow G_3 &= 5x^2 + 22x^6 \text{ and } G_3^* = 5.
\end{aligned}
\tag{7.2.11}
$$

As a matter of fact, for any integer $n \geq 1$,

$$
\begin{aligned}
x^2G_n = x^4\left(Q_{n-1} + x\frac{\mathrm{d}Q_{n-1}}{\mathrm{d}x}\right) + (G_{n-1} - x^2G_{n-1}^*) \\
+ 2x^4\sum_{i-0}^{n-1} Q_iG_{n-1-i}.
\end{aligned}
\tag{7.2.12}
$$

Lemma 7.2.1. *For any integer $n \geq 1$, G_n has no term with odd degree of x.*

Proof. From (7.2.9)–(7.2.11), it is easily seen that G_1, G_2 and G_3 have no term of odd degree. Furthermore, for $n \geq 3$, we proceed by induction on n. Assume it holds for all G_i, $1 \leq i \leq n-1$: there is no term of odd degree. Because of the parity, $x\frac{\mathrm{d}Q_{n-1}}{\mathrm{d}x}$ has no term of odd degree. By the assumption and (7.2.12), G_n has no term with odd degree of x. □

Although all $G_n \in \mathcal{R}\{x\}$, for $n \geq 1$ are even functions, it is still unknown whether or not G_n has a constant term for some n.

Lemma 7.2.2. *For any integer $n \geq 1$, G_n has terms with degree of x not less than 2.*

Proof. From (7.2.9)–(7.2.11), it is seen that among G_1, G_2 and G_3, the degree of x is not less than 2. Further, for integer $n \geq 3$ we proceed by induction on n. Assume all of G_i for $1 \leq i \leq n-1$, do not have a term of x whose degree is less than 2. Because of $G_{n-1}^* = \partial_x^2 G_{n-1}$, the assumption leads to $x^2|G_{n-1}$. By Lemma 7.2.1, $x^4|(G_{n-1} - x^2G_{n-1}^*)$. By (7.2.12), $x^2|G_n$. Therefore, the conclusion is drawn. □

This lemma tells us that, for any integer $n \geq 1$, $x^2 | G_n$.

Lemma 7.2.3. *For any integer $n \geq 1$, G_n has x^2 as a factor.*

Proof. This is a direct result of Lemma 7.2.2. □

Next, an upper bound of term degrees among all terms of G_n for integer $n \geq 1$ is estimated.

Lemma 7.2.4. *For any integer $n \geq 1$, G_n is a polynomial of x with degree at most $2n$.*

Proof. From (7.2.9)–(7.2.11), it is found that all of G_1, G_2 and G_3 satisfy the conclusion. In general, for $n \geq 4$, we proceed by induction on n. Assume for any integer i: $1 \leq i \leq n-1$, that G_i is a polynomial of degree at most $2i$. We prove that G_n is a polynomial of degree at most $2n$.

For a polynomial p of x, denote by $d(p)$ the degree of p. From (7.2.12),

$$\begin{aligned}
d(G_n) &= \max \left\{ \begin{array}{l} d\left(Q_{n-1} + x\dfrac{\mathrm{d}Q_{n-1}}{\mathrm{d}x} \right) + 2 \\ d(G_{n-1} - x^2 G^*_{n-1}) - 2 \\ d\left(\sum\limits_{i-0}^{n-1} Q_i G_{n-1-i} \right) + 2 \end{array} \right\}, \quad \text{by assumption,} \\
&= \max \left\{ \begin{array}{l} d\left(Q_{n-1} + x\dfrac{\mathrm{d}Q_{n-1}}{\mathrm{d}x} \right) + 2 \\ d\left(\sum\limits_{i-0}^{n-1} Q_i G_{n-1-i} \right) + 2 \end{array} \right\} \\
&= d\left(\sum_{i-0}^{n-1} Q_i G_{n-1-i} \right) + 2 \\
&= d(G_{n-1}) + 2, \quad \text{by assumption,} \\
&= 2(n-1) + 2 = 2n.
\end{aligned}$$

Therefore, the conclusion is drawn. □

This theorem enables us to evaluate a solution g determined by G_n ($n \geq 1$) of equation (7.2.3) such that all coefficients on G_n are in $\mathbb{Z}_+$.

Theorem 7.2.5. *Equation* (7.2.3), *and hence equation* (7.2.1), *is well-defined on* $\mathcal{R}\{x, y\}$.

Proof. First, from the equivalency between equation (7.2.4) and equation (7.2.3) and $G_0 = 0$ as the initial condition of equation (7.2.3), on the basis of the above lemmas, $g : G_n$, for $n \geq 0$ as determined by (7.2.8)–(7.2.11) and (7.2.12) provides a solution of equation (7.2.3) on $\mathcal{R}\{x, y\}$.

Then, by considering the uniqueness of the procedure to evaluate a solution g on $\mathcal{R}\{x, y\}$ under the initial condition of equation (7.2.3), the solution of equation (7.2.3) is the only one. □

On the basis of Theorem 7.2.5, one more structure of the solution g has to be mentioned.

Lemma 7.2.6. *For any integer $n \geq 1$, the polynomial $G_n \in \mathcal{R}_+\{x\}$.*

Proof. From (7.2.9)–(7.2.11), it is seen that all of G_1, G_2 and G_3 are in $\mathcal{R}_+\{x\}$. In general for $n \geq 4$, we proceed by induction on n. Assume for any integer i ($1 \leq i \leq n-1$), $G_i \in \mathcal{R}_+\{x\}$. We prove $G_n \in \mathcal{R}_+\{x\}$.

From Theorem 7.1.5 it can be deduced that, for any integer $n \geq 0$, $Q_n \in \mathcal{R}_+\{x\}$, and hence $x\frac{\mathrm{d}Q_n}{\mathrm{d}x} \in \mathcal{R}_+\{x\}$. Furthermore,

$$Q_{n-1} + x\frac{\mathrm{d}Q_{n-1}}{\mathrm{d}x} \in \mathcal{R}_+\{x\}.$$

By assumption, $G_{n-1} - x^2 G^*_{n-1} \in \mathcal{R}_+\{x\}$ and

$$\sum_{i-0}^{n-1} Q_i G_{n-1-i} \in \mathcal{R}_+\{x\}.$$

By (7.2.12), $G_n \in \mathcal{R}_+\{x\}$. □

Lemmas 7.2.2, 7.2.4 and 7.2.6 enable us, for any integer $n \geq 1$, to write

$$G_n = \sum_{i=1}^{n} G_{2i,n} \tag{7.2.13}$$

where $G_{2i,n} \in \mathcal{R}$.

From (7.1.22),

$$\begin{aligned} x^2\left(Q_{n-1} + \frac{\mathrm{d}Q_{n-1}}{\mathrm{d}x}\right) &= x^2\left(\sum_{m=1}(2m+1)Q_{2m,n-1}x^{2m}\right) \\ &= \sum_{m=2}^{n}(2m-1)Q_{2m-2,n-1}x^{2m}. \end{aligned} \tag{7.2.14}$$

Because of $G^*_n = \partial^2_x G_n = G_{2,n}$, for $n \geq 1$,

$$\begin{aligned} G_{n-1} - x^2 G^*_{n-1} &= \sum_{m=2}^{n-1} G_{2m,n-1}x^{2m} \\ &= x^2 \sum_{m=1}^{n-2} G_{2m,n-1}x^{2m}. \end{aligned} \tag{7.2.15}$$

From (7.1.22) and (7.1.23),

$$\begin{aligned}\sum_{i=0}^{n-1} Q_i G_{n-1-i} &= \sum_{\substack{l\le t\le (n-1-i)+l\\ 0\le l\le i\\ 0\le i\le n-1}} Q_{2l,i} G_{2(t-l),n-1-i} x^{2t}\\ &= \sum_{i=0}^{n-1}\Big(\sum_{\substack{0\le l\le t\\ 0\le t\le n-i-2}} + \sum_{\substack{t-(n-i-1)\le l\le i\\ n-i-1\le t\le n-1}}\Big) Q_{2l,i} G_{2(t-l),n-1-i} x^{2t}\\ &= \sum_{i=0}^{n-1}\sum_{t=0}^{i} \psi_{2t,i} x^{2t}, \quad \psi_{2l,i} \text{ in (7.2.17) below,}\\ &= \sum_{t=0}^{n-1} \Psi_{2t,n-1} x^{2t}\end{aligned}$$

where

$$\Psi_{2t,n-1} = \sum_{i=t}^{n-1} \psi_{2t,i} \tag{7.2.16}$$

and

$$\psi_{2t,i} = \begin{cases} \sum_{l=0}^{t} Q_{2l,i} G_{2(t-l),n-1-i}, & \text{when } 0 \le t \le n-i-2;\\ \sum_{l=t-(n-i-1)}^{i} Q_{2l,i} G_{2(t-l),n-1-i}, & \text{when } n-i-1 \le t \le n-1. \end{cases} \tag{7.2.17}$$

Therefore,

$$x^2 [f_{\text{0-quad}} g]_{n-1} = \sum_{m=1}^{n} \Psi_{2(m-1),n-1} x^{2m}. \tag{7.2.18}$$

Theorem 7.2.7. *Let $g = f_{\tilde{1}\text{-quad}}$ determined by $P_n = \partial_y^n f_{\tilde{1}\text{-quad}} \in \mathcal{R}_+\{x\}$, for $n \ge 0$ be the solution of equation* (7.2.1). *Then*

$$P_n = \begin{cases} 0, & \textit{when } n = 0;\\ x^2, & \textit{when } n = 1;\\ 5x^4, & \textit{when } n = 2;\\ 5x^2 + 22x^6, & \textit{when } n = 3;\\ \sum_{m=1}^{n} P_{2m,n} x^{2m}, & \textit{when } n \ge 4, \end{cases} \tag{7.2.19}$$

where

$$P_{2m,n} = \begin{cases} P_{2,n-1} + 2\Psi_{0,n-1}, & \textit{when } m = 1;\\ (2m-1) Q_{2(m-1),n-1} + P_{2m,n-1} + 2\Psi_{2(m-1),n-1}, & \textit{when } 2 \le m \le n-2;\\ (2n-3) Q_{2(n-2),n-1} + 2\Psi_{2(n-2),n-1}, & \textit{when } m = n-1;\\ (2n-1) Q_{2(n-1),n-1} + 2\Psi_{2(n-1),n-1}, & \textit{when } m = n. \end{cases} \tag{7.2.20}$$

Proof. When $n = 0, 1, 2$ and 3, the results are, respectively, clear from the initiation of equation (7.2.1), (7.2.9), (7.2.10) and (7.2.11). For $n \ge 4$, by (7.2.12),

$$P_n = x^2\left(Q_{n-1} + \frac{dQ_{n-1}}{dx}\right) + \frac{P_{n-1} - x^2 P^*_{n-1}}{x^2} + 2x^2 \sum_{i=0}^{n-1} Q_i G_{n-1-i}$$
$$= \sum_{m=2}^{n} (2m-1) Q_{2(m-1),n-1} x^{2m} + \sum_{m=1}^{n-2} P_{2m,n-1} x^{2m}$$
$$+ \sum_{m=1}^{n} 2\Psi_{2(m-1),n-1} x^{2m}.$$

After rearrangement, the conclusion is drawn. □

This theorem enables us to get the solution of equation (7.2.1) in the form of a sum with all terms positive.

Example 1. From Ren H-Liu YP [65] (1999), one finds the equation for f and g,

$$\begin{cases} x^4 yzf^2 + (z - x^2)f + (x^2 - z - x^2 zf^*) = 0; \\ g = 2x^2 yzfg + x^2 z \dfrac{\partial(xf)}{\partial x} + x^{-2} z(g - x^2 g^*); \\ f|_{x=y=z=0} = 1, \quad g|_{x=y=z=0} = 0, \end{cases} \tag{7.2.21}$$

where $f^* = \partial_x^2 f$ and $g^* = \partial_x^2 g$.

Because $\partial_z^n g \in \mathcal{R}\{x, y\}$, $n \geq 1$, are polynomials of x and y, the issue of being well-defined of this partial differential equation and the solution in the form of a finite sum with all terms positive can be addressed by following the procedure described in the context of this section. Some explicisions derived from Lagrange inversion are very simple. However, the result obtained in this section is very favorable for use of a computer. Moreover, the inversion can also be treated as a functional from the function space to itself. On this topic, it is absolutely necessary to do further research.

Example 2. Consider the system of equations about g and f

$$\begin{cases} g = \dfrac{x^4 y(f + x\frac{\partial f}{\partial x}) - x^4 y^2(1 + g^\diamond)}{x^2 - y - 2x^4 yf}; \\ f = \dfrac{x^4 yf^2 - x^* yf^* + x^2 - y}{x^2 - y}; \\ f|_{x=y=0} = 1, \quad g|_{x=y=0} = 0, \end{cases} \tag{7.2.22}$$

where $f^* = \partial_x^2 f$ and $g^\diamond = \partial_x^4 g$.

The issue of being well-defined of equation (7.2.22) can be addressed in a similar manner to the context in this section.

Lemma 7.2.8. *For any integer $n \geq 0$,*

$$\left.\begin{array}{ll} \text{when } n = 0 (\text{mod } 2), & P_{2,n} \\ \text{when } n = 1 (\text{mod } 2), & P_{4,n} \end{array}\right\} = 0.$$

Proof. The proof is similar to the proof of Lemma 7.1.6. On account of (7.2.12), we proceed by induction. □

Lemma 7.2.9. *For any integer $n \geq 1$, $P_{2(n-1),n} = 0$.*

Proof. The proof is similar to the proof of Lemma 7.1.7. On account of (7.2.12), we proceed by induction. □

Lemma 7.2.10. *For any integer $n \geq 1$, $P_{4,n} = P_{2,n+1}$ in* (7.2.20).

Proof. We proceed on the basis of Lemma 7.2.8 and Lemma 7.2.9, by employing (7.2.12), inductively, or directly evaluating by (7.2.20). □

Via Lemma 7.2.10, the equivalence between equation (7.2.22) and equation (7.2.1 can be directly derived.

Example 3. Root-isomorphic classification of near-quadrangulations by size and root-face valency on projective plane. The solution $g = f_{\tilde{1}\text{-quad}}$ of equation (7.2.1) provides the classes of near-quadrangulations by size and root-face valency on projective plane.

In Figure 7.2.1, we have the classes of near-quadrangulations on projective plane by sizes: 1–3. For example, $1a$ represents that the quadrangulations of size 1 have only 1 class and the root-face valency of them is 2. This is $P_1 = x^2$. Near-quadrangulations of size 2 on projective plane have $4b + 1c$, *i. e.*, $4 + 1 = 5$ classes. Because of them all coming with root-face valency 4, $P_2 = (4+1)x^4 = 5x^4$. Near-quadrangulations of size 3 on the projective plane have two parts. One part of them with root-valency 2 has $2d + 2e + 1f$, *i. e.*, $2 + 2 + 1 = 5$ classes. The other part of them with root-valency 6 has $6g + 3h + 6i + 6j + 1k$, *i. e.*, $3 \times 6 + 3 + 1 = 22$ classes. Therefore, $P_3 = 5x^2 + 22x^6$.

7.3 Quadrangulations on torus

Consider the partial differential system of equations for f, g and h

$$\begin{cases} x^4 y\left(z\dfrac{\partial g}{\partial z}\right)\Big|_{z=x} = x^2(1-2x^2yh)f - yf_{i_x\geq 4}; \\ x^3zy\delta_{z,x}(uh|_{x=u}) = (x^2 - 2x^4yh)g - yg_{i_x\geq 3}; \\ x^4yh^2 + (y-x^2)h - x^2yh_{2_x} + x^2 - y = 0; \\ f|_{x=y=0} = 0; \quad g|_{x=z=y=0} = 1; \quad h|_{x=y=0} = 1 \end{cases} \tag{7.3.1}$$

where $f_{i_x\geq 4}$ and $g_{i_x\geq 3}$ are, respectively, obtained by deleting the terms of degrees for x not greater than 4 and 3 from the functions f and g.

This is equation (25) in Introduction, which is meaningful in a classification for quadrangulations on the torus.

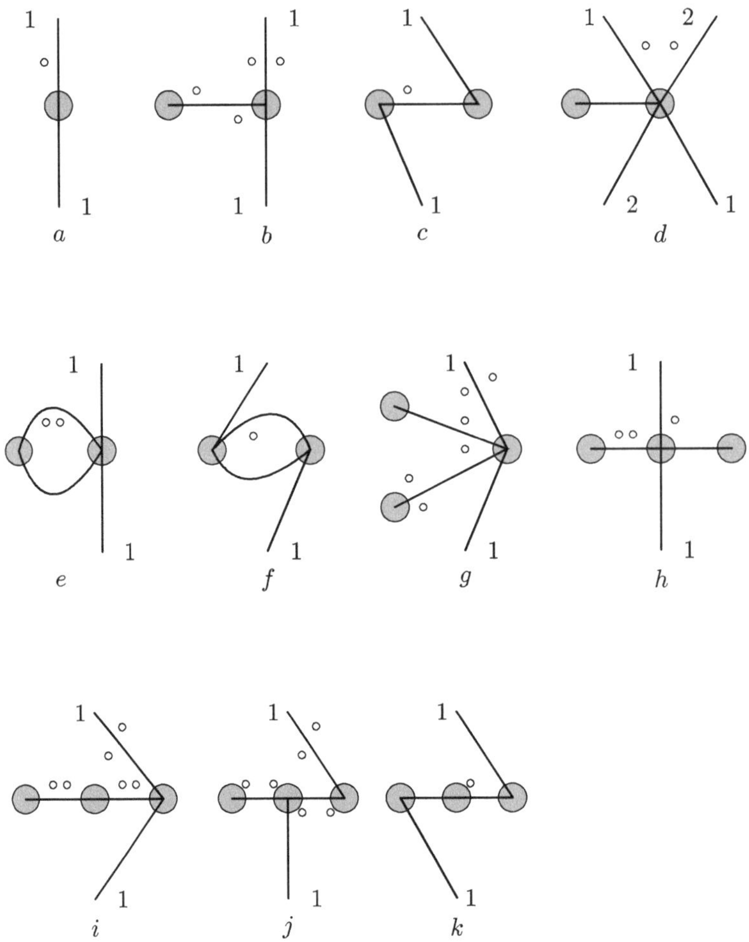

Figure 7.2.1: Classes of near-quadrangulations on projective plane with sizes 1–3.

We proceed on the basis of the equivalence between the system composed of the second and third equations of equation (7.3.1) and the system equation (7.1.27). Their solutions $g = f_{\text{mis-1}}$ and $h = f_{\text{0-quad}}$, are, respectively given by (7.1.2)–(7.1.4) and (7.1.35)–(7.1.36). It is only necessary to consider the first of equation (7.3.1).

For convenience, equation (7.3.1) is transformed into a suitable equivalence:

$$\begin{aligned} f &= x^2y\left(z\frac{\partial g}{\partial z}\right)\bigg|_{z=x} + 2x^2yhf + x^{-2}yf_{i_x\geq 4} \\ &= x^2y\left(z\frac{\partial f_{\text{mis-1}}}{\partial z}\right)\bigg|_{z=x} + 2x^2yf_{\text{0-quad}}f + x^{-2}yf_{i_x\geq 4}. \end{aligned} \tag{7.3.2}$$

Let f be determined by $[f]_n\partial_y^n f = F_n$, for $n \geq 0$. We proceed on the basis of equation (7.3.2). Because y has a factor on the right hand side,

$$y^0: \quad [f]_0 = 0 \Longrightarrow F_0 = 0, \text{ hence } F_0|_{i_x \geq 4} = 0. \tag{7.3.3}$$

This is the initial condition of equation (7.3.1): $f|_{x=y=0} = 0$.

For any integer $n \geq 1$,

$$\begin{aligned} y^n: \quad [f]_n &= x^2\left[z\frac{\partial f_{\text{mis-1}}}{\partial z}\right]_{n-1}\bigg|_{z=x} + 2x^2[f_{\text{0-quad}}f]_{n-1} \\ &\quad + x^{-2}[f_{i_x\geq 4}]_{n-1} \\ \Longrightarrow \quad F_n &= x^2\left(z\frac{\partial O_{n-1}}{\partial z}\right)\bigg|_{z=x} + 2x^2\sum_{i=0}^{n-1} F_i Q_{n-1-i} \\ &\quad + x^{-2}F_{n-1}|_{i_x\geq 4}. \end{aligned} \tag{7.3.4}$$

By employing Theorem 7.1.4 and Theorem 7.1.10,

$$\begin{aligned} F_1 &= x^2\left(z\frac{\partial O_0}{\partial z}\right)\bigg|_{z=x} + 2x^2Q_0F_0 + x^{-2}F_0|_{i_x\geq 4} \\ &= x^2(0) + 2x^2(0) + x^{-2}(0) = 0, \quad F_1|_{i_x\geq 4} = 0, \end{aligned} \tag{7.3.5}$$

$$\begin{aligned} F_2 &= x^2\left(z\frac{\partial O_1}{\partial z}\right)\bigg|_{z=x} + 2x^2(Q_0F_1 + Q_1F_0) \\ &\quad + x^{-2}F_1|_{i_x\geq 4} \\ &= x^2(x^4) + 2x^2(0) + x^{-2}(0) = x^4, \\ &\quad \text{hence } F_2|_{i_x\geq 4} = x^4, \end{aligned} \tag{7.3.6}$$

$$\begin{aligned} F_3 &= x^2\left(z\frac{\partial O_2}{\partial z}\right)\bigg|_{z=x} + 2x^2(Q_0F_2 + Q_1F_1 + Q_2F_0) \\ &\quad + x^{-2}F_2|_{i_x\geq 4} \\ &= x^2(8x^4) + 2x^2(x^4) + x^{-2}(x^4) = x^2 + 10x^6, \\ &\quad \text{hence } F_3|_{i_x\geq 4} = 10x^6, \end{aligned} \tag{7.3.7}$$

$$\begin{aligned} F_4 &= x^2\left(z\frac{\partial O_3}{\partial z}\right)\bigg|_{z=x} + 2x^2(Q_0F_3 + Q_1F_2 \\ &\quad + Q_2F_1 + Q_3F_0) + x^{-2}F_3|_{i_x\geq 4} \\ &= x^2(10x^6 + 8x^6 + 12x^6 + 8x^6 + 10x^6 + 3x^2) \\ &\quad + 2x^2(x^2 + 10x^6) + x^{-2}(10x^6) \\ &= (48x^8 + 3x^4) + (20x^8 + 2x^4) + 10x^4, \\ &\quad \text{hence } F_4 = 15x^4 + 68x^8. \end{aligned} \tag{7.3.8}$$

Via evaluating the root-isomorphic classes of near-quadrangulation of at most 4 edges on sphere, the results are the same as those obtained by (7.3.5)–(7.3.8).

Lemma 7.3.1. *For any integer $n \geq 1$, F_n is a polynomial of x with maximum degree not greater than $2n$ without term of odd degree and a constant term on $\mathcal{R}_+[x]$.*

Proof. From (7.3.5)–(7.3.8), the conclusion is true for $1 \le n \le 4$. In general, we proceed by induction on n for $n \ge 5$. Assume for any integer i $(1 \le i \le n-1)$, F_i is a polynomial of degree $2i$ without a term of odd degree and a constant term, to prove the conclusion for $i = n$. In (7.3.4), it is found that O_n and Q_n are a polynomial of x with degree $2n$ without term of odd degree and constant term in $\mathcal{R}_+[x]$. From the assumption and that the minimum degree is x in $F_{n-1}|_{i_x \ge 4}$ is at least 4, F_n is a polynomial of x with degree $2n$ without term of odd degree and constant term in $\mathcal{R}_+[x]$. This is the conclusion. □

Based on this lemma, F_n, $n \ge 1$, can be expressed as

$$F_n = \sum_{m=1}^{n} B_{2m,n} x^{2m}, \quad B_{m,n} \in \mathcal{R}_+. \tag{7.3.9}$$

Thus,

$$x^{-2} F_n|_{i_x \ge 4} = \sum_{m=2}^{n} B_{2m,n} x^{2m-2} = \sum_{m=1}^{n-1} B_{2(m+1),n} x^{2m}. \tag{7.3.10}$$

By (7.3.4),

$$\begin{aligned} F_n = {} & x^2 \left(z \frac{\partial O_{n-1}}{\partial z} \right)\bigg|_{z=x} + 2x^2 \sum_{i=0}^{n-1} F_i Q_{n-1-i} \\ & + \sum_{m=1}^{n-2} B_{2(m+1),n-1} x^{2m}. \end{aligned} \tag{7.3.11}$$

Theorem 7.3.2. *The system of partial differential equations* (7.3.1) *is well-defined on* $\mathcal{R}_+\{x, y\}$.

Proof. Because a function determined by (7.3.11) satisfies equation (7.3.2), from equivalency, this function provides a solution of equation (7.3.1).

Furthermore, from the uniqueness of the procedure to evaluating the function $(F_n, n \ge 0)$ by (7.3.11) on $\mathcal{R}_+[x]$ based on the initiation, this solution is the only one. □

On the basis of this theorem, let the solution of equation (7.3.1) be $f = f_{\text{1-nq}}$, $g = f_{\text{crq}}$ and $h = f_{\text{0-nq}}$, then, for any integer $n \ge 0$,

$$\partial_x^n f_{\text{1-nq}} = F_n, \quad \partial_x^n f_{\text{crq}} = O_n \quad \text{and} \quad \partial_x^n f_{\text{0-nq}} = Q_n \tag{7.3.12}$$

are, respectively, determined by (7.3.11), Theorem 7.1.10 and Theorem 7.1.4.

Theorem 7.3.3. *For the solution f of the partial differential system of equations equation* (7.3.1), *write $\partial_x^n f_{\text{1-nq}} = T_n$, then, for any integer $n \ge 1$, T_n has the form of a finite sum with all terms positive,*

$$\begin{aligned} T_n = {} & x^2 \left(z \frac{\partial O_{n-1}}{\partial z} \right)\bigg|_{z=x} + 2x^2 \sum_{i=0}^{n-1} T_i Q_{n-1-i} \\ & + \sum_{m=1}^{n-2} T_{2(m+1),n-1} x^{2m} \end{aligned} \tag{7.3.13}$$

where

$$T_{2(m+1),n-1} = \partial_x^{2m} T_{n-1}, \quad 1 \le m \le n-1.$$

Proof. From Theorem 7.3.2, it is known that, for any integer $n \ge 1$, $T_n = F_n$. By (7.3.11), the conclusion is drawn. □

Example 1. Root-isomorphic classes of near-quadrangulations on torus. In Figure 7.3.1, classes of near-quadrangulations with size 3 on torus are shown: $1a = x^2$ and $6b + 3c + 1d = 10x^6$, *i. e.*, $T_3 = x^2 + 10x^6(F_3)$. This is the result given in (7.3.7). These are listed in Table 7.3.1.

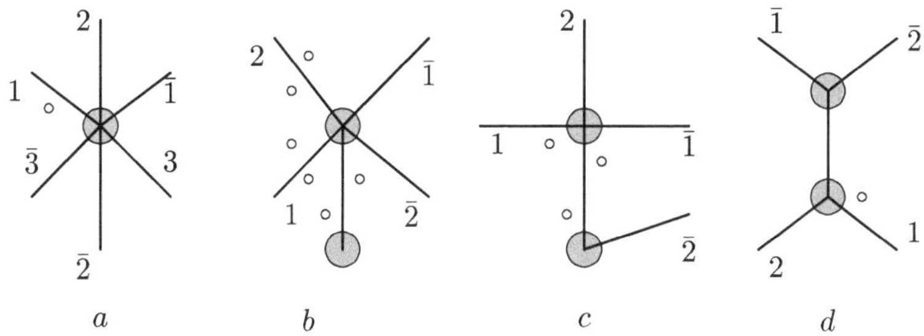

Figure 7.3.1: Classes of near-quadrangulations on torus of size 3.

Table 7.3.1: Vertices, faces and classes in Figure 7.3.1.

Order	Vertices	Faces	Classes
a	$(1,2,\gamma 1,3,\gamma 2,\gamma 3)$	$(1,3)(\gamma 1,2,\gamma 3,\gamma 2)$	$\{1,\beta 1,3,\beta 3\}$
b	$(3,1,2,\gamma 1,\gamma 2)(\gamma 3)$	$(1,\gamma 2,\gamma 1,2,3,\gamma 3)$	$\{3,\alpha 3\},\{1,\beta 1\},\{\alpha 1,\gamma 2\},$ $\{2,\beta 1\},\{\alpha 2,\gamma 1\},\{\gamma 2,\beta 3\}$
c	$(3,1,2,\gamma 1)(\gamma 2,\gamma 3)$	$(1,3,\gamma 2,\gamma 1,2,\gamma 3)$	$\{3,\alpha 3,2,\alpha 2\},\{1,\alpha 1,$ $\gamma 1,\beta 1\},\{\gamma 3,\beta 3,\gamma 2,\beta 2\}$
d	$(3,1,2)(\gamma 3,\gamma 1,\gamma 2)$	$(3,\gamma 1,2,\gamma 3,1,\gamma 2)$	$\{K1+K2+K3\}$

In this table, from the column of order (First, or Order), it is seen that the maps in the alphabetical order are marked by $a, b, c, \ldots$. From the column of faces (Third, or Faces), it is seen that all the faces in the face set of the map are marked by the letter at the corresponding entry of first column. The same for the column of vertices (Second, or Vertices) and the column of classes (Fourth, or Classes).

The number of root-isomorphic classes of a map is the number of hollows in the map shown in Figure 7.3.1.

For example, in the first column, all near-quadrangulations of size 3 on torus are listed as a, b, c and d.

In each entry of the second column, the first pair of parentheses shows the root-vertex with its valency.

In each entry of the third column, the first pair of parentheses shows the root-face with its valency.

In each entry of the fourth column, a pair of braces represents a root-isomorphic class in which each symbol (quadricell) produces an automorphism of the corresponding map.

In the Faces column, from the first pair of parentheses in the first entry, it is seen that the root-face valency of map a is 2 (*i. e.*, x^2). In the Classes column, from the number of brace pairs in the first entry, it is seen that the number of root-isomorphic classes is 1. Thus, $\partial_x^2 T_3 = 1$.

The first symbol (or quadricell) in the Classes column is chosen to be its representative denoted by a hollow as shown in Figure 7.3.2. The first quadricell in the root-vertex and the first quadricell in the root-face are the same. This quadricell is the root of the corresponding map. See also Table 7.3.2.

In Figure 7.3.2, root-isomorphic classes of near-quadrangulations on torus of size 4 are shown as $1a + 4b + 8c + 2d = 15x^4$, *i. e.*, $T_{4,4} = 15x^4(F_{4,4})$ which is the same as $\partial_x^4 F_4 = 15x^4$, obtained by (7.3.8).

Figure 7.3.2: Classes of near-quadrangulations on torus of size 4.

Table 7.3.2: Vertices, faces and classes in Figure 7.3.2.

Order	Vertices	Faces	Classes
a	$(4, 1, 2, 3)$ $(\gamma 4, \gamma 1, \gamma 2, \gamma 3)$	$(1, \gamma 2, 3, \gamma 4)$ $(\gamma 1, 2, \gamma 3, 4)$	$\{K1 + K2 + K3 + K4\}$
b	$(4, 1)$ $(r4, 2, 3, \gamma 1, \gamma 2, \gamma 3)$	$(1, \gamma 2, 3, r4)$ $(\gamma 1, 4, 2, \gamma 3)$	$\{4, \alpha 4, 1, \alpha 1\}, \{\gamma 4, \beta 4, \gamma 1, \beta 1\},$ $\{2, \alpha 3, \gamma 2, \beta 3\}, \{\alpha 2, 3, \beta 2, \gamma 3\}$
c	$(4, 1, 2, \gamma 1, 3, \gamma 2, \gamma 3)$ $(r4)$	$(1, 3, 4, r4)$ $(\gamma 1, 2, \gamma 3, \gamma 2)$	$\{4, \alpha 4\}, \{1, \beta 3\}, \{\alpha 1, \gamma 3\}, \{2, \beta 2\},$ $\{\alpha 2, \gamma 2\}, \{\gamma 1, \alpha 3\}, \{\beta 1, 3\}, \{\gamma 4, \beta 4\}$
d	$(4, 1, 2, \gamma 1)$ $(r4, 3, \gamma 2, \gamma 3)$	$(1, 4, 3, r4)$ $(\gamma 1, 2, \gamma 3, \gamma 2)$	$\{4, \alpha 4, 2, \alpha 2, \gamma 4, \beta 4, \gamma 2, \beta 2\},$ $\{1, \alpha 1, \gamma 1, \beta 1, 3, \alpha 3, \gamma 3, \beta 3\}$

7.4 Quadrangulations on Klein bottle

Consider the partial differential system of equations about f, g, h and p

$$\begin{cases} x^4y\left(x\frac{\partial g}{\partial x} + \left[z\frac{\partial h}{\partial z}\right]_{z=x}\right) = x^2f - x^4y(g + g^2 + 2pf) - y(f - x^2\partial_x^2 f); \\ x^4y\left(p + x\frac{\partial p}{\partial x}\right) = x^2(x^2 - y - 2x^4yp)g + x^2y\partial_x^2 g; \\ \frac{x^3zy}{z-x}\delta_{z,x}(up|_{x=u}) = x^2(1 - 2x^2yp)h - y(h - x^2\partial_x^2 h); \\ x^4yp^2 + (y - x^2)p - x^2y\partial_x^2 p + x^2 - y = 0; \\ f|_{x=z=y=0} = g|_{x=z=y=0} = h|_{x=z=y=0} = 0; \quad p|_{x=z=y=0} = 1. \end{cases} \tag{7.4.1}$$

This is equation (26) in Introduction, which is meaningful in a classification for quadrangulations on the Klein bottle.

From the system of equations (7.1.2), equation (7.1.27) and (7.2.1),

$$p = f_{\text{0-nq}}, \quad h = f_{\text{crq}} \quad \text{and} \quad g = f_{\tilde{\text{1}}\text{-nq}}. \tag{7.4.2}$$

Their expressions in the form of a finite sum with all terms positive are, respectively, given by (7.1.2)–(7.1.4), (7.1.35)–(7.1.36) and (7.2.19)–(7.2.20). Thus, it is only necessary to determine f via $F_n = \partial_y^n f = [f]_n$, $n \geq 1$.

First, the first equation in (7.4.1) is transformed into one of its equivalences:

$$\begin{aligned} f = {} & x^2y\left(x\frac{\partial g}{\partial x} + \left[z\frac{\partial h}{\partial z}\right]_{z=x}\right) + x^2y(g + g^2 + 2pf) \\ & + x^{-2}y(f - x^2\partial_x^2 f). \end{aligned} \tag{7.4.3}$$

Since y is a factor of the part on the right hand side with the equal sign,

$$[f]_n = F_0 = 0. \tag{7.4.4}$$

This is the initial condition to equation (7.4.3) from equation (7.4.1).

For any integer $n \geq 1$,

$$\begin{aligned} [f]_n = {} & x^2\left[x\frac{\partial g}{\partial x} + \left(z\frac{\partial h}{\partial z}\right)_{z=x}\right]_{n-1} + x^2[g + g^2 + 2pf]_{n-1} \\ & + x^{-2}[f - x^2\partial_x^2 f]_{n-1}, \text{ by (7.4.2)}, \\ = {} & x^2\left(x\frac{\partial P_{n-1}}{\partial x} + \left(z\frac{\partial O_{n-1}}{\partial z}\right)_{z=x}\right) \\ & + x^2\left(P_{n-1} + \sum_{i=0}^{n-1} P_iP_{n-1-i} + 2\sum_{i=0}^{n-1} Q_iF_{n-1-i}\right) \\ & + x^{-2}(F_{n-1} - x^2\partial_x^2 F_{n-1}). \end{aligned} \tag{7.4.5}$$

On the basis of this, by (7.2.19), (7.1.35) and (7.1.2),

$$\begin{aligned}F_1 &= x^2\left(x\frac{\partial P_0}{\partial x} + \left(z\frac{\partial O_0}{\partial z}\right)_{z=x}\right)\\&\quad + x^2(P_0 + P_0P_0 + 2Q_0F_0)\\&\quad + x^{-2}(F_0 - x^2\partial_x^2F_0)\\&= 0,\ \text{hence } F_1 - x^2\partial_x^2F_1 = 0,\end{aligned}\tag{7.4.6}$$

$$\begin{aligned}F_2 &= x^2\left(x\frac{\partial P_1}{\partial x} + \left(z\frac{\partial O_1}{\partial z}\right)_{z=x}\right)\\&\quad + x^2(P_1 + 2P_0P_1 + 2(Q_0F_1 + Q_1F_0)\\&\quad + x^{-2}(F_1 - x^2\partial_x^2F_1)\\&= x^2(2x^2 + x^2) + x^2(x^2) = 4x^4,\\&\quad \text{hence } F_2 - x^2\partial_x^2F_2 = 0 = 4x^4,\end{aligned}\tag{7.4.7}$$

$$\begin{aligned}F_3 &= x^2\left(x\frac{\partial P_2}{\partial x} + \left(z\frac{\partial O_2}{\partial z}\right)_{z=x}\right)\\&\quad + x^2\left(P_2 + \sum_{i=0}^{2} P_iP_{2-i} + 2\sum_{i=0}^{2} Q_iF_{2-i}\right)\\&\quad + x^{-2}(F_2 - x^2\partial_x^2F_2)\\&= x^2(20x^4 + 3x^4 + 2x^4 + 3x^4)\\&\quad + x^2(5x^4 + x^4 + 8x^4) + x^{-2}(4x^4)\\&= 4x^2 + 42x^6,\\&\quad \text{hence } F_3 - x^2\partial_x^2F_3 = 42x^6,\end{aligned}\tag{7.4.8}$$

$$\begin{aligned}F_4 &= x^2\left(x\frac{\partial P_3}{\partial x} + \left(z\frac{\partial O_3}{\partial z}\right)_{z=x}\right)\\&\quad + x^2\left(P_3 + \sum_{i=0}^{3} P_iP_{2-i} + 2\sum_{i=0}^{3} Q_iF_{2-i}\right)\\&\quad + x^{-2}(F_3 - x^2\partial_x^2F_3)\\&= x^2(13x^2 + 180x^6)\\&\quad + x^2(13x^2 + 124x^6)\\&\quad + x^{-2}(42x^6)\\&= 68x^4 + 304x^8,\\&\quad \text{hence } F_4 - x^2\partial_x^2F_4 = 68x^4 + 304x^8.\end{aligned}\tag{7.4.9}$$

By directly classifying near-quadrangulations of size 4 on the Klein bottle into root-isomorphic classes via joint trees shown in Liu YP [46] (2003, pp. 14–18), or [62]

(Liu, YP, 2017, pp. 197–201), it is seen that the results are the same as those in (7.4.5)–(7.4.8).

Lemma 7.4.1. *For any integer $n \geq 1$, F_n is a polynomial of x with degree at most $2n$ with neither a term of odd degree nor a constant term on $\mathcal{R}_+[x]$.*

Proof. From (7.4.6)–(7.4.9), it is found that the conclusion is true for $1 \leq n \leq 4$. For $n \geq 5$, we proceed by induction on n. Assume for any integer $1 \leq i \leq n-1$, that F_i is a polynomial of x with degree $2n$ with neither a term of odd degree nor a constant term. We prove the conclusion for $i = n$. On the basis of the last three sections, P_{n-1}, O_{n-1} and Q_{n-1} in (7.4.5) are all polynomials of degree at most $2(n-1)$ on $\mathcal{R}_+[x]$ with neither a term of odd degree nor a constant term. By the assumption and knowing that all terms in $F_{n-1}|_{i_x \geq 4}$ are with degree of x at least 4, F_n is deduced to be a polynomial of degree at most $2(n-1)+2 = 2n$ in $\mathcal{R}_+[x]$ with neither a term of odd degree nor a constant term. This is the conclusion. □

Based on this lemma, F_n, for $n \geq 1$, can be expressed in the form of

$$F_n = \sum_{m=1}^{n} K_{m,n} x^{2m}, \quad K_{m,n} \in \mathcal{R}_+. \tag{7.4.10}$$

Thus,

$$x^{-2}(F_n - x^2 \partial_x^2 F_n) = \sum_{m=2}^{n} K_{m,n} x^{2m-2} = \sum_{m=1}^{n-1} K_{m+1,n} x^{2m}. \tag{7.4.11}$$

From (7.4.5),

$$\begin{aligned} F_n &= x^2\left(x\frac{\partial P_{n-1}}{\partial x} + \left(z\frac{\partial O_{n-1}}{\partial z}\right)_{z=x}\right) \\ &\quad + x^2\left(P_{n-1} + \sum_{i=0}^{n-1} P_i P_{n-1-i} + 2\sum_{i=0}^{n-1} Q_i F_{n-1-i}\right) \\ &\quad + \sum_{m=1}^{n-1} K_{m+1,n} x^{2m}. \end{aligned} \tag{7.4.12}$$

Theorem 7.4.2. *The system of partial differential equations* (7.4.1) *is well-defined on* $\mathcal{R}_+[x,z]$.

Proof. It is easily shown that the function evaluated from (7.4.12) is a solution of equation (7.4.3), and hence equivalently of equation (7.4.1).

Furthermore, by considering the uniqueness of the evaluation procedure based on (7.4.12) from the initial condition of equation (7.4.1), the solution is the only one. □

On the basis of this theorem, let $f = f_{\tilde{2}\text{-nq}}$ $g = f_{\tilde{1}\text{-nq}}$ $h = f_{\text{crq}}$ $p = f_{\text{0-nq}}$ be consisting of the solution set of the system of equations (7.4.1), then, for any integer $n \geq 0$,

$$\partial_x^n f_{\tilde{1}\text{-nq}} = P_n, \quad \partial_x^n f_{\text{crq}} = O_n \quad \text{and} \quad \partial_x^n f_{\text{0-nq}} = Q_n \tag{7.4.13}$$

are determined by (7.4.12), Theorems 7.1.10 and 7.1.4.

Theorem 7.4.3. *In the solution set of system of equations (7.4.1), write* $\partial_x^n f_{\tilde{2}\text{-nq}} = K_n$, *then, for any integer* $n \geq 0$, K_n *has an expression in the form of a finite sum with all terms positive,*

$$\begin{aligned} K_n = {} & x^2\left(x\frac{\partial P_{n-1}}{\partial x} + \left(z\frac{\partial O_{n-1}}{\partial z}\right)_{z=x}\right) \\ & + x^2\left(P_{n-1} + \sum_{i=0}^{n-1} P_i P_{n-1-i} + 2\sum_{i=0}^{n-1} Q_i K_{n-1-i}\right) \\ & + \sum_{m=1}^{n-1} K_{m+1,n} x^{2m} \end{aligned} \tag{7.4.14}$$

where $n \geq 1$ *and* $K_0 = 0$.

Proof. From Theorem 7.4.2, it is known that, for any integer $n \geq 1$, $K_n = F_n$. By (7.4.12), the conclusion is drawn. □

Example 1. Root-isomorphic lasses of near-quadrangulations on Klein bottle. Only the case of size not greater than 3 is allowed. Because maps of size 1 only are allowed on the sphere, near-quadrangulations on the Klein bottle come with at least 2 edges. Figure 7.4.1 shows near-quadrangulations of size 2. Figures 7.4.2 and 7.4.3 show those of size 3, with, respectively, 1 and 2 vertices.

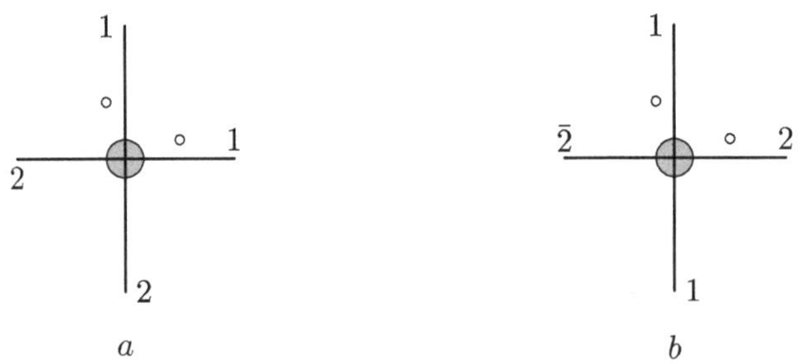

Figure 7.4.1: Classes of near-quadrangulations of size 2 on Klein bottle.

Table 7.4.1: Vertices, edges and classes in Figure 7.4.1.

Order	Vertices	Faces	Classes
a	$(1,\beta 1,2,\beta 2)$	$(1,\alpha 1,2,\alpha 2)$	$\{1,\alpha 1,2,\alpha 2\}$hen we proceed by induction $\{\beta 1,\gamma 1,\beta 2,\gamma 2\}$
b	$(1,2,\beta 1,\gamma 2)$	$(1,\alpha 2,\alpha 1,2)$	$\{1,\alpha 1,\beta 1,\gamma 1\}$hen we proceed by induction $\{2,\alpha 2,\beta 2,\gamma 2\}$

In Table 7.4.1, vertices, faces and root-isomorphic classes of near-quadrangulations of size 2 in Figure 7.4.1 are listed (*cf.* the explanation of Table 7.3.1).

As a matter of fact, all these near-quadrangulations are only quadrangulations.

In Table 7.4.2, vertices, faces and root-isomorphic classes of near-quadrangulations of size 3 and order 1 in Figure 7.4.2 are listed (*cf.* the explanation of Table 7.3.1).

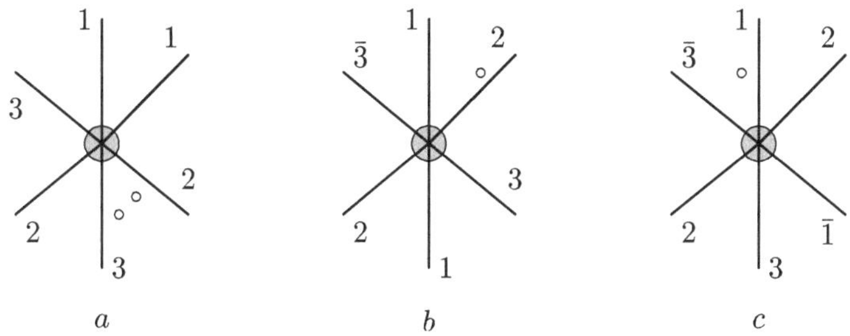

Figure 7.4.2: Classes of near-quadrangulations of size 3 and order 1 on Klein bottle.

Table 7.4.2: Vertices, edges and classes in Figure 7.4.2.

Order	Vertices	Faces	Classes
a	$(1, \beta 1, 2{,}3, \beta 2, \beta 3)$	$(1, \alpha 1, 2, \alpha 3)(3, \gamma 2)$	$\{3, \gamma 2\}$, $\{\alpha 2, \beta 3\}$
b	$(1, 2, 3, \beta 1, \beta 2, \gamma 3)$	$(1, \alpha 3, \gamma 2, 3)(\alpha 1, \beta 2)$	$\{\alpha 1, 2, \gamma 1, \beta 2\}$
c	$(1, 2, \gamma 1, 2, \beta 2, \gamma 3)$	$(1, 3)(\alpha 1, \alpha 2, \beta 3, \beta 2)$	$\{1, \beta 1, 3, \beta 3\}$

In Table 7.4.3, vertices, faces and root-isomorphic classes of near-quadrangulations of size 3 and order 2 in Figure 7.4.3 are listed (*cf.* the explanation of Table 7.3.1).

where $\{Ki\} = \{i\}, \{\alpha i\}, \{\beta i\}, \{\gamma i\}$, $i = 1, 2, 3$, and $K = \{1, \alpha, \beta, \gamma\}$ is the Klein group (see Liu YP [44]).

Example 2. Root-isomorphic classes of quadrangulations on Klein bottle (continued!).

In Table 7.4.4, vertices, faces and root-isomorphic classes of quadrangulations of size 4 and order 2 in Figure 7.4.4 are listed (*cf.* the explanation of Table 7.3.1).

In Table 7.4.5, vertices, faces and root-isomorphic classes of quadrangulations of size 4 and order 2 in Figure 7.4.5 are listed (*cf.* the explanation of Table 7.3.1).

In Table 7.4.6, vertices, faces and root-isomorphic classes of quadrangulations of size 4 and order 2 in Figure 7.4.6 are listed (*cf.* the explanation of Table 7.3.1).

In Table 7.4.7, vertices, faces and root-isomorphic classes of quadrangulations of size 4 and order 2 in Figure 7.4.7 are listed (*cf.* the explanation of Table 7.3.1).

From Figures 7.4.4–7.4.7, it is seen that quadrangulations of size 4 and order 2 on a Klein bottle have $12 + 16 + 8 + 32 = 68$ root-isomorphic classes in all.

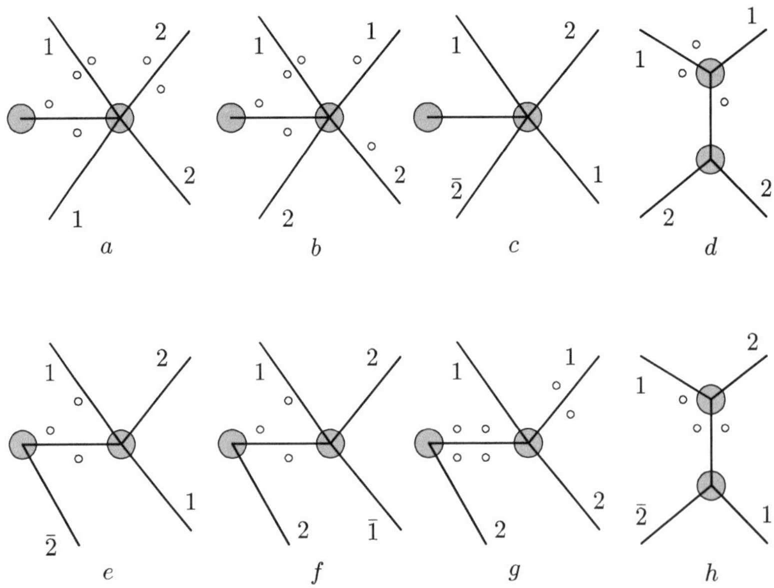

Figure 7.4.3: Classes of near-quadrangulations of size 3 and order 2 on Klein bottle.

Table 7.4.3: Vertices, edges and classes in Figure 7.4.3.

Order	Vertices	Faces	Classes
a	$(3,1,2,\beta 2,\beta 1)(\gamma 3)$	$(1,\gamma 2,\beta 2,\alpha 1,3,\gamma 3)$	$\{3,\alpha 3\},\{\gamma 3,\beta 3\},\{1,\gamma 1\},\{\alpha 1,\beta 1\},\{2,\gamma 2\},\{\alpha 2,\beta 2\}$
b	$(3,1,\beta 1,2,\beta 2)(\gamma 3)$	$(1,\alpha 1,2,\alpha 2,3,\gamma 3)$	$\{3,\alpha 3\},\{\gamma 3,\beta 3\},\{1,\gamma 2\},\{\alpha 1,\beta 2\},\{2,\gamma 1\},\{\alpha 2,\beta 1\}$
c	$(3,1,2,\beta 1,\gamma 2)(\gamma 3)$	$(1,\alpha 2,\gamma 1,2,3,\gamma 3)$	$\{K1\},\{K2\},\{K3\}$
d	$(3,1,\beta 1)(\gamma 3,2,\beta 2)$	$(1,\alpha 1,3,2,\alpha 2,\gamma 3)$	$\{3,\alpha 3,\gamma 3,\beta 3\},\{1,\gamma 1,2,\gamma 2\},\{\alpha 1,\beta 1,\alpha 2,\beta 2\}$
e	$(3,1,2,\beta 1)(\gamma 3,\gamma 2)$	$(1,\alpha 2,\beta 3,\gamma 1,2,\gamma 3)$	$\{3,\alpha 3,2,\alpha 2\},\{\gamma 3,\beta 3,\gamma 2,\beta 2\},\{1,\alpha 1,\beta 1,\gamma 1\}$
f	$(3,1,2,\gamma 1)(\gamma 3,\beta 2)$	$(1,3,\beta 2,\alpha 1,\alpha 2,\gamma 3)$	$\{3,\alpha 3,2,\alpha 2\},\{\gamma 3,\beta 3,\beta 2,\gamma 2\},\{1,\alpha 1,\gamma 1,\beta 1\}$
g	$(3,1,\beta 1,2)(\gamma 3,\beta 2)$	$(1,\alpha 1,2,\beta 3,\alpha 2,\gamma 3)$	$\{3,\alpha 2\},\{\alpha 3,2\},\{\beta 3,\gamma 2\},\{\gamma 3,\beta 2\},\{1,\gamma 1\},\{\alpha 1,\beta 1\}$
h	$(3,1,2)(\gamma 3,\beta 1,\gamma 2)$	$(1,\beta 3,\alpha 2,\gamma 1,2,\gamma 3)$	$\{3,\alpha 2,\gamma 3,\beta 2\},\{\alpha 3,2,\beta 3,\gamma 2\},\{1,\alpha 1,\beta 1,\gamma 1\}$

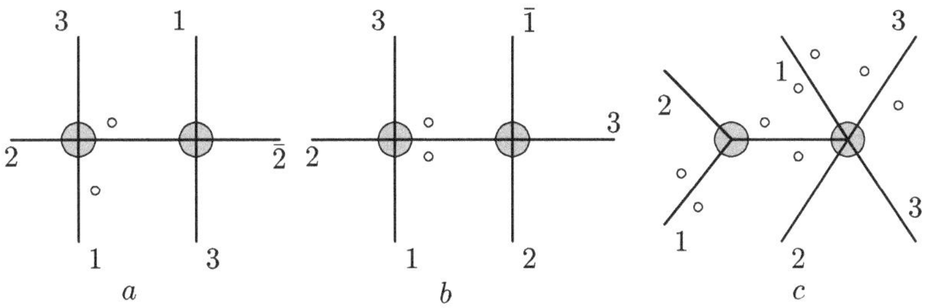

Figure 7.4.4: Classes of quadrangulations of size 4 and order 2 on Klein bottle I.

Table 7.4.4: Vertices, edges and classes in Figure 7.4.4.

Order	Vertices	Faces	Classes
a	$(4,1,2,3)$ $(\gamma4,\beta1,\gamma2,\beta3)$	$(1,\beta4,\alpha3,\gamma4)$ $(\alpha1,\gamma2,3,\beta2)$	$\{4,\alpha4,2,\alpha2,\gamma4,\beta4,\gamma2,\beta2\}$, $\{1,\alpha1,3,\alpha3,\beta1,\gamma1,\beta3,\gamma3\}$
b	$(4,1,2,3)$ $(\gamma4,\gamma1,\beta3,\beta2)$	$(1,\beta3,\alpha2,\gamma4)$ $(2,\gamma3,4,\gamma1)$	$\{4,\alpha1,2,\alpha3,\gamma4,\beta1,\beta3,\gamma2\}$, $\{\alpha4,1,\alpha2,3,\beta4,\gamma1,\gamma3,\beta2\}$
c	$(4,1,2)$ $(\gamma4,\beta1,3,\beta3,\beta2)$	$(1,\beta4,\alpha2,\gamma4)$ $(\alpha1,3,\alpha3,\beta2)$	$\{4,\alpha4\},\{\gamma4,\beta4\},\{1,\alpha2\},\{\alpha1,2\}$, $\{\beta1,\gamma2\},\{\gamma1,\beta2\},\{3,\gamma3\},\{\alpha3,\beta3\}$

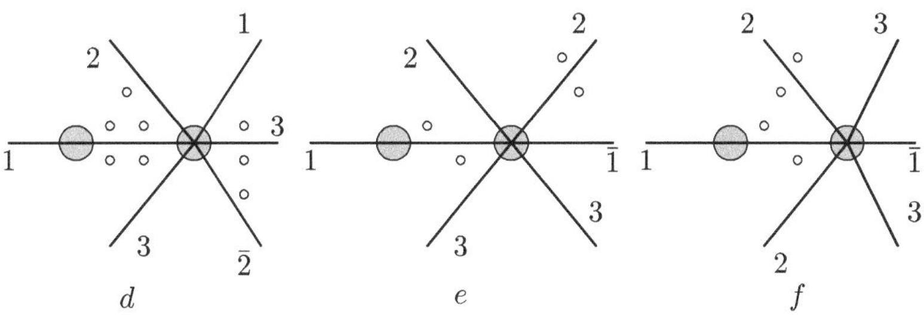

Figure 7.4.5: Classes of quadrangulations of size 4 and order 2 on Klein bottle II.

Table 7.4.5: Vertices, edges and classes in Figure 7.4.5.

Order	Vertices	Faces	Classes
d	$(4,1)$ $(\gamma4,2,\beta1,3,\gamma2,\beta3)$	$(1,\alpha2,\alpha3,\gamma4)$ $(\alpha1,3,\beta2,\beta4)$	$\{4,1\},\{\alpha4,\alpha1\},\{\beta4,\beta1\},\{\gamma4,\gamma1\}$, $\{2,\alpha2\},\{\gamma2,\beta2\},\{3,\gamma3\},\{\alpha3,\beta3\}$
e	$(4,1)$ $(\gamma4,2,\beta2,\gamma1,3,\beta3)$	$(1,3,\alpha3,\gamma4)$ $(\alpha1,\gamma2,\beta2,\beta4)$	$\{4,\alpha4,1,\alpha1\},\{\gamma4,\beta4,\gamma1,\beta1\}$, $\{2,\gamma2,3,\gamma3\},\{\alpha2,\beta2,\alpha3,\beta3\}$
f	$(4,1)$ $(\gamma4,2,3,\gamma1,\beta3,\beta2)$	$(1,\beta3,\alpha2,\gamma4)$ $(\alpha1,\alpha3,\beta2,\beta4)$	$\{4,\alpha4,1,\alpha1\},\{\gamma4,\beta4,\gamma1,\beta1\}$, $\{2,\alpha3,\beta3,\gamma2\},\{\alpha2,3,\gamma3,\beta2\}$

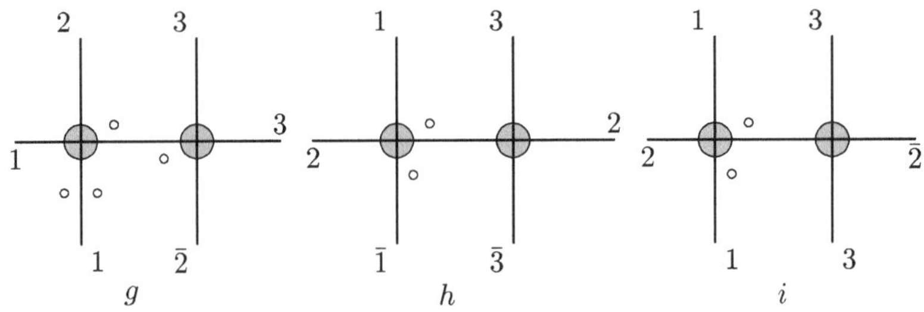

Figure 7.4.6: Classes of quadrangulations of size 4 and order 2 on Klein bottle III.

Table 7.4.6: Vertices, edges and classes in Figure 7.4.6.

Order	Vertices	Faces	Classes
g	$(4, 1, \beta 1, 2)$ $(\gamma 4, 3, \beta 3, \gamma 2)$	$(1, \alpha 1, 2, \gamma 4)$ $(\alpha 2, \gamma 3, \beta 3, \beta 4)$	$\{4, \alpha 2, \gamma 4, \beta 2\}, \{\alpha 4, 2, \beta 4, \gamma 2\},$ $\{1, \gamma 1, 3, \gamma 3\}, \{\alpha 1, \beta 1, \alpha 3, \beta 3\}$
h	$(4, 1, 2, \gamma 1)$ $(\gamma 4, 3, \beta 2, \gamma 3)$	$(1, 4, 3, \gamma 4)$ $(\alpha 1, \alpha 2, \gamma 3, \beta 2)$	$\{4, \alpha 4, 2, \alpha 2, \gamma 4, \beta 4, \beta 2, \gamma 2\},$ $\{1, \alpha 1, \gamma 1, \beta 1, 3, \alpha 3, \gamma 3, \beta 3\}$
i	$(4, 1, 2, \beta 1)$ $(\gamma 4, 3, \gamma 2, \beta 3)$	$(1, \alpha 2, \alpha 3, \gamma 4)$ $(\alpha 1, 4, 3, \beta 2)$	$\{4, \alpha 4, 2, \alpha 2, \gamma 4, \beta 4, \gamma 2, \beta 2\},$ $\{1, \alpha 1, \beta 1, \gamma 1, 3, \alpha 3, \beta 3, \gamma 3\}$

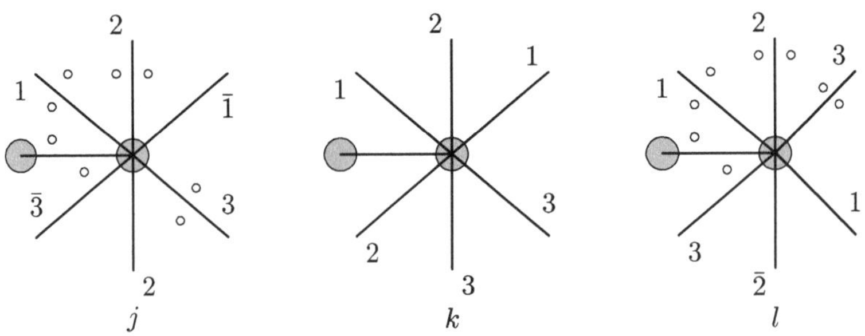

Figure 7.4.7: Classes of quadrangulations of size 4 and order 2 on Klein bottle IV.

Table 7.4.7: Vertices, edges and classes in Figure 7.4.7.

Order	Vertices	Faces	Classes
j	$(4, 1, 2, \gamma 1, 3, \beta 2, \gamma 3)$ $(\gamma 4)$	$(1, 3, 4, \gamma 4)$ $(\alpha 1, \alpha 2, \gamma 3, \beta 2)$	$\{4, \alpha 4\}, \{\gamma 4, \beta 4\}, \{1, \beta 3\}, \{\alpha 1, \gamma 3\},$ $\{2, \gamma 2\}, \{\alpha 2, \beta 2\}, \{3, \beta 1\}, \{\alpha 3, \gamma 1\}$
k	$(4, 1, 2, \beta 1, 3, \beta 3, \beta 2)$ $(r4)$	$(1, \alpha 2, 4, \gamma 4)$ $(\alpha 1, 3, \alpha 3, \beta 2)$	$\{K1\}, \{K2\}, \{K3\}, \{K4\}$
l	$(4, 1, 2, 3, b1, r2, b3)$ $(r4)$	$(1, \alpha 3, 4, \gamma 4)$ $(\alpha 1, \gamma 2, 3, \beta 2)$	$\{4, \alpha 4\}, \{\gamma 4, \beta 4\}, \{1, \gamma 3\}, \{\alpha 1, \beta 3\},$ $\{2, \beta 2\}, \{\alpha 2, \gamma 2\}, \{3, \gamma 1\}, \{\alpha 3, \beta 1\}$

7.5 Surface loopless model

Consider the partial differential equation for $f \in \mathcal{R}\{x, y\}$

$$\begin{cases} axy\left(2y\frac{\partial f}{\partial y} - x\frac{\partial f}{\partial x}\right) = (1 - xyf|_{x=1})f - 1; \\ f|_{x=0,y=0} = 1, \end{cases} \tag{7.5.1}$$

where $a \in \mathcal{R}_+$, $a \neq 0$. This is equation (27) in Introduction when $c = d = 1$ because it is meaningful in a classification for loopless maps on orientable surfaces and whole (orientable and non-orientable) surfaces according as $a = 1$ and $a = 2$.

Equation (7.5.1) is transformed on $\mathcal{R}\{x, y\}$ into one of the equivalences as

$$f = axy\left(2y\frac{\partial f}{\partial y} - x\frac{\partial f}{\partial x}\right) + xyf|_{x=1}f + 1. \tag{7.5.2}$$

For convenience, let us write

$$f = \sum_{n\geq 0} F_n y^n, \quad F_n = [f]_n \in \mathcal{R}\{x\}. \tag{7.5.3}$$

From the initial condition of equation (7.5.1),

$$[f|_{x=0}]_0 = 1. \tag{7.5.4}$$

On the basis of (7.5.3), for any integer $n \geq 0$,

$$[f|_{x=1}]_n = F_n|_{x=1}, \quad \left[y\frac{\partial f}{\partial y}\right]_n = nF_n \quad \text{and} \quad \left[x\frac{\partial f}{\partial x}\right]_n = x\frac{\mathrm{d}F_n}{\mathrm{d}x}. \tag{7.5.5}$$

Let $H_n = F_n|_{x=1}$, for $n \geq 0$, then from (7.5.3) and (7.5.5) we have, for any integer $n \geq 0$,

$$[f|_{x=1}f]_n = \sum_{i=0}^{n} H_i F_{n-i}. \tag{7.5.6}$$

We proceed on the basis of (7.5.3)–(7.5.6). From (7.5.2),

$$\begin{aligned} y^0: \quad & [f]_0 = 1 \Longrightarrow F_0 = 1; \\ & \Longrightarrow F_0 = 1,\ H_0 = 1, \end{aligned} \tag{7.5.7}$$

$$\begin{aligned} y^1: \quad & [f]_1 = ax\left[2y\frac{\partial f}{\partial y} - x\frac{\partial f}{\partial x}\right]_0 + x[f|_{x=1}f]_0, \\ & \text{by (7.5.5)–(7.5.6)}, \\ & = x(H_0F_0) = x \\ & \Longrightarrow F_1 = x,\ H_1 = 1, \end{aligned} \tag{7.5.8}$$

$$
\begin{aligned}
y^2: \quad [f]_2 &= ax\left[2y\frac{\partial f}{\partial y} - x\frac{\partial f}{\partial x}\right]_1 + x[f|_{x=1}f]_1, \\
&\quad \text{by (7.5.5)–(7.5.6)}, \\
&= ax(2x - x) + x(H_0F_1 + H_1F_0), \\
&\quad \text{by (7.5.5)–(7.5.8)}, \\
&= ax(2x - x) + x(1 + x) = x + (a+1)x^2 \\
&\Longrightarrow F_2 = x + (a+1)x^2,\ H_2 = a + 2,
\end{aligned} \tag{7.5.9}
$$

$$
\begin{aligned}
y^3: \quad [f]_3 &= ax\left[2y\frac{\partial f}{\partial y} - x\frac{\partial f}{\partial x}\right]_2 + x[f|_{x=1}f]_2, \\
&\quad \text{by (7.5.5)–(7.5.6)}, \\
&= ax\left(2(2F_2) - x\frac{\mathrm{d}F_2}{\mathrm{d}x}\right) + x\sum_{i=0}^{2} H_iF_{2-i}, \\
&\quad \text{by (7.5.5)–(7.5.9)}, \\
&= ax(3x + 2(a+1)x^2) + x((a+2) \\
&\quad + 2x + (a+1)x^2) \\
&\Longrightarrow \begin{cases} F_3 = (a+2)x + (3a+2)x^2 + (2a^2+3a+1)x^3, \\ H_3 = 2a^2 + 7a + 5, \end{cases}
\end{aligned} \tag{7.5.10}
$$

and, for any integer $n \geq 4$,

$$
\begin{aligned}
y^n: \quad [f]_n &= ax\left[2y\frac{\partial f}{\partial y} - x\frac{\partial f}{\partial x}\right]_{n-1} + x[f|_{x=1}f]_{n-1}, \\
&\quad \text{by (7.5.5)–(7.5.6)}, \\
&\Longrightarrow F_n = ax\left(2(n-1)F_{n-1} - x\frac{\mathrm{d}F_{n-1}}{\mathrm{d}x}\right) \\
&\quad + x\sum_{i=0}^{n-1} H_iF_{n-1-i}.
\end{aligned} \tag{7.5.11}
$$

Lemma 7.5.1. *For any integer $n \geq 1$, F_n is a polynomial of x with degree n without constant term.*

Proof. From (7.5.8)–(7.5.10), for $n = 1$, 2 and 3, the conclusion is true. For $n \geq 4$ in general, we proceed by induction on n. Assume F_i is a polynomial of x with degree without a constant term for any integer i: $n > i \geq 1$. We prove the conclusion for $i = n$. Because x is a factor of the part on the right hand side of (7.5.11), F_n has no constant term. Denote by d the degree of a polynomial of x. From (7.5.11),

$$
\begin{aligned}
d(F_n) &= 1 + d(F_{n-1}), \quad \text{by the assumption}, \\
&= 1 + (n-1) = n.
\end{aligned}
$$

Therefore, the conclusion is proved. □

Based on Lemma 7.5.1, F_n can be expressed in the form of

$$F_n = \sum_{i=1}^{n} F_{i,n} x^i \tag{7.5.12}$$

where $F_{i,n} \in \mathcal{R}$, for $n \geq i \geq 1$.

Lemma 7.5.2. *In (7.5.11), the polynomial*

$$2(n-1)F_{n-1} - x\frac{\mathrm{d}F_{n-1}}{\mathrm{d}x} \geq 0$$

for $n \geq 2$ if, and only if, $F_{n-1} \geq 0$.

Proof. From (7.5.12),

$$\begin{aligned} 2(n-1)F_{n-1} - x\frac{\mathrm{d}F_{n-1}}{\mathrm{d}x} &= 2(n-1)\sum_{i=1}^{n-1} F_{i,n-1}x^i - \sum_{i=1}^{n-1} iF_{i,n-1}x^i \\ &= \sum_{i=1}^{n-1}(2(n-1)-i)F_{i,n-1}x^i. \end{aligned}$$

Because of $n \geq 2$, $2(n-1) - i > 0$ for any i: $1 \leq i \leq n$. This leads to the conclusion. □

This lemma enables us to write

$$\Lambda_{n-1} = 2(n-1)F_{n-1} - x\frac{\mathrm{d}F_{n-1}}{\mathrm{d}x}$$

as a polynomial of x with degree not greater than $n-1$ determined only by F_i $(1 \leq i \leq n-1)$ such that, for any integer $1 \leq i \leq n-1$,

$$\begin{aligned} \Lambda_{i,n-1} &= \partial_x^i \Lambda_{n-1} (\in \mathcal{R}_+) \\ &= (2n-i-2)F_{i,n-1}. \end{aligned}$$

Lemma 7.5.3. *For any integer $n \geq 1$, $F_n \in \mathcal{R}_+[x]$ (i. e., the set of all polynomials of x with every coefficient non-negative).*

Proof. Based on Lemmas 7.5.1–7.5.2 and $a \in \mathcal{R}_+$, we proceed by induction on n, $F_{i,n} = \partial_x^i F_n \in \mathcal{R}_+$ for $1 \leq i \leq n$ and then $F_n \in \mathcal{R}_+[x]$ is found. □

This lemma shows the probability of equation (7.5.1) to have a solution whose coefficient can be expressed as either a finite sum of positive terms or it is summation-free.

Theorem 7.5.4. *Equation* (7.5.1) *is well-defined on $\mathcal{R}_+\{x,y\}$.*

Proof. Let f_{nl}, $\partial_y^n f_{\mathrm{nl}} = F_n$, for $n \geq 0$, be evaluated by (7.5.7)–(7.5.11). Because f_{nl} satisfies equation (7.5.2) and equation (7.5.2) is equivalent to equation (7.5.1), f_{nl} is also a solution of equation (7.5.1). From Lemma 7.5.3, $f_{\mathrm{nl}} \in \mathcal{R}_+\{x,y\}$.

By considering the procedure to evaluate f_{nl} from (7.5.7) to (7.5.11) on $\mathcal{R}_+\{x,y\}$, the uniqueness of f_{nl} is seen for the initial condition of equation (7.5.1). This solution is the only one. □

On the basis of the three lemmas mentioned above, we are allowed to construct the solution of equation (7.5.1) in a form preferable to automatic production.

Theorem 7.5.5. *The solution $f = f_{\mathrm{nl}}$ of equation (7.5.1) on $\mathcal{R}_+[x, y\}$ are determined by $F_n = \partial_y^n f_{\mathrm{nl}}$, for $n \geq 0$, in the form of*

$$\begin{cases} \partial_y^n f_{\mathrm{nl}} = x(F_{n-1} + a\Lambda_{n-1} + \Sigma_{n-1}), & \text{when } n \geq 1; \\ \partial_y^0 f_{\mathrm{nl}} = 1, & \text{when } n = 0, \end{cases} \tag{7.5.13}$$

where

$$\begin{cases} \Lambda_{n-1} = \sum_{i=1}^{n-1} (2n - i - 2) F_{i,n-1} x^i; \\ \Sigma_{n-1} = \sum_{i=1}^{n-1} F_i|_{x=1} F_{n-1-i}. \end{cases} \tag{7.5.14}$$

Proof. On the basis of Theorem 7.5.4 and Lemma 7.5.2 with their proofs, via rearrangements, (7.5.13) is then found. □

Because F_{n-1}, Λ_{n-1} and Σ_{n-1} all are of the form of a finite sum with all terms positive in $\mathcal{R}_+[x]$, the F_n, for $n \geq 1$ are all of the form of a finite sum with all terms positive in $\mathcal{R}_+[x]$.

In addition, it is also seen that all coefficients in F_n, for $n \geq 1$, are in $\mathbb{Z}_+$.

Example 1. Given root-vertex valency and size, determine the root-isomorphic classes of loopless maps on orientable surfaces. Consider the equation

$$\begin{cases} xy\left(2y\frac{\partial f}{\partial y} - x\frac{\partial f}{\partial x}\right) = (1 - xyf|_{x=1})f - 1; \\ f|_{x=0.y=0} = 1, \end{cases} \tag{7.5.15}$$

found in Liu YP [48] (p. 211, (8.4.8)). Its solution is the enufunction of loopless rooted maps on orientable surfaces with root-vertex valency and size as parameters in which the coefficient of $x^m y^n$ in $\mathbb{Z}_+$ provides the number of root-isomorphic classes of such maps with root-vertex valency m and size n.

When $a = 1$, equation (7.5.1) becomes equation (7.5.15). Hence, the solution of equation (7.5.15) is just (7.5.13) and (7.5.14) in the case of $a = 1$.

Theorem 7.5.6. *Let $\phi \in \mathcal{R}\{x, y\}$ be determined by $\Phi_n = \partial_y^n \phi$, for $n \geq 0$, be the enufunction of loopless rooted maps on orientable surfaces with root-vertex valency $m(x^m)$ and size $n(y^n)$ as two parameters, then*

$$\begin{cases} \Phi_n = x(\Phi_{n-1} + \Lambda'_{n-1} + \Sigma'_{n-1}), & n \geq 1; \\ \Pi_0 = 1, & n = 0, \end{cases}$$

where

$$\begin{cases} \Lambda'_{n-1} = \sum_{i=1}^{n-1} (2n-i-2)\Phi_{i,n-1}x^i; \\ \Sigma'_{n-1} = \sum_{i=1}^{n-1} \Phi_i|_{x=1}\Phi_{n-1-i}. \end{cases}$$

Proof. This is a result of Theorem 7.5.5 when $a = 1$. □

Example 2. Given root-vertex valency and size, determine the root-isomorphic classes of loopless maps on all (orientable and non-orientable) surfaces. Consider the equation

$$\begin{cases} 2xy\left(2y\frac{\partial f}{\partial y} - x\frac{\partial f}{\partial x}\right) = (1 - xyf|_{x=1})f - 1; \\ f|_{x=0,y=0} = 1, \end{cases} \tag{7.5.16}$$

found in Liu YP [48] (p. 213, (8.4.17)).

Its solution is the enufunction of loopless rooted maps on all surfaces with root-vertex valency and size as parameters in which the coefficient of $x^m y^n$ in $\mathbb{Z}_+$ provides the number of root-isomorphic classes of such maps with root-vertex valency m and size n.

When $a = 2$, equation (7.5.1) becomes equation (7.5.16). Hence, the solution of equation (7.5.16), is just (7.5.13) and (7.5.14) in the case of $a = 2$.

From the two examples, we see the reason that equation (7.5.1) is called a *surface loopless model.*

7.6 Surface endless model

Consider the equation for $f \in \mathcal{R}\{x,y\}$ as

$$\begin{cases} ax^3y\frac{\partial f}{\partial x} = \left(1 - ax^2y + \frac{xy}{1-x}\right)f - \frac{x^2y}{1-x}f|_{x=1} - xy - 1; \\ f|_{x=0,y=0} = 1, \end{cases} \tag{7.6.1}$$

where $a \in \mathcal{R}_+$, $a \neq 0$. This is equation (28) in Introduction when $b = c = d = 1$ because it is meaningful in a classification for endless maps on orientable surfaces and whole (orientable and non-orientable) surfaces according as $a = 1$ and $a = 2$.

On $\mathcal{R}\{x,y\}$, equation (7.6.1) is transformed into a suitable one of its equivalences:

$$f = ax^3y\frac{\partial f}{\partial x} + ax^2yf + \frac{xy}{1-x}(xf|_{x=1} - f) + xy + 1. \tag{7.6.2}$$

Let $f \in \mathcal{R}\{x,y\}$ be determined by $F_n = [f]_n = \partial_y^n f \in \mathcal{R}\{x\}$, $n \geq 0$. This is

$$f = \sum_{n\geq 0} F_n y^n, \quad F_n = [f]_n \in \mathcal{R}\{x\}. \tag{7.6.3}$$

By the initial condition of equation (7.6.1),

$$[f|_{x=0}]_0 = 1. \tag{7.6.4}$$

From (7.6.3), it is seen that, for $n \geq 0$,

$$[f|_{x=1}]_n = F_n|_{x=1} \quad \text{and} \quad \left[x\frac{\partial f}{\partial x}\right]_n = x\frac{\mathrm{d}F_n}{\mathrm{d}x}. \tag{7.6.5}$$

Write $H_n = F_n|_{x=1}$, $n \geq 0$. By (7.6.5),

$$H_n = \sum_{i\geq 0} \partial_x^i F_n \tag{7.6.6}$$

for $n \geq 0$.

On the basis of (7.6.4)–(7.6.6). By (7.6.2),

$$\begin{aligned} y^0: \quad & [f]_0 = 1 \text{ (all terms but constant term with } y \\ & \qquad \text{on the right hand side of equation (7.6.2));} \\ & \Longrightarrow F_0 = 1, \quad H_0 = 1, \end{aligned} \tag{7.6.7}$$

$$\begin{aligned} y^1: \quad [f]_1 &= ax^2\left[x\frac{\partial f}{\partial x}\right]_0 + ax^2[f]_0 + \frac{x[xf|_{x=1} - f]_0}{1-x} + x, \\ & \quad \text{by (7.6.5)–(7.6.6),} \\ &= ax^2 - x + x = ax^2 \\ & \Longrightarrow F_1 = ax^2, \quad H_1 = a, \end{aligned} \tag{7.6.8}$$

$$\begin{aligned} y^2: \quad [f]_2 &= ax^2\left[x\frac{\partial f}{\partial x}\right]_1 + ax^2[f]_1 + \frac{x[xf|_{x=1} - f]_1}{1-x}, \\ & \quad \text{by (7.6.5)–(7.6.6),} \\ &= 2a^2x^4 + a^2x^4 + ax^2 = ax^2 + 3a^2x^4 \\ & \Longrightarrow F_2 = ax^2 + 3a^2x^4, \quad H_2 = a + 3a^2, \end{aligned} \tag{7.6.9}$$

and, for any integer $n \geq 3$,

$$\begin{aligned} y^n: \quad [f]_n &= ax^2\left[x\frac{\partial f}{\partial x}\right]_{n-1} + ax^2[f]_{n-1} \\ & \quad + \frac{x[xf|_{x=1} - f]_{n-1}}{1-x}, \\ & \quad \text{by (7.6.5)–(7.6.6),} \\ & \Longrightarrow F_n = ax^3\frac{\mathrm{d}F_{n-1}}{\mathrm{d}x} + ax^2F_{n-1} \\ & \quad + \frac{x(xH_{n-1} - F_{n-1})}{1-x}. \end{aligned} \tag{7.6.10}$$

Lemma 7.6.1. *For any integer $n \geq 1$, F_n is a polynomial in x with terms of degrees between 2 and $2n$.*

Proof. From (7.6.8)–(7.6.9), the lemma is true for $n = 1$ and 2. For $n \geq 3$ in general, we proceed by induction on n. Assume, for any integer i ($n - 1 > i \geq 1$), that F_i is a polynomial of x with all terms of degrees between 2 and $2i$. We prove the lemma for $i = n$. Let $d(p)$ is the degree of polynomial p of x. Because x^2 is a factor on the right hand side of (7.6.10), the minimum degree of all terms of F_n is not less than 2. By (7.6.10),

$$\begin{aligned} d(F_n) &= 2 + d(F_{n-1}), \quad \text{by the assumption,} \\ &= 2 + 2(n-1) \leq 2n. \end{aligned}$$

Therefore, the lemma is true. □

On the basis of this lemma, it is seen that, for $n \geq 1$, F_n has the form

$$F_n = \sum_{i=2}^{2n} F_{i,n} x^i \quad \text{and} \quad H_n = \sum_{i=2}^{2n} F_{i,n}. \tag{7.6.11}$$

This lemma shows that, for any integer n given, F_n is of the form of a finite sum of terms not greater than $2n$.

Lemma 7.6.2. *For any integer $n \geq 1$,*

$$(1-x)|(xH_n - f_n) \quad \textit{and} \quad \frac{xH_n - F_n}{1-x} \geq 0. \tag{7.6.12}$$

Proof. From (7.6.11),

$$\begin{aligned} \frac{xH_n - F_n}{1-x} &= \frac{\sum_{i=2}^{2n} F_{i,n}(x - x^i)}{1-x} \\ &= x \sum_{i=2}^{2n} F_{i,n} \left(\frac{1 - x^{i-1}}{1-x} \right). \end{aligned} \tag{7.6.13}$$

Because of $(1-x)|(1-x^{i-1})$ for any integer $i \geq 2$, the first conclusion is true. Because of

$$\frac{1 - x^{i-1}}{1-x} = \sum_{j=0}^{i-2} x^j,$$

for any integer $i \geq 2$, the second conclusion is true. □

From the proof of the lemma, it is seen that

$$\frac{xH_n - F_n}{1-x} = \sum_{\substack{j+1 \leq i \leq 2n \\ 1 \leq j \leq 2n-2}} F_{i,n} x^j. \tag{7.6.14}$$

Lemma 7.6.3. *For any integer $n \geq 1$, $F_n \in \mathcal{R}_+[x]$.*

Proof. Because $a \in \mathcal{R}_+$, on the basis of two lemmas mentioned above, (7.6.10) and the inductive principle lead to $F_n \in \mathcal{R}_+[x)$ if, and only if,

$$(xH_n - F_n)/(1-x) \in \mathcal{R}_+[x).$$

By (7.6.13), the conclusion is drawn. □

As a matter of fact, the conclusion of this lemma can be strengthened to all coefficients of polynomial F_n being in $\mathbb{Z}_+$ whenever $a \in \mathcal{Z}_+$.

Theorem 7.6.4. *Equation* (7.6.1) *is well-defined on* $\mathcal{R}_+[x,y\}$.

Proof. Let $f = f_{\rm nd}$, $\partial_y^n f_{\rm nd} = F_n$, $n \geq 0$. They are evaluated from (7.6.7)–(7.6.10). Because of $f_{\rm nd}$ satisfying the equation (7.6.2), the equivalence between equation (7.6.2) and equation (7.6.1) shows that $f_{\rm nd}$ is a solution of equation (7.6.1) as well. From Lemma 7.6.3, $f_{\rm nd} \in \mathcal{R}_+[x,y\}$.

By considering the procedure to get $f_{\rm nd}$ by employing (7.6.7)–(7.6.10) on $\mathcal{R}_+[x,y\}$, it is seen that $f_{\rm nd}$ is unique for the initial condition. Therefore, $f_{\rm nd}$ is the only solution of equation (7.6.1). □

This theorem shows that it only is necessary to investigate $f_{\rm nd}$ for exploiting more useful structures on any general solution of equation (7.6.1).

Theorem 7.6.5. *The solution* $f_{\rm nd}$ *of equation* (7.6.1) *on* $\mathcal{R}_+[x,y\}$ *obeys an expression of the form of a finite sum with all terms positive like, for any integer* $n \geq 0$,

$$[f_{\rm nd}]_n = \begin{cases} 1, & \text{when } n = 0; \\ ax^2, & \text{when } n = 1; \\ ax^2\Lambda_{n-1} + x\Pi_{n-1}, & \text{when } n \geq 2, \end{cases} \tag{7.6.15}$$

where

$$\begin{cases} \Lambda_{n-1} = \sum\limits_{i=2}^{2n-2} (i+1)F_{i,n-1}x^i; \\ \Pi_{n-1} = \sum\limits_{\substack{i+1\leq j\leq 2n-2 \\ 1\leq i\leq 2n-4}} F_{j,n-1}x^i. \end{cases} \tag{7.6.16}$$

Proof. By substituting (7.6.11) and (7.6.14) into (7.6.10), (7.6.15) can be found after rearrangement. □

Example 1. Root-isomorphic classes of Endless maps on orientable surfaces with root-vertex valency and size given. Consider the equation for $f \in \mathcal{R}\{x,y\}$

$$\begin{cases} x^3y\dfrac{\partial f}{\partial x} = \left(1 - x^2y + \dfrac{xy}{1-x}\right)f - \dfrac{x^2y}{1-x}f|_{x=1} - xy - 1; \\ f|_{x=0,y=0} = 1. \end{cases} \tag{7.6.17}$$

A special case of the first in equation (7.6.17) might be seen in [48] (2008, p. 180, equation (7.5.12)). One of its solution is the enufunction of determining the number of root-isomorphic classes of end-cutless maps on all orientable surfaces with root-vertex valency and size given.

When $a = 1$, equation (7.6.1) becomes equation (7.6.17). Because of it being well-defined, the solution of equation (7.6.17) is just the case of $a = 1$ in the solution of equation (7.6.15).

Theorem 7.6.6. *The solution $f = f_{\mathrm{elo}} \in \mathcal{R}_+[x, y\}$ of equation (7.6.17) determined by $O_n = \partial_y^n f_{\mathrm{elo}}$, for $n \geq 0$, is of the form of a finite sum with all terms positive,*

$$O_n = \begin{cases} 1, & \text{when } n = 0; \\ x^2, & \text{when } n = 1; \\ x^2\Lambda_{n-1} + x\Pi_{n-1}, & \text{when } n \geq 2, \end{cases} \tag{7.6.18}$$

where

$$\begin{cases} \Lambda_{n-1} = \sum\limits_{i=2}^{2n-2} (i+1) O_{i,n-1} x^i; \\ \Pi_{n-1} = \sum\limits_{i=1}^{2n-4} \left(\sum\limits_{j=i+1}^{2n-2} O_{j,n-1} \right) x^i. \end{cases} \tag{7.6.19}$$

Proof. This is a direct result of Theorem 7.6.5 □

Example 2. Root-isomorphic classes of endless maps on all (orientable and non-orientable) surfaces with root-vertex valency and size given. Consider the equation for $f \in \mathcal{R}\{x, y\}$

$$\begin{cases} 2x^3 y \dfrac{\partial f}{\partial x} = \left(1 - 2x^2 y + \dfrac{xy}{1-x}\right) f - \dfrac{x^2 y}{1-x} f|_{x=1} - xy - 1; \\ f|_{x=0, y=0} = 1. \end{cases} \tag{7.6.20}$$

When $a = 2$, equation (7.6.1) becomes equation (7.6.20). From the fact of it being well-defined, the solution of equation (7.6.20) is just the case of $a = 2$ in the solution of equation (7.6.1).

Theorem 7.6.7. *The solution $f = f_{\mathrm{ela}} \in \mathcal{R}_+[x, y\}$ of equation (7.6.17) as determined by $A_n = \partial_y^n f_{\mathrm{ela}}$, for $n \geq 0$, is in the form of a finite sum with all terms positive,*

$$A_n = \begin{cases} 1, & \text{when } n = 0; \\ 2x^2, & \text{when } n = 1; \\ 2x^2\Lambda_{n-1} + x\Pi_{n-1}, & \text{when } n \geq 2, \end{cases} \tag{7.6.21}$$

where

$$\begin{cases} \Lambda_{n-1} = \sum_{i=2}^{2n-2} (i+1)A_{i,n-1}x^i; \\ \Pi_{n-1} = \sum_{i=1}^{2n-4} \left(\sum_{j=i+1}^{2n-2} A_{j,n-1} \right) x^i. \end{cases} \tag{7.6.22}$$

Proof. This is a direct result of Theorem 7.6.5. □

The two examples above suggest us to call equation (7.6.1) a *surface endless model.*

7.7 Surface Euler model

Consider the equation for $f \in \mathcal{R}\{x, y\}$

$$\begin{cases} 2ax^4y\dfrac{\partial f}{\partial x^2} = \left(1 - ax^2y + \dfrac{x^2y}{1-x^2}\right)f - \dfrac{x^2y}{1-x^2}f|_{x=1} - 1; \\ f|_{x=0,y=0} = 1, \end{cases} \tag{7.7.1}$$

where $a \in \mathbb{Z}_+$, $a \neq 0$. This is equation (29) in Introduction when $b = c = d = 1$ because it is meaningful in a classification for Eulerian maps on orientable surfaces and whole (orientable and non-orientable) surfaces according as $a = 1$ and $a = 2$.

Equation (7.7.1) is on $\mathcal{R}\{x^2, y\}$ (because of only x^2 occurring in the equation!) $\subseteq \mathcal{R}\{x, y\}$ transformed into the suitable one of equivalences as

$$f = 1 + ax^2y\left(f + 2x^2\frac{\partial f}{\partial x^2}\right) + \frac{x^2y}{1-x^2}(f|_{x=1} - f). \tag{7.7.2}$$

Let $z = x^2$. For $f \in \mathcal{R}\{z, y\}$ instead of $f \in \mathcal{R}\{x^2, y\} \subseteq \mathcal{R}\{x, y\}$, let f be determined by $F_n = [f]_n = \partial_y^n f \in \mathcal{R}\{z\}$, for $n \geq 0$, *i. e.*,

$$f = \sum_{n\geq 0} F_n y^n, \quad F_n = [f]_n \in \mathcal{R}\{z\}. \tag{7.7.3}$$

From the initial condition of equation (7.7.1),

$$[f|_{x=0}]_0 = 1. \tag{7.7.4}$$

By (7.7.3), it is seen that, for $n \geq 0$,

$$[f|_{z=1}]_n = F_n|_{z=1} \quad \text{and} \quad \left[z\frac{\partial f}{\partial z}\right]_n = z\frac{\mathrm{d}F_n}{\mathrm{d}z}. \tag{7.7.5}$$

Write $H_n = F_n|_{z=1}$, for $n \geq 0$. By (7.7.5), it is known that, for $n \geq 0$,

$$H_n = \sum_{i\geq 0} \partial_z^i F_n. \tag{7.7.6}$$

On the basis of (7.7.4)–(7.7.6). By (7.7.2),

$$
\begin{aligned}
y^0: \quad & [f]_0 = 1(\text{all terms but constant term on right} \\
& \quad \text{side of equation (7.7.2) with } y \text{ as factor}); \\
& \Longrightarrow F_0 = 1(\text{the initiation of equation (7.7.1)}), \\
& \quad H_0 = 1,
\end{aligned} \tag{7.7.7}
$$

$$
\begin{aligned}
y^1: \quad & [f]_1 = az\left[f + 2z\frac{\partial g}{\partial z}\right]_0 + \frac{z[f|_{x=1} - f]_0}{1-z}, \\
& \quad \text{by (7.7.5)–(7.7.6)}, \\
& \quad = ax^2(1+0) + 0 = az \\
& \Longrightarrow F_1 = az, \quad H_1 = a,
\end{aligned} \tag{7.7.8}
$$

$$
\begin{aligned}
y^2: \quad & [f]_2 = az\left[f + 2z\frac{\partial g}{\partial z}\right]_1 + \frac{z[f|_{z=1} - f]_1}{1-z}, \\
& \quad \text{by (7.7.5)–(7.7.6)}, \\
& \quad = az(az + 2az) + \frac{z(a - az)}{1-z} = 3a^2z^2 + az \\
& \Longrightarrow F_2 = 3a^2z^2 + az, \quad H_2 = 3a^2 + a,
\end{aligned} \tag{7.7.9}
$$

and, for any integer $n \geq 3$,

$$
\begin{aligned}
y^n: \quad & [f]_n = az\left[f + 2z\frac{\partial g}{\partial z}\right]_{n-1} + \frac{z[f|_{z=1} - f]_{n-1}}{1-z}, \\
& \quad \text{by (7.7.5)–(7.7.6)}, \\
& \Longrightarrow F_n = az\left(F_{n-1} + 2z\frac{\mathrm{d}F_{n-1}}{\mathrm{d}z}\right) \\
& \quad + \frac{z}{1-z}(H_{n-1} - F_{n-1}).
\end{aligned} \tag{7.7.10}
$$

Lemma 7.7.1. *For any integer $n \geq 1$, F_n is a polynomial of x with degree of each term at least 1 and at most n.*

Proof. On the basis of (7.7.8)–(7.7.9). For $n = 1$ and 2, the conclusion is true. For $n \geq 3$ in general, we proceed by induction. Assume, for any integer i: $n > i \geq 1$, F_i is a polynomial of z with the degree of each term at least 1 and at most i. We prove the case of $i = n$. Because z is a factor on the right hand side of (7.7.10), F_n has no constant term with degree of each term at least 1. Let $d(p)$ be the degree of a polynomial of z. From (7.7.10),

$$
\begin{aligned}
d(F_n) &= 2 + d(F_{n-1}), \quad \text{by the assumption,} \\
&= 2 + 2(n-1) = 2n.
\end{aligned}
$$

Therefore, the conclusion is drawn. □

On the basis of this lemma, we are allowed to write F_n for $n \geq 1$ in the form of

$$F_n = \sum_{i=1}^{n} F_{i,n} z^i \quad \text{and} \quad H_n = \sum_{i=1}^{n} F_{i,n}. \tag{7.7.11}$$

Lemma 7.7.2. *For any integer $n \geq 1$,*

$$(1-z)|(H_n - f_n), \quad \textit{and} \quad \frac{H_n - F_n}{1-z} \geq 0 \tag{7.7.12}$$

if, and only if, $F_n \in \mathcal{R}_+[z]$.

Proof. From (7.7.11),

$$\frac{H_n - F_n}{1-z} = \frac{\sum_{i=1}^{n} F_{i,n}(1-z^i)}{1-z}. \tag{7.7.13}$$

Because of $(1-z)|(1-z^i)$ for $i \geq 1$, the first conclusion is true. Because of

$$\frac{1-z^i}{1-z} = \sum_{j=0}^{i-1} z^j \geq 0$$

for $i \geq 1$, the second conclusion is true. □

In the proof of this lemma, it is seen that

$$\frac{H_n - F_n}{1-z} = \sum_{\substack{j+1 \leq i \leq n \\ 1 \leq j \leq n-1}} F_{i,n} z^j. \tag{7.7.14}$$

Lemma 7.7.3. *For any integer $n \geq 1$, $F_n \in \mathcal{R}_+[z]$.*

Proof. We proceed on the basis of the two lemmas above. Because $a \in \mathbb{Z}_+$, (7.7.10) and induction principle show that $F_n \in \mathcal{R}_+[z)$ if, and only if, $H_n - F_n/1 - z \in \mathcal{R}_+[z)$. From (7.7.13), the conclusion is drawn. □

This lemma enables us to evaluate the solution of equation (7.7.1)) in the form of a sum with all terms positive.

Theorem 7.7.4. *Equation* (7.7.1) *is well-defined on $\mathcal{R}_+[z,y\}$.*

Proof. The f_{eu} in which $\partial_y^n f_{eu} = F_n$, for $n \geq 0$, are determined by (7.7.7)–(7.7.10). Because the f_{nd} satisfy equation (7.7.2), the equivalence between equation (7.7.2) and equation (7.7.1) shows that f_{eu} is a solution of equation (7.7.1) as well. From Lemma 7.7.3, $f_{eu} \in \mathcal{R}_+[z,y\}$.

By considering the procedure for getting f_{eu} by (7.7.7)–(7.7.10), it is seen that f_{eu} is unique from the initial condition of equation (7.7.1) on $\mathcal{R}_+[z,y\}$. This solution is the only one on $\mathcal{R}_+[z,y\}$. □

This theorem enables us to evaluate the solution of equation (7.7.1) in a form favorable to the use of computers

Theorem 7.7.5. *The solution f_{eu} of equation (7.7.1) on $\mathcal{R}_+[z,y\}(z = x^2!)$ determined by $F_n = [f_{\mathrm{eu}}]_n = \partial_y^n f_{\mathrm{eu}}$ obeys an expression of the form of a finite sum with all terms positive as, for any integer $n \geq 0$,*

$$[f_{\mathrm{eu}}]_n = \begin{cases} 1, & \text{when } n = 0; \\ ax^2, & \text{when } n = 1; \\ 3a^2x^4 + ax^2, & \text{when } n = 2; \\ \sum_{m=1}^{n} A_{2m,n-1}x^{2m}, & \text{when } n \geq 3, \end{cases} \tag{7.7.15}$$

where for $2 \leq m \leq n-1$,

$$A_{2m,n-1} = \begin{cases} \sum_{i=1}^{n-1} F_{2i,n-1}, & \text{when } m = 1; \\ (2m-1)aF_{2(m-1),n-1} + \sum_{i=m}^{n-1} F_{2i,n-1}, & \text{when } 2 \leq m \leq n-1; \\ (2n-1)aF_{2(n-1),n-1}, & \text{when } m = n. \end{cases} \tag{7.7.16}$$

Proof. By substituting (7.7.11) and (7.7.13) into (7.7.10), $[f_{\mathrm{eu}}]_n$ is obtained for $n \geq 3$ after rearrangement. For $0 \leq n \leq 2$, the result follows from (7.7.7)–(7.7.9). □

Example 1. Root-isomorphic classes of Euler maps on orientable surfaces with root-vertex valency and size given. We have the partial differential equation

$$\begin{cases} 2x^4y\dfrac{\partial f}{\partial x^2} = \left(1 - x^2y + \dfrac{x^2y}{1-x^2}\right)f - \dfrac{x^2y}{1-x^2}f|_{x=1} - 1; \\ f|_{x=0,y=0} = 1, \end{cases} \tag{7.7.17}$$

whose first case appears in Liu YP [48] (2008, Section 6.5). Its solution is the enufunction of Euler rooted maps on orientable surfaces with root-vertex valency and size as two parameters.

Attention has to be paid to the fact that when $a = 1$, equation (7.7.1) becomes equation (7.7.17). Thus, a solution of equation (7.7.17) is a specific case in the solution of equation (7.7.15) or equation (7.7.16) when $a = 1$.

Theorem 7.7.6. *The solution f_{euo} of equation (7.7.17) on $\mathcal{R}_+[z,y\}$ $(z = x^2!)$ as determined by $O_n = [f_{\mathrm{euo}}]_n = \partial_y^n f_{\mathrm{euo}}$ has an expression in the form of a finite sum with all terms positive as for any integer $n \geq 0$,*

$$O_n = \begin{cases} 1, & \text{when } n = 0; \\ x^2, & \text{when } n = 1; \\ 3x^4 + x^2, & \text{when } n = 2; \\ \sum_{m=1}^{n} A_{2m,n-1}x^{2m}, & \text{when } n \geq 3, \end{cases} \tag{7.7.18}$$

where for $2 \leq m \leq n-1$,

$$A_{2m,n-1} = \begin{cases} \sum_{i=1}^{n-1} O_{2i,n-1}, & \text{when } m = 1; \\ (2m-1)O_{2(m-1),n-1} + \sum_{i=m}^{n-1} O_{2i,n-1}, & \text{when } 2 \leq m \leq n-1; \\ (2n-1)O_{2(n-1),n-1}, & \text{when } m = n. \end{cases} \tag{7.7.19}$$

Proof. This is the case $a = 1$ of (7.7.15) and (7.7.16). □

Example 2. Root-isomorphic classes of Euler maps on all (orientable and non-orientable) surfaces with root-valency and size given. We have the partial differential equation

$$\begin{cases} 4x^4 y \dfrac{\partial f}{\partial x^2} = \left(1 - 2x^2 y + \dfrac{x^2 y}{1 - x^2}\right) f - \dfrac{x^2 y}{1 - x^2} f|_{x=1} - 1; \\ f|_{x=0,y=0} = 1, \end{cases} \tag{7.7.20}$$

whose first equation is a specific case of the first of equation (7.7.1) when $a = 2$. This solution is just the enufunction of Euler rooted maps on all (orientable and non-orientable) surfaces with root-vertex valency and size as two parameters.

Attention has to be paid to the fact that when $a = 2$, equation (7.7.1) becomes equation (7.7.20). Thus, a solution of (7.7.20) is the specific case in the solution of equation (7.7.1) when $a = 2$. This is why equation (7.7.1) is called a *surface Euler model.*

Theorem 7.7.7. *The solution* f_{eua} *of equation* (7.7.1) *on* $\mathcal{R}_+[z, y\}$ ($z = x^2!$) *as determined by* $Q_n = [f_{\text{eua}}]_n = \partial_y^n f_{\text{eua}}$ *has an expression in the form of a finite sum with all terms positive as, for any integer* $n \geq 0$,

$$Q_n = \begin{cases} 1, & \text{when } n = 0; \\ 2x^2, & \text{when } n = 1; \\ 12x^4 + 2x^2, & \text{when } n = 2; \\ \sum_{m=1}^{n} A_{2m,n-1} x^{2m}, & \text{when } n \geq 3, \end{cases} \tag{7.7.21}$$

where for $2 \leq m \leq n-1$,

$$A_{2m,n-1} = \begin{cases} \sum_{i=1}^{n-1} Q_{2i,n-1}, & \text{when } m = 1; \\ 2(2m-1)Q_{2(m-1),n-1} + \sum_{i=m}^{n-1} O_{2i,n-1}, & \text{when } 2 \leq m \leq n-1; \\ 2(2n-1)Q_{2(n-1),n-1}, & \text{when } m = n. \end{cases} \tag{7.7.22}$$

Proof. The case $a = 2$ of (7.7.15) and (7.7.16). □

7.8 Notes

7.8.1. The sphere is a foundation of general surfaces. In Example 4 in Section 7.1, a surface with some boundaries is paid attention to. As a matter of fact, lack-1 face can be seen as a boundary. By considering the root-face also as a boundary, a cylinder is obtained. Thus, the classification of maps on surfaces can be transformed into that on the sphere.

7.8.2. In Ren H-Liu YP [67], the root-isomorphic classes of near-triangulations on cylinder are discussed. Although an explicision is given, because of the complication for calculation, no final result is treated. However, by an application of the theory presented in this book, an expression in the form of a finite sum with all terms positive can be given favorable to computers. If a map has all vertices, but two, of even valency, then it is called *Euler trace*, or *universal map*. In Zhang YL-Liu YP-Cai JL [90], a cubic equation is provided. Its being well-defined and its solution in the form of a finite sum with all terms positive can also be established by the theory in this book.

7.8.3. In Cai JL-Liu YP [6], a quadratic equation of three variables can, by taking $y = z$, be transformed into equation (7.1.1). Naturally, the solution of three variables in the form of a sum with all terms positive can be derived in a similar way to Section 7.1.

7.8.4. The root-isomorphic classification of maps is initiated from Tutte WT [80] (1962) for planar triangulations. By following this method, Brown WT [5] (1965) investigated such a classification for quadrangulations on the sphere and on a disc (or a sphere with a boundary as the simplest surface with a hole). A sphere with two holes is called a *pan-cylinder* in Liu YP [53] (2012). For a quadrangulation on a pan-cylinder, if each boundary is treated as a face, then it becomes a lack-1 face quadrangulation (see Example 4 in Section 7.1) on the sphere. A lack-1 quadrangulation (Section 7.1) can also be dealt with as a map with all faces quadrangles but with two exceptions. The dual of a lack-1 quadrangulation is a type of a *universal map* (or *Euler walk*) because all vertices have even valencies except for two vertices. A closed Euler walk is called an *Euler tour*. In Liu YP [55] (1979), one sees that an Euler tour is applied to determining the maximum genus of a graph.

7.8.5. In Ren H-Liu YP [68] (2001), root-isomorphic classes are discussed on a torus to determine the enufunctions with up to 4 variables for 4 parameters. Although explicisions are found with certain complication, most of them are in the form of summations with terms alternate, not yet all only positive. In Ren H-Liu YP [72] (2002) and Hao RX-Liu YP [9] (2002) etc., this is seen as well.

7.8.6. On projective plane, see Ren H-Liu YP [66] (2000), [70] (2002), [71] (2002), Ren H-Deng M-Liu YP [73] (2005), Li ZX-Liu YP [10] (2002), Xu Y-Liu YP [89] (2007), Liu WZ-Liu YP [11] etc. All functions encountered can be expressed in the form of a finite sum with all terms positive.

7.8.7. Equation system (7.4.1) in Section 7.4 as for the dual of 4-regular maps on a Klein bottle looks very much simpler than those in Ren H-Liu YP [69] (2001). And then, all 68, instead of 67 in [69], quadrangulations on a Klein bottle are shown in Example 2 of Section 7.4.

7.8.8. On the relationship between systems of differential equations in Sections 7.2–7.4 and combinatorial enumerations, see Liu YP [53] (2012), where details are given.

7.8.9. On the equations in Sections 7.5–7.7, see Liu YP [54] (2012). Some numerical results for a size not greater than 10 in Section 7.5 might be found in Pan LY-Liu YP [64] (2013).

8 Tree equations

8.1 Planted model

Consider the equation

$$\begin{cases} a_2 \int\limits_y \dfrac{y^2 f}{1 - a_3 y f} = f - a_1 y_1; \\ f|_{\mathbf{y}=\mathbf{0}} = a_0, \end{cases} \tag{8.1.1}$$

where $\mathbf{a} = (a_0, a_1, a_2, a_3) \in \mathbb{Z}_+^4$, $f = f(\mathbf{y}) \in \mathcal{R}\{\mathbf{y}\}$ and $\mathbf{y} = (y_1, y_2, y_3, \ldots)$. This is equation (30) in Introduction when a_0, a_1, a_2 and a_3 are, respectively, replaced by d, b, a and c.

The first equation of equation (8.1.1) has appeared in [18, 20] (Liu YP, 1985) when $a_1 = a_2 = 1$, *i. e.*,

$$\int\limits_y \frac{y^2 f}{1 - y f} = f - y_1. \tag{8.1.2}$$

Because a solution of equation (8.1.2) holds for planted trees for $a_0 = 0$, equation (8.1.1) is called of *planted tree model*.

In equation (8.1.1), let $f_{\mathbf{a}} = f_{\mathbf{a}}(\mathbf{y}) \in \mathcal{R}\{\mathbf{y}\}$, $\mathbf{y} = (y_1, y_2, y_3, \ldots)$, be a solution, then it is of the form

$$f_{\mathbf{a}} = \sum_{n \ge 0} F_{\mathbf{a}[n]} \tag{8.1.3}$$

where $F_{\mathbf{a}[n]}$ is a homogeneous polynomial of $\mathbf{y}$ with degree n. Then we have

$$F_{\mathbf{a}[n]} = \sum_{\mathbf{n} \in \mathcal{N}} F_{\mathbf{a},\mathbf{n}} \mathbf{y}^{\mathbf{n}} \tag{8.1.4}$$

where $\mathcal{N} = \{\mathbf{n} \mid \mathbf{n} \ge \mathbf{0}, |\mathbf{n}| = n\}$, and $|\mathbf{n}|$ is the sum of all components in $\mathbf{n}$.

Because of

$$\frac{1}{1 - a_3 y f} = \sum_{m \ge 0} (a_3 y f)^m, \tag{8.1.5}$$

we have

$$\frac{y^2 f}{1 - a_3 y f} = \sum_{m \ge 1} a_3^m y^{m+1} f^m.$$

Equivalently, equation (8.1.1) becomes

$$\begin{cases} f = a_1 y_1 + a_2 \sum\limits_{m \ge 1} a_3^m y_{m+1} f^m; \\ F_{\mathbf{a}[0]} = a_0. \end{cases} \tag{8.1.6}$$

https://doi.org/10.1515/9783110625837-008

In equation (8.1.6), for integer $i \geq 1$, let

$$f_{\mathbf{a}}^{i} = \sum_{n\geq 0} F_{\mathbf{a}[n]}^{[i]} \tag{8.1.7}$$

where $F_{\mathbf{a}[n]}^{[i]}$ is a homogeneous polynomial of $\mathbf{y}$ with degree n in $f_{\mathbf{a}}^{i}$.

From the initial condition of equation (8.1.6) and equation (8.1.3), because there is no constant term on the right hand side of the first equation in equation (8.1.6), for $i \geq 1$, we have

$$F_{\mathbf{a}[0]}^{[i]} = a_0^i = 0. \tag{8.1.8}$$

Since $f_{\mathbf{a}}^{1} = f_{\mathbf{a}}$, we have $F_{\mathbf{a}[n]}^{[1]} = F_{\mathbf{a}[n]}$, for $n \geq 0$. Furthermore, from equation (8.1.7), for $n \geq 0$ and $i \geq 1$,

$$F_{\mathbf{a}[n]}^{[i]} = \begin{cases} 0; & \text{when } n = 0; \\ \sum_{j=1}^{n-1} F_{\mathbf{a}[j]} F_{\mathbf{a}[n-j]}^{[i-1]}, & \text{otherwise.} \end{cases} \tag{8.1.9}$$

Observation 8.1.1. *For any integers $i \geq 2$ and $n \geq 1$, $F_{\mathbf{a}[n]}^{[i]}$ are determined only by $F_{\mathbf{a}[j]}$, $1 \leq j \leq n-1$.*

Proof. From (8.1.9), by induction, the fact can be seen. □

We consider the fact that $\mathbf{y}^{\mathbf{n}}$ for nonnegative integral vector $\mathbf{n} = (n_1, n_2, n_3, \ldots)$ can be seen as a partition of a set with n elements into n_i subsets of cardinality i, $i \geq 1$. For this reason, f as shown in (8.1.3) is called a *partition function.*

Observation 8.1.2. *For any integer $n = |\mathbf{n}|$ and $\pi(\mathbf{n})(= n_1 + 2n_2 + 3n_3 + \cdots) \geq n \geq 1$, the partition functions $F_{\mathbf{a}[n]}$ are all polynomials.*

Proof. From the nonnegative integrality, for $\pi(\mathbf{n}) \geq n \geq 1$, $F_{\mathbf{a}[n]}$ involves a finite number of y_i, $i \geq 1$, and hence, as well, a finite number of possibilities to form a term of degree n. □

On the basis of equation (8.1.6), $F_{\mathbf{a}[n]}$, $0 \leq m \leq 4$, are evaluated in what follows.

Because of $a_0 = 0$, $F_0 = 0$.

For $n = 1$, from (8.1.8), it is seen that

$$\sum_{m\geq 1} a_3^m y_{m+1} F_{\mathbf{a}[0]}^{[m]} = 0$$

and hence

$$\begin{aligned} F_{\mathbf{a}[1]} &= a_1 y_1 + a_2 \sum_{i\geq 1} a_3^i y_{i+1} F_{\mathbf{a}[0]}^{[i]} \\ &= a_1 y_1. \end{aligned} \tag{8.1.10}$$

Therefore, from (8.1.9), we have

$$F^{[i]}_{\mathbf{a}[1]} = 0 \ (i \geq 2). \tag{8.1.11}$$

For $n = 2$, from (8.1.10), we have

$$\sum_{m\geq 2} a_3^m y_{m+1} F^{[m]}_{\mathbf{a}[1]} = 0$$

and hence

$$\begin{aligned} F_{\mathbf{a}[2]} &= a_2\Big(a_3 y_2 F^{[1]}_{\mathbf{a}[1]} + \sum_{i\geq 2} a_3^i y_{i+1} F^{[i]}_{\mathbf{a}[0]}\Big) \\ &= a_1 a_2 a_3 y_1 y_2. \end{aligned} \tag{8.1.12}$$

For $n = 3$, from (8.1.12), we have

$$\sum_{m\geq 3} a_3^m y_{m+1} F^{[m]}_{\mathbf{a}[2]} = 0$$

and hence

$$\begin{aligned} F_{\mathbf{a}[3]} &= a_2\Big(a_3^2 y_3 F^{[2]}_{\mathbf{a}[2]} + a_3 y_2 F^{[1]}_{\mathbf{a}[2]} + \sum_{i\geq 3} a_3^i y_{i+1} F^{[i]}_{\mathbf{a}[0]}\Big) \\ &= a_2(a_3^2 y_3 F^{[2]}_{\mathbf{a}[2]} + a_3 y_2 F^{[1]}_{\mathbf{a}[2]}) \\ &= a_2(a_3^2 y_3 (a_1^2 y_1^2) + a_3 y_2 (a_1 a_2 a_3 y_1 y_2)) \\ &= a_1^2 a_2 a_3^2 y_3 y_1^2 + a_1 a_2^2 a_3 y_1 y_2^2. \end{aligned} \tag{8.1.13}$$

For $n = 4$, from (8.1.13), we have

$$\sum_{m\geq 4} a_3^m y_{m+1} F^{[m]}_{\mathbf{a}[2]} = 0$$

and hence

$$\begin{aligned} F_{\mathbf{a}[4]} &= a_2(a_3^3 y_4 F^{[3]}_{\mathbf{a}[3]} + a_3^2 y_3 F^{[2]}_{\mathbf{a}[3]} + a_3 y_2 F^{[1]}_{\mathbf{a}[3]}) \\ &\quad + a_2 \sum_{i\geq 4} a_3^i y_{i+1} F^{[i]}_{\mathbf{a}[0]} \\ &= a_2(a_3^3 y_4 F^{[3]}_{\mathbf{a}[3]} + a_3^2 y_3 F^{[2]}_{\mathbf{a}[3]} + a_3 y_2 F^{[1]}_{\mathbf{a}[3]}). \end{aligned} \tag{8.1.14}$$

On account of (8.1.10), (8.1.12) and (8.1.13), from (8.1.9),

$$\begin{cases} F^{[1]}_{\mathbf{a}[3]} = a_1^2 a_2 a_3^2 y_3 y_1^2 + a_1 a_2^2 a_3 y_1 y_2^2 \\ \qquad\quad = a_1 a_2 (a_1 a_3^2 y_3 y_1^2 + a_2 a_3 y_1 y_2^2); \\ F^{[2]}_{\mathbf{a}[3]} = 2F_{\mathbf{a}[1]} F_{\mathbf{a}[2]} = 2a_1^2 a_2 a_3 y_1^2 y_2; \\ F^{[3]}_{\mathbf{a}[3]} = (F_{\mathbf{a}[1]})^3 = a_1^3 y_1^3. \end{cases}$$

By (8.1.14), it turns out that

$$\begin{aligned} F_{\mathbf{a}[4]} &= a_2(a_3^3y_4(a_1^3y_1^3) + a_3^2y_3(2a_1^2a_2y_1^2y_2)) \\ &\quad + a_2(a_3y_2a_1a_2(a_1y_3y_1^2 + a_2y_1y_2^2)) \\ &= a_1^3a_2a_3^3y_1^3y_4 + 2a_1^2a_2a_3^2y_1^2y_2y_3 \\ &\quad + a_1^2a_2a_3y_1^2y_2y_3 + a_1^2a_2^2a_3y_1y_2^3 \\ &= a_1^3a_2a_3^3y_1^3y_4 + a_1^2a_2a_3(2a_1a_3+1)y_1^2y_2y_3 \\ &\quad + a_1a_2^3a_3y_1y_2^3. \end{aligned} \tag{8.1.15}$$

Lemma 8.1.3. *For any integer $n \geq 1$, $F_{\mathbf{a}[n]}$ is a homogeneous polynomial of $\mathbf{y}_n = (y_1, y_2, \ldots, y_n)$ with degree n.*

Proof. For $n \geq 4$, from (8.1.9), (8.1.10), (8.1.12), (8.1.13) and (8.1.15), the conclusion is seen. In fact, for $n \geq 1$, from (8.1.9), we have $F_{\mathbf{a}[n-1]}^{[i-1]} = 0, i \geq n+1$. Because of $y_i F_{\mathbf{a}[n-1]}^{[i-1]} = 0$ for all $i \geq n+1$, from equation (8.1.6), it is seen that $F_{\mathbf{a}[n]}$ is independent of all y_i, $i \geq n_1$, and hence we have the conclusion. □

This lemma tells us that $F_{\mathbf{a},n}$ is independent of y_i, $i \geq n+1$, for any positive integer n. On this basis, from equation (8.1.6), we have the following.

Lemma 8.1.4. *Equation* (8.1.6), *and hence equation* (8.1.1), *is equivalent to the system of equations:*

$$F_{\mathbf{a}[n]} = \begin{cases} 0, & \text{when } n = 0; \\ a_1 y_1, & \text{when } n = 1; \\ a_2 \sum_{i=1}^{n-1} a_3^i y_{i+1} F_{\mathbf{a}[n-1]}^{[i]}, & \text{when } n \geq 1. \end{cases} \tag{8.1.16}$$

Proof. This is a direct result of Lemma 8.1.3. □

Now, we are allowed to present our main result of this section.

Theorem 8.1.5. *On $\mathcal{R}\{\mathbf{y}\}$, equation* (8.1.1) *is well-defined if, and only if, $a_0 = 0$.*

Proof. The necessity is obvious from what was mentioned above. Then it is only necessary to show that the system of equations (8.1.16) has, and is the only one to have, a solution for the sufficiency.

We proceed by induction on $n \geq 1$. We evaluate $F_{\mathbf{a}[1]}, F_{\mathbf{a},2}, \ldots, F_{\mathbf{a}[n]}, \ldots$ in this order. First, from (8.1.10), (8.1.12), (8.1.13) and (8.1.15), we have seen that $F_{\mathbf{a}[1]}$, $F_{\mathbf{a},2}$, $F_{\mathbf{a}[3]}$ and $F_{\mathbf{a}[4]}$ each is determined by those already done before. Then, to determine $F_{\mathbf{a}[n]}$ for $n \geq 5$ whenever $F_{\mathbf{a}[i]}$, $i \leq n-1$, we use the assumption. Since all terms on the right hand side of the last equation in equation (8.1.16) are known, $F_{\mathbf{a}[n]}$ is determined as well. The conclusion is drawn. □

8.2 Solution planted

Although equation (8.1.16) provides a method to evaluate the solution of equation (8.1.6) and hence equation (8.1.1) in its own right, it is still necessary to observe the structure of the solution for certain distinct purposes.

Observation 8.2.1. *In the solution evaluated from equation* (8.1.16), *the terms of* $\mathbf{y}^{\mathbf{i}}$ *have a factor* $a_1^{i_1}a_2^{|\mathbf{i}|-i_1}$.

Proof. On the basis of equation (8.1.16), we proceed by induction on $n \geq 1$. When $n = 1$, because of $F_{\mathbf{a}[1]} = a_1y_1$, the result for $F_{\mathbf{a}[1]}$ and then for $F^{[i]}_{\mathbf{a}[i]}$ $(i \geq 2)$, is easily checked to be true.

For $n \geq 2$, assume that, for any $i \leq n-1$, the $F_{\mathbf{a}[i]}$, and then $F^{[j]}_{\mathbf{a}[i]}$ $(j \geq 1)$ as well, satisfy the result. Because of all $F^{[j]}_{\mathbf{a}[i]}$ $(1 \leq i \leq n-1, j \geq 1)$ obey the result, by considering the additivity of the result, $F_{\mathbf{a}[n]}$ satisfies the result as well. □

On the basis of Lemma 8.1.3, we are allowed to write the homogeneous polynomial $F_{\mathbf{a}[n]}$ of degree n as

$$F_{\mathbf{a}[n]} = \sum_{|\mathbf{i}_n|=n} F_{\mathbf{a}[\mathbf{i}_n]}\mathbf{y}^{\mathbf{i}_n} \tag{8.2.1}$$

where $\mathbf{y}_n = (y_1, y_2, \ldots, y_n)$ and $\mathbf{i}_n = (i_1, i_2, \ldots, i_n)$. For convenience, let us denote by

$$\pi(F_{\mathbf{i}_n}) = \sum_{j=1}^{n} ji_j = \mathbf{n}\mathbf{i}_n^{\mathrm{T}} \tag{8.2.2}$$

its size where $\mathbf{n} = (1, 2, \ldots, n)$.

Observation 8.2.2. *For two homogeneous polynomials A and B, we have* $\pi(AB) = \pi(A) + \pi(B)$.

Proof. Let $\mathbf{i}$ and $\mathbf{j}$ be a power vector of a term in, respectively, A and B. Since $\mathbf{i}+\mathbf{j}$ is a power vector of AB, we have $\pi(AB) = \mathbf{n}(\mathbf{i}+\mathbf{j})^{\mathrm{T}} = \mathbf{n}\mathbf{i}^{\mathrm{T}} + \mathbf{n}\mathbf{j}^{\mathrm{T}} = \pi(A) + \pi(B)$. This is the conclusion. □

Lemma 8.2.3. *For any integers* $n \geq 1$ *and* k, $1 \leq k \leq n$, *if* $F^{[k]}_{\mathbf{a}[\mathbf{i}_n]} > 0$, *then*

$$\pi(F_{\mathbf{a}[\mathbf{i}_n]}) = 2n - k. \tag{8.2.3}$$

Proof. We proceed by induction on n. For $n = 1$, we have $k = 1$. Because of $F^{[1]}_{\mathbf{a}[1]} = F_{\mathbf{a}[1]} = a_1y_1$ in the second line of (8.1.16), we have $\pi(F^{[1]}_{\mathbf{a}[1]}) = \pi(F_{\mathbf{a}[1]}) = 1 \times 1 = 2 \times 1 - 1 = 2n - 1$.

For $n \geq 2$, let us assume $\pi(F^{[j]}_{\mathbf{a}[i]}) = 2i - j$ for $1 \leq j \leq i$ and $i \leq n-1$ to evaluate $\pi(F^{[k]}_{\mathbf{a}[n]})$ for $1 \leq k \leq n$. On the basis of the last line in (8.1.16), we have to evaluate $\pi(F^{[i]}_{\mathbf{a}[n-1]})$ for $1 \leq i \leq n-1$. Because of all these results together with the assumption, we have $\pi(F^{[i]}_{\mathbf{a}[n-1]}) = 2(n-1) - i$ for $1 \leq i \leq n-1$. On account of

$$\pi(y_{i+1}F^{[i]}_{\mathbf{a}[n-1]}) = (i+1) + (2n-2-i) = 2n-1,$$

which is independent of i, from the last line in (8.1.16), we have

$$\pi(F_{\mathbf{a}[n]}) = \pi(y_{i+1}F^{[i]}_{\mathbf{a}[n-1]}) = 2n-1.$$

Then we have $F^{[i]}_{\mathbf{a}[n]}$, $2 \le i \le n$, by the second line of (8.1.9). Because of

$$\begin{aligned}\pi(F_{\mathbf{a}[j]}F^{[i-1]}_{\mathbf{a}[n-j]}) &= (2j-1) + (2(n-j) - (i-1)) \\ &= 2n-i,\end{aligned}$$

which is independent of j, we have

$$\pi(F^{[i]}_{\mathbf{a}[n]}) = \pi(F_{\mathbf{a}[j]}F^{[i-1]}_{\mathbf{a}[n-j]}) = 2n-i.$$

The conclusion is drawn. □

The case of $k = 1$ holds independently from a fundamental point of view.

Corollary 8.2.4. *For any integer $n \ge 1$, $\pi(F_{\mathbf{a}[n]}) = 2n-1$.*

Proof. This is the case of $k = 1$ in Lemma 8.2.3. □

Let $\mathcal{I}_n = \{\mathbf{i}_n \ge \mathbf{0} \mid \mathbf{n}\mathbf{i}_n^{\mathrm{T}} = 2n-1\}$. From Corollary 8.2.4, we have

$$F_{\mathbf{a}[n]} = \sum_{\mathbf{i}_n \in \mathcal{I}_n} F_{\mathbf{a}[\mathbf{i}_n]}\mathbf{y}^{\mathbf{i}_n}. \tag{8.2.4}$$

Furthermore, for integer k, $2 \le k \le n$, from (8.1.9), we have

$$F^{[k]}_{\mathbf{a}[n]} = \sum_{\mathbf{i}_n \in \mathcal{I}_n} F^{[k]}_{\mathbf{a}[\mathbf{i}_n]}\mathbf{y}^{\mathbf{i}_n}. \tag{8.2.5}$$

Of course, when $k = 1$, then $F^{[1]}_{\mathbf{a}[n]} = F_{\mathbf{a}[n]}$ and $F^{[1]}_{\mathbf{a}[\mathbf{i}_n]} = F_{\mathbf{a}[\mathbf{i}_n]}$. This is (8.2.4).

Theorem 8.2.5. *Equation (8.1.1) for $a_0 = 0$ has its solution in $\mathcal{R}\{\mathbf{y}\}$ determined by*

$$F_{\mathbf{a}[n]} = \begin{cases} 0, & \text{when } n = 0; \\ a_1y_1, & \text{when } n = 1; \\ a_1a_2a_3y_1y_2, & \text{when } n = 2; \\ a_1^2a_2a_3^2y_1y_2^2 + a_1a_2^2a_3y_1^2y_2, & \text{when } n = 3; \\ a_1^3a_2a_3^3y_1^3y_4 + a_1^2a_2a_3(2a_1a_3+1)y_1^2y_2y_3 + a_1a_2^3a_3y_1y_2^3, & \text{when } n = 4; \\ a_2a_3y_2F_{\mathbf{a}[n-1]} + a_2\sum_{i=3}^{n-1}a_3^{i-1}y_iF^{[i-1]}_{\mathbf{a}[n-1]} + a_1^{n-1}a_2y_1^{n-1}y_n, & \text{when } n \ge 5. \end{cases} \tag{8.2.6}$$

Proof. When $n = 0$, it is determined by the initial condition of equation (8.1.1) for $a_0 = 0$.

When $n = 1, 2, 3$ and 4, the results follow from (8.1.10), (8.1.12), (8.1.13) and (8.1.10).

When $n \geq 5$, because of $F^{[1]}_{\mathbf{a}[n-1]} = F_{\mathbf{a}[n-1]}$ and $F^{[n-1]}_{\mathbf{a}[n-1]} = a_1^{n-1} y_1^{[n-1]}$, from (8.1.16), we have

$$\begin{aligned}
F_{\mathbf{a}[n]} &= a_2 \sum_{i=1}^{n-1} a_3^i y_{i+1} F^{[i]}_{\mathbf{a}[n-1]}, \text{ by } i \Leftarrow i+1, \\
&= a_2 \sum_{i=2}^{n} a_3^{i-1} y_i F^{[i-1]}_{\mathbf{a}[n-1]}, \text{ by } F^{[1]}_{\mathbf{a}[n-1]} = F_{\mathbf{a}[n-1]}, \\
&= a_2 a_3 y_2 F_{\mathbf{a}[n-1]} + a_2 \sum_{i=3}^{n} a_3^{i-1} y_i F^{[i-1]}_{\mathbf{a}[n-1]}, \\
&\quad \text{by } F^{[n-1]}_{\mathbf{a}[n-1]} = a_1^{n-1} a_3^{n-1} y_1^{n-1}, \\
&= a_2 a_3 y_2 F_{\mathbf{a}[n-1]} + a_2 \sum_{i=3}^{n-1} a_3^{i-1} y_i F^{[i-1]}_{\mathbf{a}[n-1]} \\
&\quad + a_1^{n-1} a_2 a_3^{n-1} y_1^{n-1} y_n.
\end{aligned}$$

Because of Lemma 8.1.4, this is what we want to prove. □

On account of $a_0, a_1, a_2 \in \mathbb{R}_+$, from this theorem, by considering that all of $F_{\mathbf{a}[k]}$, $1 \leq k \leq 4$, are of the form of a sum of positive terms and that, for any integer $n \geq 5$, $F_{\mathbf{a}[n]}$ is of the form of a sum of positive terms whenever $F_{\mathbf{a},k}$ is so for any integer $k \leq n-1$, it is seen that all $F_{\mathbf{a}[n]}$ are of the form of a sum of positive terms.

8.3 Restrictions planted

Now, we observe the special case of equation (8.1.1) when $a_0 = 0$ and $a_1 = a_2 = a_3 = 1$. This is the equation

$$\begin{cases} \int\!\!\int_y \dfrac{y^2 f}{1 - yf} = f - y_1; \\ f|_{\mathbf{y}=\mathbf{0}} = 0. \end{cases} \tag{8.3.1}$$

This is an equation satisfied by the enumerating function, denoted by $f = F_0 + F_1 + F_2 + \cdots$, of rooted plane planted trees with non-rooted vertex partition vector as parameter.

Theorem 8.3.1. *Equation* (8.3.1) *has, and is the only one to have, the solution determined by*

$$F_n = \begin{cases} 0, & \text{when } n = 0; \\ y_1, & \text{when } n = 1; \\ y_1 y_2, & \text{when } n = 2; \\ y_1 y_2^2 + y_1^2 y_2, & \text{when } n = 3; \\ y_1 y_2^2 + 3y_1^2 y_2 y_3 + y_1^3 y_4, & \text{when } n = 4; \\ y_2 F_{n-1} + \sum_{i=3}^{n-1} y_i F_{n-1}^{[i-1]} + y_1^{n-1} y_n, & \text{when } n \geq 5. \end{cases} \tag{8.3.2}$$

Proof. On the basis of Lemma 8.1.4 and Theorem 8.1.5, the conclusion is drawn by taking $a_1 = a_2 = a_3 = 1$ and by substituting F_n, F_{n-1} and $F_{n-1}^{[i-1]}(3 \leq i \leq n-1)$ for $F_{\mathbf{a}[n]}$, $F_{\mathbf{a}[n-1]}$ and $F_{\mathbf{a}[n-1]}^{[i-1]}$ $(3 \leq i \leq n-1)$, respectively, in (8.2.6). □

Let $F_{\mathbf{i}}$ be the coefficient of the term with $\mathbf{y}^{\mathbf{i}}$ in f, then

$$F_n = \sum_{\substack{|\mathbf{i}|=n \\ \mathbf{i}>\mathbf{0}}} F_{\mathbf{i}} \mathbf{y}^{\mathbf{i}}$$

where $F_{\mathbf{i}}$ is the number of non-isomorphic classes of rooted planted plane trees each of which is with the partition vector $\mathbf{i}$, *i. e.*, i_j non-rooted vertices of valency $j \geq 1$.

In Figures 8.1.1 and 8.1.2 of Liu [59], all distinct isomorphic classes of rooted planted plane trees with order $n+1$ for $n = 1, 2, 3$ and 4 show the correctness of (8.3.2).

According to Liu [49], it is seen that

$$F_{\mathbf{i}} = \frac{(n-1)!}{\mathbf{i}!}. \tag{8.3.3}$$

Theorem 8.3.2. *The solution of equation* (8.3.1) *is determined by*

$$F_n = \sum_{\mathbf{i} \in \mathcal{I}_n} \frac{(n-1)!}{\mathbf{i}!} \mathbf{y}^{\mathbf{i}} \tag{8.3.4}$$

where $\mathcal{I}_n = \{\mathbf{i} \geq \mathbf{0} \,|\, |\mathbf{i}| = n, \pi(\mathbf{i}) = \mathbf{n}\mathbf{i}^{\mathrm{T}} = 2n-1\}$, *for* $n \geq 0$.

Proof. We proceed on the basis of Theorems 8.1.1 and 8.2.5 for $a_0 = 0$ and $a_1 = a_2 = a_3 = 1$. From (8.3.3), the conclusion is drawn. □

Lemma 8.3.3. *Equation* (8.3.1) *is equivalent to*

$$\begin{cases} f = y_1 + \sum_{m \geq 1} y_{m+1} f^m; \\ f|_{\mathbf{y}=\mathbf{0}} = 0, \end{cases} \tag{8.3.5}$$

in $\mathcal{R}\{\mathbf{y}\}$.

Proof. By considering that equation (8.1.1) is equivalent to equation (8.1.6), the conclusion is the case of $a_0 = 0$ and $a_1 = a_2 = a_3 = 1$. □

Let $\mathbf{t} = (t_1, t_2, t_3, \ldots) = (f, f^2, f^3, \ldots)$, then from Liu [20] (or Liu [18]), by considering that only one way is available to generate a plane tree of rooted valency $i \geq 2$ via producing a new vertex valency $j \geq i-1$ from a plane tree of valency j, it is shown that the solution $f = t_1$ of equation (8.3.1) is just the first entry of a solution of the infinite dimensional vector equation

$$(\mathbf{I} - \mathbf{Y})\mathbf{t}^{\mathrm{T}} = y_1 \mathbf{e}_1^{\mathrm{T}} \tag{8.3.6}$$

where $\mathbf{e}_1 = (1, 0, 0, 0, \ldots)$, $\mathbf{I}$ is the unit matrix of infinite dimension and $\mathbf{Y} = (y_{i,j})_{i\geq 1, j\geq 1}$ with

$$y_{i,j} = \begin{cases} y_{j-i+2}, & \text{when } i \leq j+1; \\ 0, & \text{when } i \geq j+2. \end{cases} \tag{8.3.7}$$

In fact, $t_i = f^i$, $i \geq 2$, is the enumerating function of plane trees with rooted valency i for the partition vector of non-rooted vertices as parameter.

Because of

$$(\mathbf{I} - \mathbf{Y})^{-1} = \sum_{n\geq 0} \mathbf{Y}^n \tag{8.3.8}$$

in $\mathcal{R}\{\mathbf{y}\}$, equation (8.3.6) has a solution

$$\mathbf{t}^{\mathrm{T}} = y_1 \Big(\sum_{n\geq 0} \mathbf{Y}^n \Big) \mathbf{e}_1^{\mathrm{T}}. \tag{8.3.9}$$

In what follows, our aim is the determination of $\mathbf{Y}^n = (y_{i,j}^{[n]})_{i\geq 1, j\geq 1}$.

Lemma 8.3.4. *For any integer $n \geq 2$, $y_{i,j}^{[n]} = 0$ $(i \geq n+2, 1 \leq j \leq i-n-1)$.*

Proof. When $n = 2$, from (8.3.7), for any integer $i \geq 3$, in the ith row of Y, $y_{i,j} = 0$, $1 \leq j \leq i-2$ and for any integer $j \geq 1$, in the jth column Y, $y_{i,j} = 0$, $i \geq j+2$. Because of

$$y_{i,j}^{[2]} = \sum_{k\geq 1} y_{i,k} y_{k,j}$$

for $i, j \geq 1$, we have, for any $i \geq 4$, $1 \leq j \leq i-3$,

$$\begin{aligned} y_{i,j}^{[2]} &= \sum_{k=i-1}^{j+1} y_{i,k} y_{k,j}, \quad \text{from } i \geq j+3, \\ &= \sum_{k=j+2}^{j+1} y_{i,k} y_{k,j}, \quad \text{from } j+2 > j+1, \\ &= 0. \end{aligned}$$

This is the case for $n = 2$ in the conclusion.

For any integer $n \geq 3$, assume that, for $2 \leq l \leq n-1$, we have $y_{i,j}^{[l]} = 0$, $i \geq l+2$, $1 \leq j \leq i-l-1$. We prove the case for $l = n$. From (8.3.7), for any $i \geq 3$, in the ith row of Y, $y_{i,j} = 0$, $1 \leq j \leq i-2$ and from the assumption, for any $j \geq 1$, in the jth column of Y, $y_{i,j} = 0$, $i \geq j + (n-1) + 1$. On account of the associative law for matrix multiplication, for any $i, j \geq 1$,

$$y_{i,j}^{[n]} = \sum_{k \geq 1} y_{i,k} y_{k,j}^{[n-1]}.$$

For any $i \geq n+2$, $1 \leq j \leq i-n-1$, because of all entries which are 0 in $[y_{i,1}, y_{i,i-2}]$ (others not 0!) on the ith row of Y and all entries which are 0 in $[y_{j+n,j}, y_{\infty,j})$ (the others not 0!) on the jth column of Y^{n-1}, we have

$$\begin{aligned} y_{i,j}^{[n]} &= \sum_{k \geq 1} y_{i,k} y_{k,j}^{[n-1]}, \\ &\quad \text{by } y_{i,k} = 0,\ k \in [1, i-2], \\ &= \sum_{k \in \Lambda} y_{i,k} y_{k,j}^{[n-1]}, \\ &\quad \text{by } \Lambda = [i-n, \infty) \cap [1, n+j-1] = \emptyset, \\ &= 0. \end{aligned}$$

This is the case for $n \geq 3$ in the conclusion. □

This lemma enables us to write $Y^n = (y_{i,j}^{[n]})_{i \geq 1, j \geq 1}$ as

$$y_{i,j}^{[n]} = \begin{cases} \sum_{k=1}^{n+j-1} y_{i,k} y_{k,j}^{[n-1]}, & \text{when } 1 \leq i \leq 2, j \geq 1; \\ \sum_{k=i-1}^{n+j-1} y_{i,k} y_{k,j}^{[n-1]}, & \text{when } i \geq 3, j-i \geq -n; \\ 0, & \text{otherwise, } i.\,e., i \geq 3, j-i \leq -(n+1), \end{cases} \tag{8.3.10}$$

for $n \geq 2$.

Lemma 8.3.5. *For any integer $n \geq 2$, $y_{i,j}^{[n]}$, $i \geq 1$, $j \geq 1$, are determined only by $\{y_{i,1}^{[n]}$, $2 \leq i \leq n+1\}$ and $\{y_{1,j}^{[n]}$, $j \geq 1\}$.*

Proof. When $n = 1$, from (8.3.7), it is seen that, for any integer $s \geq 1$, $y_{i+s,j+s} = y_{(j+s)-(i+s)+2} = y_{j-i+2} = y_{i,j}$. For convenience, such a property of a matrix is called *slope translational*. It can be seen that the product of two slope translational matrices is still slope translational. Hence, Y^2 is slope translational, *i. e.*, for any integer $s \geq 0$, $y_{i+s,j+s}^{[2]} = y_{i,j}^{[2]}$. Furthermore, for any integer $n \geq 3$, Y^n is slope translational, *i. e.*, for any integer $s \geq 0$, $y_{i+s,j+s}^{[n]} = y_{i,j}^{[n]}$. From Lemma 8.3.4, the conclusion is drawn. □

Based on this lemma, for any integer $n \geq 1$, let us write

$$y_k^{[n]} = \begin{cases} y_{n-k+2,1}^{[n]}, & \text{when } 1 \leq k \leq n; \\ y_{1,n-k+2}^{[n]}, & \text{when } k \geq n+1. \end{cases} \tag{8.3.11}$$

Conversely, for any integer $i, j \geq 1$,

$$y_{i,j}^{[n]} = \begin{cases} y_{j-i+2}^{[n]}, & \text{when } j - i \geq -n; \\ 0, & \text{when } j - i \leq -(n+1). \end{cases} \tag{8.3.12}$$

Attention. (1) Because of $y_{i,j}^{[1]} = y_{i,j}$, when $n = 1$, (8.3.12) becomes (8.3.7).

(2) For any integer $n \geq 1$, $y_1^{[n]} = y_1^n$ and for $n \geq 2$, $y_2^{[n]} = ny_1^{n-1}y_2$.

From (8.3.9),

$$\begin{aligned} \tau_1 &= y_1\Big(\sum_{n\geq 0} y_{1,1}^{[n]}\Big), \quad \text{by (8.3.12);} \\ &= y_1\Big(\sum_{n\geq 0} y_2^{[n]}\Big), \end{aligned} \tag{8.3.13}$$

where $y_0^{[0]} = 1$.

Lemma 8.3.6. *For F_n, $n \geq 1$, in the solution of equation* (8.1.1) *with $a_0 = 0$ and $a_1 = a_2 = 1$, we have*

$$F_n = y_1 \sum_{i=1}^{n-1} y_{i+1} y_{n-i}^{[n-2]}. \tag{8.3.14}$$

Proof. Since each nonzero entry of Y^n, $n \geq 1$, is a homogeneous polynomial of degree n, all $y_i^{[n]}$, $i \geq 1$, are homogeneous polynomials of degree n. Because τ_1 is a solution of equation (8.3.1), it is seen that $y_1 y_{n-i}^{[n-2]}$ for $1 \leq i \leq n-1$ in (8.3.13) is a part of F_{n-1}, *i. e.*, a part of τ_1 is a homogeneous polynomial of degree $n-1$. From the uniqueness of the solution, $F_n = y_1 y_{n-1}^{[n-1]}$. From (8.3.10) and (8.3.12), the conclusion can easily be drawn. □

Theorem 8.3.7. *For any integer $n \geq 1$, we have*

$$\partial_{\mathbf{y}}^{\mathbf{i}} F_n = \frac{(n-1)!}{\mathbf{i}!} \tag{8.3.15}$$

where $n = |\mathbf{n}|$ and $\partial_{\mathbf{y}}^{\mathbf{i}}$ is the coefficient operator for extracting the coefficient of the term with $\mathbf{y}^{\mathbf{i}}$.

Proof. From Theorem 8.3.1 and Theorem 8.3.2, the conclusion is drawn. □

Example 1. Root-isomorphic classes of planted trees with vertex partition vector as parameters. On the basis of (8.3.2), one finds that, for any integer n, F_n provides the polynomial of $\mathbf{y}$ whose coefficient of $\mathbf{y}^{\mathbf{n}}$ is the number of root-isomorphic classes of planted trees with the vertex partition vector $\mathbf{n} = (n_1, n_2, n_3, \ldots)$ where n_i is the number of non-rooted vertices of valency $i \geq 1$ and order $n+1$ and size $n = |\mathbf{n}|$.

Figure 8.3.1 provides the root-isomorphic classes of planted trees with vertex partition vectors as $a = (1)$, $b = (1,1)$, $c = (2,0,1)$ and $d = (1,2,0)$ of orders 2, 3 and 4. For

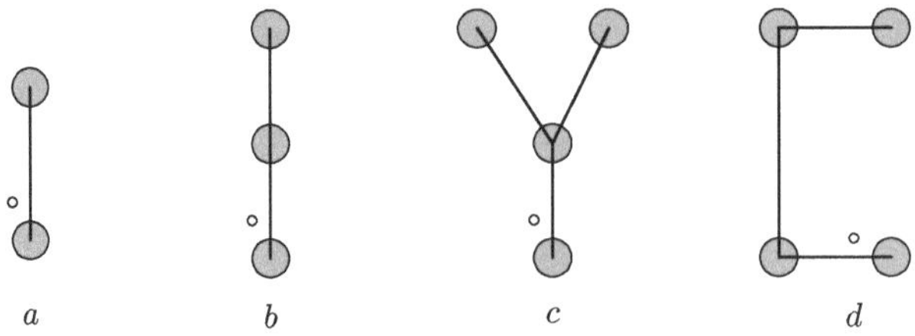

Figure 8.3.1: Classes of planted trees with vertex partition of order 2–4.

example, a has 1 class, *i. e.*, $F_1 = y_1$; b has 1 class, *i. e.*, $F_2 = y_1y_2$; each of c and d has 1 class, *i. e.*, $F_3 = y_1^2y_3 + y_1^3$.

Figure 8.3.2 provides the root-isomorphic classes of planted trees with vertex partition vectors as $a = (3,0,0,1)$, $b = (2,1,1,0)$, and $c = (1,3,0,0)$ of order 5. Here, a has 1 class, b has 3 classes and c has 1 class, *i. e.*, $F_4 = y_1^3y_4 + 3y_1^2y_2y_3 + y_1y_2^3$.

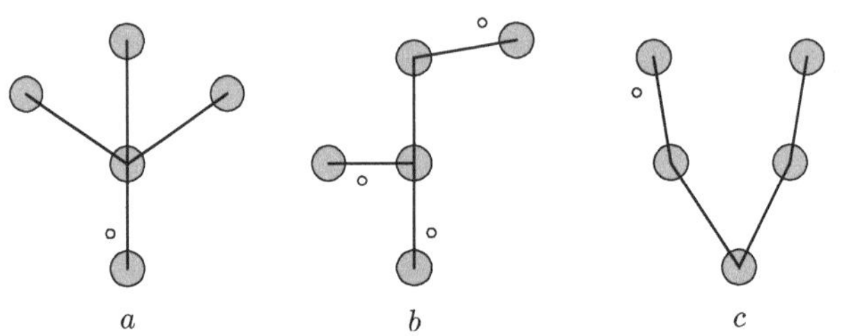

Figure 8.3.2: Classes of planted trees with vertex partition of order 5.

8.4 Explicit expressions planted

Let us go back to equation (8.1.6) which is equivalent to equation (8.1.1) for $a_0 = 0$.

Now, one might be motivated to establish an equation system of infinite dimension,

$$\mathbf{t}_{\mathbf{a}}^{\mathrm{T}} = a_2\mathbf{Y}_{\mathbf{a}}\mathbf{t}_{\mathbf{a}}^{\mathrm{T}} + a_1y_1\mathbf{e}_1^{\mathrm{T}} \tag{8.4.1}$$

where $\mathbf{a} = (a_1, a_2, a_3) \in \mathbb{Z}_+^3$, $\mathbf{t}_{\mathbf{a}} = (t_{\mathbf{a}1}, t_{\mathbf{a}2}, t_{\mathbf{a}3}, \ldots)$, $\mathbf{e}_1 = (1, 0, 0, \ldots)$ and

$$y_{\mathbf{a}[i,j]} = \begin{cases} a_3^j y_{j-i+2}, & \text{when } -1 \le j - i; \\ 0, & \text{otherwise, i. e., } j - i \le -2, \end{cases} \tag{8.4.2}$$

for integers $i, j \ge 1$.

We have

$$\mathbf{Y_a} = \mathbf{Y}[a_3, a_3^2, a_3^3, \ldots]$$

where $[a_3, a_3^2, a_3^3, \ldots] = (a_3, a_3^2, a_3^3, \ldots)\mathbf{I}$, a diagonal matrix of infinite dimension, and $\mathbf{Y} = (y_{i,j})_{i,j\geq 1}$ is given by (8.3.7).

Observation 8.4.1. *If* $\mathbf{t}_a$ *is a solution of equation* (8.4.1), *then* $t_{\mathbf{a}1}$ *is the solution of equation* (8.1.1) *for* $a_0 = 0$.

Proof. Because the first equation in equation (8.4.1) is just equation (8.1.6) and hence equation (8.1.1) for $a_0 = 0$. This is the conclusion. □

This observation tells us that an explicision of the solution of equation (8.1.1) for $a_0 = 0$ can be extracted from a solution of equation (8.4.1).

Theorem 8.4.2. *Equation* (8.4.1) *is well-defined for* $\mathbf{a} \in \mathbb{Z}_+^3$ *and* $a_1 a_2 a_3 \neq 0$.

Proof. We proceed on the basis of the equivalence between equation (8.4.1) and

$$(I - a_2\mathbf{Y_a})\mathbf{t}_\mathbf{a}^\mathrm{T} = a_1 y_1 \mathbf{e}_1^\mathrm{T}.$$

Because the inverse of $I - a_2\mathbf{Y_a}$ is

$$(I - a_2\mathbf{Y_a})^{-1} = \sum_{i\geq 0}(a_2\mathbf{Y_a})^i,$$

the conclusion is drawn. □

This theorem and Observation 8.4.1 enable us to determine the solution of equation (8.1.1) via that of equation (8.4.1).

Observation 8.4.3. *When integer* $n = |\mathbf{n}|$ *is given, then the solution* $\mathbf{t}_a$ *of equation* (8.4.1) *is only dependent on some* y_i *for* $1 \leq i \leq n$.

Proof. Because of the maximum valency of a planted tree of order $n+1$ is n, the conclusion follows. □

This observation enables us to restrict equation (8.4.1) in terms of dimension n for integer n given.

Lemma 8.4.4. *For integer* $n \geq 1$ *given, the infinite dimensional system equation* (8.4.1) *is equivalent to the following equation system in dimension* $n-1$:

$$(I_{(n-1)\times(n-1)} - a_2\mathbf{Y}_{\mathbf{a}((n-1)\times(n-1))})\mathbf{t}_{\mathbf{a}(n-1)}^\mathrm{T} = a_1 y_1 \mathbf{e}_{1(n-1)}^\mathrm{T} \tag{8.4.3}$$

where $\mathbf{t}_{\mathbf{a}(n-1)} = (t_{\mathbf{a}(1)}, t_{\mathbf{a}(2)}, \ldots, t_{\mathbf{a}(n-1)})$ *and* $\mathbf{Y}_{\mathbf{a}((n-1)\times(n-1))}$ *are, respectively, all on* $\mathcal{R}\{\mathbf{y}_{n-1}\}^{n-1}$ *and* $\mathcal{R}\{\mathbf{y}_{n-1}\}^{(n-1)\times(n-1)}$.

Proof. This is a direct result of Observation 8.4.3. □

Now, we are allowed to describe an explicision of the solution of equation (8.4.1) and then that of equation (8.1.1).

Theorem 8.4.5. *The solution $t_{\mathbf{a}(1)}$ of equation* (8.4.1) *for integer $n \geq 1$ given has an explicision*

$$t_{\mathbf{a}1(n-1)} = a_1 y_1 \left(1 + \sum_{i=1}^{n-1} a_2^i y_{\mathbf{a}[1,1]}^{[i]} \right) \tag{8.4.4}$$

where $y_{\mathbf{a}[1,1]}^{[1]} = y_{\mathbf{a}[1,1]}$ is determined by (8.4.2) *and $y_{\mathbf{a}[1,1]}^{[i]}$ is the entry at* (1, 1) *of the squared matrix $\mathbf{Y}_{\mathbf{a}[(n-1)\times(n-1)]}^i$ for $2 \leq i \leq n-1$.*

Proof. We proceed on the basis of Lemma 8.4.4. Because

$$\begin{aligned} t_{\mathbf{a}1(n-1)} &= a_1 y_1 [(I_{(n-1)\times(n-1)} - a_2 \mathbf{Y}_{\mathbf{a}((n-1)\times(n-1))})^{-1}]_{\leq n-1} \mathbf{e}_{1(n-1)}^{\mathrm{T}} \\ &= a_1 y_1 \left(1 + \sum_{i=1}^{n-1} a_2^i y_{\mathbf{a}[1,1]}^{[i]} \right), \end{aligned}$$

the conclusion is drawn. □

Example 1. When $n = 4$. We proceed on the basis of Theorem 8.4.5. Because of the coefficient matrix with dimension 3

$$\mathbf{Y}_{\mathbf{a}(3\times3)} = \begin{pmatrix} a_3 y_2 & a_3^2 y_3 & a_3^3 y_4 \\ a_3 y_1 & a_3^2 y_2 & a_3^3 y_3 \\ 0 & a_3^2 y_1 & a_3^3 y_2 \end{pmatrix}$$

we have $y_{\mathbf{a}[1,1]}^{[1]} = a_3 y_2$, $y_{\mathbf{a}[1,1]}^{[2]} = a_3^2 y_2^2 + a_3^3 y_1 y_3$ and $y_{\mathbf{a}[1,1]}^{[3]} = a_3^3 y_2^3 + (2+a_3) a_3^4 y_1 y_2 y_3 + a_3^6 y_1^2 y_4$.

By employing (8.4.4),

$$\begin{aligned} t_{\mathbf{a}1(3)} = a_1 y_1 & \\ &+ a_1 a_2 a_3 y_1 y_2 \\ &+ a_1 a_2^2 (a_3^2 y_1 y_2^2 + a_3^3 y_1^2 y_3) \\ &+ a_1 a_2^3 (a_3^3 y_1 y_2^3 + (2 + a_3) a_3^4 y_1^2 y_2 y_3 + a_3^6 y_1^3 y_4). \end{aligned}$$

For $a_1 = a_2 = a_3 = 1$, this result can be checked in Figure 8.3.1 and Figure 8.3.2.

In what follows, a summation-free explicision of the solution of equation (8.1.1) for $a_0 = 0$ and $a_3 = 1$ is extracted in a direct manner.

Because of the only occurrence of $a_1 y_1$ and $a_2 y_i$, $i \geq 2$ in the equation, f is a function of $a_1 y_1$ and $a_2 y_i$, $i \geq 2$. This enables us to introduce the substitution

$$z_j = \begin{cases} a_1 y_1, & \text{when } j = 1; \\ a_2 y_j, & \text{when } j \geq 2, \end{cases} \tag{8.4.5}$$

for integer $j \geq 1$.

By substituting z_j for a_1y_1 ($j = 1$) and a_2y_j ($j \geq 2$) in equation (8.1.6), the result is

$$\begin{cases} f = z_1 + \sum_{m\geq 1} a_3^m z_{m+1} f^m; \\ f|_{\mathbf{z}=\mathbf{0}} = 0 \end{cases} \tag{8.4.6}$$

in $\mathcal{R}\{\mathbf{z}\}$. This is just equation (8.3.5) when $a_3 = 1$.

Theorem 8.4.6. *The solution of equation* (8.1.1) *for* $a_3 = 1$ *is determined by the explicit expression*

$$\partial_{\mathbf{y}}^{\mathbf{i}} F_{\mathbf{a},|\mathbf{i}|} = \frac{a_1^{i_1} a_2^{|\mathbf{i}|-i_1}(|\mathbf{i}|-1)!}{\mathbf{i}!} \tag{8.4.7}$$

for $|\mathbf{i}| = n \geq 1$ *where* $\mathbf{i} = (i_1, i_2, i_3, \ldots)$.

Proof. On the basis of Theorem 8.3.7. From (8.4.1) and (8.4.2), the conclusion is drawn. □

8.5 General model

Second, consider the equation

$$\begin{cases} a_2 x \int_y y\delta_{x,y}(uf|_{x=u}) = f - a_1; \\ f|_{x=0,\mathbf{y}=\mathbf{0}} = a_0, \end{cases} \tag{8.5.1}$$

where $\mathbf{a} = (a_0, a_1, a_2) \in \mathbb{Z}_+^3$, $f = f(x, \mathbf{y}) \in \mathcal{R}\{x, \mathbf{y}\}$, $\mathbf{y} = (y_1, y_2, y_3, \ldots)$ and

$$\delta_{x,y}(uf|_{x=u}) = \frac{xf(x) - yf(y)}{x - y},$$

called the *straight difference* of the function $uf(u)$ between x and y.

Because a solution of equation (8.5.1) in the case $a_0 = a_1 = a_3 = 1$ given in [18, 20] (Liu YP, 1985) holds for general plane trees, this equation is called in *general tree model*.

This is equation (31) in Introduction when a_0, a_1 and a_2 are, respectively, replaced by d, c and a.

For convenience, let $F_{\mathbf{a}[m]}(\mathbf{y}) = \partial_x^m f$, $m \geq 0$, and $\mathrm{id}(\mathbf{y}) = |\mathbf{n}|$, where $\mathbf{n} = (n_1, n_2, n_3, \ldots)$, *i. e.*, the power vector of $\mathbf{y}$, $|\mathbf{n}| = n_1 + n_2 + n_3 + \cdots$. For any integer $n \geq 0$, write

$$F_{\mathbf{a}[m,n]} = F_m(\mathbf{y})|_{\mathrm{id}(\mathbf{y})=n} \left(= \sum_{|\mathbf{n}|=n} F_{\mathbf{a}[m,\mathbf{n}]} \mathbf{y}^{\mathbf{n}} \right),$$

or $F_{\mathbf{a}[m,n]}$ is the sum of terms with $\mathbf{y}^{\mathbf{n}}$, $|\mathbf{n}| = n$ in $F_{\mathbf{a}[m]}(\mathbf{y})$, a polynomial of degree n.

Therefore, it is seen that

$$f = \sum_{m\geq 0} F_{\mathbf{a}[m]}(\mathbf{y})x^m. \tag{8.5.2}$$

Because

$$\begin{aligned}\delta_{x,y}(uf|_{x=u}) &= \frac{xf - yf|_{u=y}}{x-y}\\ &= \frac{x\sum_{m\geq 0} F_{\mathbf{a}[m]}(\mathbf{y})x^m - y\sum_{m\geq 0} F_{\mathbf{a}[m]}(\mathbf{y})y^m}{x-y}\\ &= \sum_{m\geq 0} F_{\mathbf{a}[m]}(\mathbf{y})\frac{x^{m+1}-y^{m+1}}{x-y},\end{aligned}$$

$$x^{m+1} - y^{m+1} = (x-y)\left(\sum_{i=0}^{m} x^i y^{m-i}\right),$$

and

$$\begin{aligned}\delta_{x,y}(uf|_{x=u}) &= \sum_{m\geq 0} F_{\mathbf{a}[m]}(\mathbf{y})\left(\sum_{i=0}^{m} x^i y^{m-i}\right)\\ &= \sum_{\mathbf{i}\geq 0}\sum_{m\geq i} F_{\mathbf{a}[m]}(\mathbf{y})y^{m-i}x^i\\ &= \sum_{\substack{i\geq m\\ m\geq 0}} F_{\mathbf{a}[i]}(\mathbf{y})y^{i-m}x^m,\end{aligned}$$

we have

$$\int_y y\delta_{x,y}(uf|_{x=u}) = \sum_{\substack{i\geq m\\ m\geq 0}} F_{\mathbf{a}[i]}(\mathbf{y})y_{i-m+1}x^m.$$

Then the first line of equation (8.5.1) becomes

$$f = a_1 + a_2 \sum_{\substack{i\geq m-1\\ m\geq 1}} F_{\mathbf{a}[i]}(\mathbf{y})y_{i-m+2}x^m. \tag{8.5.3}$$

Theorem 8.5.1. *Equation* (8.5.1) *is equivalent to*

$$\begin{cases} f = a_1 + a_2 \sum_{\substack{i\geq m-1\\ m\geq 1}} F_{\mathbf{a}[i]}(\mathbf{y})y_{i-m+2}x^m;\\ f|_{\mathbf{y}=\mathbf{0}\Rightarrow x=0} = a_0,\end{cases} \tag{8.5.4}$$

in $\mathcal{R}\{x,\mathbf{y}\}$.

Proof. By considering that (8.5.3) is equivalent to the first line of equation (8.1.1) on $\mathcal{R}\{x,\mathbf{y}\}$, the conclusion is drawn. □

This theorem enables us then only to discuss on equation (8.5.4) instead of equation (8.1.1) without indication.

Observation 8.5.2. *When $a_0 \neq a_1$, equation* (8.5.4) *is not consistent.*

Proof. Because no solution exists for satisfying the initial condition of equation (8.5.4) when $a_0 \neq a_1$, the conclusion is drawn. □

This enables us only to investigate the case of $a_0 = a_1$ for equation (8.5.4) or equation (8.5.1).

On the basis of the observation, because

$$\sum_{i\geq m} F_{\mathbf{a}[i]}(\mathbf{y})y_{i-m+1} = \sum_{i\geq m-1} F_{\mathbf{a}[i]}(\mathbf{y})y_{i-m+2},$$

equation (8.5.4) becomes

$$\begin{cases} f = a_0 + a_2 \sum\limits_{\substack{i\geq m-1\\ m\geq 1}} F_{\mathbf{a}[i]}(\mathbf{y})y_{i-m+2}x^m; \\ f|_{\mathbf{y}=\mathbf{0}\Rightarrow x=0} = a_0. \end{cases} \tag{8.5.5}$$

Based on equations (8.5.5) and (8.5.2), we have

$$x^0: \quad F_{\mathbf{a}[0]}(\mathbf{y}) = a_0,\ \text{i.e., the initial condition of equation (8.5.5);} \tag{8.5.6}$$

$$\begin{aligned} x^1: \quad F_{\mathbf{a}[1]}(\mathbf{y}) &= a_2 \sum_{i\geq 0} y_{i+1}F_{\mathbf{a}[i]}(\mathbf{y}) \\ &= a_2(y_1F_{\mathbf{a}[0]}(\mathbf{y}) + y_2F_{\mathbf{a}[1]}(\mathbf{y}) + y_3F_{\mathbf{a}[2]}(\mathbf{y}) \\ &\quad + y_4F_{\mathbf{a}[3]}(\mathbf{y}) + \cdots), \\ &\quad \text{by } F_{\mathbf{a}[0]}(\mathbf{y}) = a_0, \\ &= a_2a_0y_1 + a_2(y_2F_{\mathbf{a}[1]}(\mathbf{y}) + y_3F_{\mathbf{a}[2]}(\mathbf{y}) \\ &\quad + y_4F_{\mathbf{a}[3]}(\mathbf{y}) + \cdots); \end{aligned} \tag{8.5.7}$$

$$\begin{aligned} x^2: \quad F_{\mathbf{a}[2]}(\mathbf{y}) &= a_2 \sum_{i\geq 1} y_iF_{\mathbf{a}[i]}(\mathbf{y}) \\ &= a_2(y_1F_{\mathbf{a}[1]}(\mathbf{y}) + y_2F_{\mathbf{a}[2]}(\mathbf{y}) + y_3F_{\mathbf{a}[3]}(\mathbf{y}) \\ &\quad + y_4F_{\mathbf{a}[4]}(\mathbf{y}) + \cdots); \end{aligned} \tag{8.5.8}$$

$$\begin{aligned} x^3: \quad F_{\mathbf{a}[3]}(\mathbf{y}) &= a_2 \sum_{i\geq 2} y_{i-1}F_{\mathbf{a}[i]}(\mathbf{y}) \\ &= a_2(y_1F_{\mathbf{a}[2]}(\mathbf{y}) + y_2F_{\mathbf{a}[3]}(\mathbf{y}) + y_3F_{\mathbf{a}[4]}(\mathbf{y}) \\ &\quad + y_4F_{\mathbf{a}[5]}(\mathbf{y}) + \cdots); \end{aligned} \tag{8.5.9}$$

$$
\begin{aligned}
x^4: \quad F_{\mathbf{a}[4]}(\mathbf{y}) &= a_2 \sum_{i \geq 3} y_{i-2} F_{\mathbf{a}[i]}(\mathbf{y}) \\
&= a_2 (y_1 F_{\mathbf{a}[3]}(\mathbf{y}) + y_2 F_{\mathbf{a}[4]}(\mathbf{y}) + y_3 F_{\mathbf{a}[5]}(\mathbf{y}) \\
&\quad + y_4 F_{\mathbf{a}[6]}(\mathbf{y}) + \cdots);
\end{aligned} \tag{8.5.10}
$$

and so forth.

Let $\mathbf{f} = (F_{\mathbf{a}[1]}, F_{\mathbf{a}[2]}, F_{\mathbf{a}[3]}, \ldots)$. Then, from (8.1.6)–(8.1.10), equation (8.5.5) becomes

$$
\mathbf{f}^{\mathrm{T}} = a_0 a_2 y_1 \mathbf{e}_1^{\mathrm{T}} + \mathbf{Y}_{\mathbf{a}} \mathbf{f}^{\mathrm{T}} \tag{8.5.11}
$$

where $\mathbf{Y}_{\mathbf{a}} = a_2 \mathbf{Y}$ and $\mathbf{Y}$ is just given in equation (8.3.6).

Theorem 8.5.3. *Equation* (8.5.5) *is equivalent to equation* (8.5.11) *in* $\mathcal{R}\{\mathbf{y}\}$.

Proof. Since each entry of $\mathbf{Y}$ is in $\mathcal{R}\{\mathbf{y}\}$, we have $\mathbf{Y} \in \mathcal{R}\{\mathbf{y}\}$. Then the conclusion can be proved. □

This theorem enables us to determine $\mathbf{f}$ from equation (8.5.11) for a solution of equation (8.5.5) and then equation (8.5.1).

Theorem 8.5.4. *Equation* (8.5.5) *is well-defined in* $\mathcal{R}\{\mathbf{y}\}$.

Proof. By the equivalent transformation on $\mathcal{R}\{\mathbf{y}\}$, equation (8.5.5) becomes

$$
(I - a_2 \mathbf{Y})^{-1} \mathbf{f}^{\mathrm{T}} = a_0 a_2 y_1 \mathbf{e}_1^{\mathrm{T}}. \tag{8.5.12}
$$

By considering

$$
\Big(I + \sum_{i \geq 1} (a_2 \mathbf{Y})^i\Big)(I - a_2 \mathbf{Y}) = I = (I - a_2 \mathbf{Y})\Big(I + \sum_{i \geq 1} (a_2 \mathbf{Y})^i\Big),
$$

it is seen that

$$
I + \sum_{i \geq 1} (a_2 \mathbf{Y})^i = (I - a_2 \mathbf{Y})^{-1}.
$$

Hence,

$$
\mathbf{f}_{\mathbf{a}}^{\mathrm{T}} = a_0 a_2 y_1 \Big(I + \sum_{i \geq 1} (a_2 \mathbf{Y})^i\Big) \mathbf{e}_1^{\mathrm{T}} \tag{8.5.13}
$$

is a solution of equation (8.5.5). Because of the uniqueness of the inverse of the coefficient matrix, the solution is the only one. □

8.6 Solution general

On the basis of (8.5.7)–(8.5.10), by the equivalent transformations on $\mathcal{R}\{x, \mathbf{y}\}$,

$$\begin{aligned} x^1: \quad F_{\mathbf{a}[1]}(\mathbf{y}) &= \sum_{i\geq 0, i\neq 1} \frac{a_2 y_{i+1}}{1-a_2y_2} F_{\mathbf{a}[i]}(\mathbf{y}) \\ &= \frac{a_2}{1-a_2y_2} a_0(y_1 + y_3 F_{\mathbf{a}[2]}(\mathbf{y}) \\ &\quad + y_4 F_{\mathbf{a}[3]}(\mathbf{y}) + \cdots); \end{aligned} \tag{8.6.1}$$

$$\begin{aligned} x^2: \quad F_{\mathbf{a}[2]}(\mathbf{y}) &= \sum_{i\geq 1, i\neq 2} \frac{a_2 y_i}{1-a_2y_2} F_{\mathbf{a}[i]}(\mathbf{y}) \\ &= \frac{a_2}{1-a_2y_2} (y_1 F_{\mathbf{a}[1]}(\mathbf{y}) + y_3 F_{\mathbf{a}[3]}(\mathbf{y}) \\ &\quad + y_4 F_{\mathbf{a}[4]}(\mathbf{y}) + \cdots); \end{aligned} \tag{8.6.2}$$

$$\begin{aligned} x^3: \quad F_{\mathbf{a}[3]}(\mathbf{y}) &= \sum_{i\geq 2, i\neq 3} \frac{a_2 y_{i-1}}{1-a_2y_2} F_{\mathbf{a}[i]}(\mathbf{y}) \\ &= \frac{a_2}{1-a_2y_2} (y_1 F_{\mathbf{a}[2]}(\mathbf{y}) + y_3 F_{\mathbf{a}[4]}(\mathbf{y}) \\ &\quad + y_4 F_{\mathbf{a}[5]}(\mathbf{y}) + \cdots); \end{aligned} \tag{8.6.3}$$

$$\begin{aligned} x^4: \quad F_{\mathbf{a}[4]}(\mathbf{y}) &= \sum_{i\geq 3, i\neq 4} \frac{a_2 y_{i-2}}{1-a_2y_2} F_{\mathbf{a}[i]}(\mathbf{y}) \\ &= \frac{a_2}{1-a_2y_2} (y_1 F_{\mathbf{a}[3]}(\mathbf{y}) + y_3 F_{\mathbf{a}[5]}(\mathbf{y}) \\ &\quad + y_4 F_{\mathbf{a}[6]}(\mathbf{y}) + \cdots); \end{aligned} \tag{8.6.4}$$

and for any integer $m \geq 5$,

$$\begin{aligned} x^m: \quad F_{\mathbf{a}[m]}(\mathbf{y}) &= \frac{a_2 y_1}{1-a_2y_2} F_{\mathbf{a}[m-1]}(\mathbf{y}) \\ &\quad + \sum_{i\geq m+1} \frac{a_2 y_{i-m+2}}{1-a_2y_2} F_{\mathbf{a}[i]}(\mathbf{y}). \end{aligned} \tag{8.6.5}$$

In what follows, relevant results are explained for determining all $F_{\mathbf{a}[m,n]}$ when any integers $m \geq 1$ and $n \geq 1$ are given. Let $\min(F_m(\mathbf{y}))$ ($\max(F_m(\mathbf{y}))$) be the minimum (maximum) degree of powers of $\mathbf{y}$ among all nonzero terms in $F_m(\mathbf{y})$.

Observation 8.6.1. *For any integer $m \geq 3$, $F_{\mathbf{a}[m]}(\mathbf{y})$ is independent of $F_{\mathbf{a}[i]}(\mathbf{y})$, $3 \leq i \leq m-2$.*

Proof. From (8.6.3), it is seen that $F_{\mathbf{a}[3]}(\mathbf{y})$ is independent of $F_{\mathbf{a}[3-2]}(\mathbf{y}) = F_{\mathbf{a}[1]}(\mathbf{y})$. From (8.6.4), it is seen that $F_{\mathbf{a}[4]}(\mathbf{y})$ is independent of $F_{\mathbf{a}[1]}(\mathbf{y})$ and $F_{\mathbf{a}[2]}(\mathbf{y})$. For $m \geq 5$, from (8.6.5), it is seen that $F_{\mathbf{a}[m]}(\mathbf{y})$ depends only on $F_{\mathbf{a}[m-1]}(\mathbf{y})$ and $F_{\mathbf{a}[i]}(\mathbf{y})$, $i \geq m+1$. This is the conclusion. □

According to (8.6.5), this observation enables us to see that all entries of $\mathbf{Y} = (y_{i,j})_{i,j\geq 1}$ in (8.5.12), $y_{i,j} = 0$ for any integer $i \geq 3$ and $1 \leq j \leq i-2$.

Observation 8.6.2. *For any integer $m \geq 2$, $\min(F_{\mathbf{a}[m]}) = \min(F_{\mathbf{a}[m-1]}) + 1$.*

Proof. On the basis of (8.5.13), it is seen that all nonzero entries are of the same degree $i \geq 2$ in Y^i and that $\min(a_0 a_1 y_1 y_{i,1}^{[i-1]}) = \min(a_0 a_1 y_1 y_1^{i-1}) = \min(y_1^i) = i$. Because of

$$\begin{cases} F_{\mathbf{a}[m]} = \sum\limits_{j\geq i-1} a_0 a_2^{j+1} y_1 y_{i,1}^{[j]}, & \text{when } i \geq 2; \\ F_{\mathbf{a}[1]} = a_0 a_2 y_1 \Big(1 + \sum\limits_{j\geq 1} a_2^j y_{1,1}^{[j]}\Big), \end{cases} \tag{8.6.6}$$

we have $\min(F_{[m]}) = \min(y_1 y_{\mathbf{a}[m,1]}^{i-1})$. On account of $\min(F_{\mathbf{a}[1]}) = \min(y_1) = 1$, $\min(F_{\mathbf{a}[m]}) = \min(F_{\mathbf{a}[m-1]}) + 1$ is done. □

In fact, this observation tells us that, for any integer $m \geq 1$, $\min(F_{[m]}) = m$ whenever $F_{\mathbf{a}[1]} = 1$. Certainly, it is true because of $\min(F_{\mathbf{a}[1]}) = \min(y_1) = 1$.

If another parameter $n = \mathrm{id}(\mathbf{y}) = |\mathbf{n}|$ where $\mathbf{n}$ is the power vector of $\mathbf{y}$ is considered, then, for any integer $m \geq 1$, we have

$$F_{\mathbf{a}[m]} = \sum_{n\geq 1} F_{\mathbf{a}[m,n]}. \tag{8.6.7}$$

Observation 8.6.3. *For any two integers m and $n(\geq m \geq 1)$, $F_{\mathbf{a}[m,n]}$ is independent of y_i for $i \geq n-m+2$.*

Proof. From (8.6.5),

$$F_{\mathbf{a}[m]}(\mathbf{y}) = a_2 \sum_{i\geq m-1} y_{i-m+2} F_{\mathbf{a}[i]}(\mathbf{y}).$$

For any integer $n \geq m-1$,

$$\begin{aligned} F_{\mathbf{a}[m,n]} &= a_2 \sum_{i\geq m-1} y_{i-m+2} F_{\mathbf{a}[i,n-1]}, \\ &\quad \text{by Observation 8.6.1,} \\ &= a_2 \sum_{i=m-1}^{n-1} y_{i-m+2} F_{\mathbf{a}[i,n-1]}. \end{aligned} \tag{8.6.8}$$

Therefore, $F_{\mathbf{a}[m,n]}$ is independent of y_i, $i \geq (n-1) - m + 2 + 1 = n - m + 2$. This is just the conclusion. □

This observation reveals the existence of a finite $\max(F_{\mathbf{a}[m]})$ for $m \geq 1$.

Lemma 8.6.4. *Given integer $m \geq 1$. For any integer $n \geq 1$, if $n \leq m-1$, then $F_{\mathbf{a}[m,n]} = 0$.*

Proof. Since $\min(F_{\mathbf{a}[m]}(\mathbf{y})) = m$ from Observation 8.6.2, the conclusion is soon done. □

This lemma provides the minimum m for $F_{\mathbf{a}[m,n]} \neq 0$.

Lemma 8.6.5. *For any integers $m, n \geq 1$, $F_{\mathbf{a}[m,n]} = a_0 a_2^m y_1^m$ if, and only if, $n = m$.*

Proof. When $m = n = 1$, from (8.6.8), $F_{\mathbf{a}[m,n]} = a_2 y_1 F_{\mathbf{a}[0,0]} = a_0 a_2 y_1$. When $m = n \geq 2$, assume that $F_{\mathbf{a}[j,j]} = a_0 a_2^j y_1^j$ for $1 \leq j \leq m-1$. We proceed by induction on $m = n$, from (8.6.8), we have

$$\begin{aligned} F_{\mathbf{a}[m,m]} &= a_2 y_1 F_{\mathbf{a}[m-1,m-1]}, \quad \text{by the assumption,} \\ &= a_2 y_1 (a_0 a_2^{m-1} y_1^{m-1}) \\ &= a_0 a_2^m y_1^m. \end{aligned}$$

This is the conclusion. □

On the basis of what was mentioned above, we are allowed to evaluate $F_{\mathbf{a}[m,n]}$ for $n \geq 1$. On account of Lemma 8.6.4, $F_{\mathbf{a}[m,n]}$ are determined in the order of n from smaller to greater as shown in what follows for $n \leq 4$

$n = 1$: $F_{\mathbf{a}[1,1]}$;
$n = 2$: $F_{\mathbf{a}[1,2]}, F_{\mathbf{a}[2,2]}$;
$n = 3$: $F_{\mathbf{a}[1,3]}, F_{\mathbf{a}[2,3]}, F_{\mathbf{a}[3,3]}$;
$n = 4$: $F_{\mathbf{a}[1,4]}, F_{\mathbf{a}[2,4]}, F_{\mathbf{a}[3,4]}, F_{\mathbf{a}[4,4]}$; etc.

By employing (8.6.8), we start from the initial condition: $F_{\mathbf{a}[0,0]} = a_0$.

$$\begin{aligned}
n = 1:\quad & F_{\mathbf{a}[1,1]} = a_2 y_1 F_{\mathbf{a}[0,0]} = a_0 a_2 y_1; \\
n = 2:\quad & F_{\mathbf{a}[1,2]} = a_2 (y_2 F_{\mathbf{a}[1,1]}) = a_0 a_2^2 y_1 y_2, \\
& F_{\mathbf{a}[2,2]} = a_2 y_1 F_{\mathbf{a}[1,1]} = a_2 y_1 (a_0 a_2 y_1) = a_0 a_2^2 y_1^2; \\
n = 3:\quad & F_{\mathbf{a}[1,3]} = a_2 (y_2 F_{\mathbf{a}[1,2]} + y_3 F_{\mathbf{a}[2,2]}) = a_0 a_2^3 (y_1 y_2^2 + y_1^2 y_3), \\
& F_{\mathbf{a}[2,3]} = a_2 (y_1 F_{\mathbf{a}[1,2]} + y_2 F_{\mathbf{a}[2,2]}) = a_0 a_2^3 y_1^2 y_2 + y_1^3, \\
& F_{\mathbf{a}[3,3]} = a_2 y_1 a_0 a_2^2 y_1^2 = a_0 a_2^3 y_1^3; \\
n = 4:\quad & F_{\mathbf{a}[1,4]} = a_2 (y_2 F_{\mathbf{a}[1,3]} + y_3 F_{\mathbf{a}[2,3]} + y_4 F_{\mathbf{a}[3,3]}) = a_0 a_2^4 (y_1 y_2^3 + 3 y_1^2 y_2 y_3 + y_1^3 y_4), \\
& F_{\mathbf{a}[2,4]} = a_2 (y_1 F_{\mathbf{a}[1,3]} + y_2 F_{\mathbf{a}[2,3]} + y_3 F_{\mathbf{a}[3,3]}) = a_0 a_2^4 (3 y_1^2 y_2^2 + 2 y_1^3 y_3), \\
& F_{\mathbf{a}[3,4]} = a_2 (y_1 F_{\mathbf{a}[2,3]} + y_2 F_{\mathbf{a}[3,3]}) = a_0 a_2^4 (2 y_1^3 y_2 + y_1^3 y_2) = 3 a_0 a_2^4 y_1^3 y_2, \\
& F_{\mathbf{a}[4,4]} = a_2 (y_1 F_{\mathbf{a}[3,3]}) = a_0 a_2^4 y_1^4; \text{ etc.}
\end{aligned}$$

Then we are allowed to provide the solution in the form of a finite sum of positive terms.

Theorem 8.6.6. *The solution of equation* (8.5.1) *is determined by*

$$F_{\mathbf{a}[m,n]} = \begin{cases} a_0, & \textit{when } m = 0, n = 0; \\ a_0 a_2 y_1, & \textit{when } m = 1, n = 1; \\ a_2 \sum_{i=m-1}^{n-1} y_{i-m+2} F_{\mathbf{a}[i,n-1]}, & \textit{when } 1 \leq m \leq n, n \geq 2; \\ 0, & \textit{otherwise}. \end{cases} \tag{8.6.9}$$

Proof. When $n = 0$ and hence $m = 0$, the result is from the initial condition of equation (8.5.1). For $1 \leq n \leq 4$, $F_{\mathbf{a}[m,h]}$ are found in the above calculation. For $n \geq 5$, the results are from (8.6.8). □

8.7 Restrictions general

Now, we are go back equation (8.5.1) in the cases of $a_0 = a_1 = a_2 = 1$. This is the equation

$$\begin{cases} x \int_y y\delta_{x,y}(uf|_{x=u}) = f - 1; \\ f|_{x=0,\mathbf{y}=\mathbf{0}} = 1. \end{cases} \tag{8.7.1}$$

This is the enumerating equation for determining the number of asymmetric plane trees via the partition vector of non-rooted vertices as parameter. It looks that the equation for the first time occurs in Liu [32] (1989) with an equivalent form. However, on the enumeration of asymmetric plane trees is from Tutte [85] (1964).

For convenience, for a function $f \in \mathcal{R}\{x, \mathbf{y}\}$, let $[f]_m = \partial_x^m f$ and $\langle f \rangle_n = \partial_{\mathbf{y}}^n f$ where $n = |\mathbf{n}|$ and $\mathbf{n}$ is a power vector of $\mathbf{y}$. Then f is determined by $F_{m,n} = \langle [f]_m \rangle_n$ for all $m \geq 0$ and $\mathbf{n} \geq \mathbf{0}$.

Theorem 8.7.1. *The solution of equation* (8.7.1) *is determined by*

$$F_{m,n} = \begin{cases} 1, & \text{when } m = 0, n = 0; \\ y_1, & \text{when } m = 1, n = 1; \\ \sum_{i=m-1}^{n-1} y_{i-m+2} F_{i,n-1}, & \text{when } 1 \leq m \leq n, n \geq 2; \\ 0, & \text{otherwise}. \end{cases} \tag{8.7.2}$$

Proof. This is a special case of Theorem 8.6.6 with $a_0 = a_1 = a_2 = 1$. □

One might like to see what happens in an enumeration of general plane trees for m, n smaller. The reader is referred to the examples in Liu [59] (2015, pp. 211–227).

For any y^n, let $\pi(\mathbf{y}^{\mathbf{n}}) = \pi$, write

$$\pi(\mathbf{n}) = \sum_{i=1}^{n} i n_i. \tag{8.7.3}$$

Lemma 8.7.2. *For any integer $m, n \geq 1$, $F_{m,n}$ has $|\mathcal{N}_{m,n}|$ terms, where*

$$\mathcal{N}_{m,n} = \{\mathbf{n} \geq \mathbf{0} \mid |\mathbf{n}| = n, \pi(\mathbf{n}) = 2n - m\}. \tag{8.7.4}$$

Proof. First, to check that, for $1 \leq n \leq 4$, in $F_{m,n}$ obtained in Section 8.6, each $\mathbf{n}$ satisfy $\pi(\mathbf{n}) = 2n - m$. These enable us to write $\pi(F_{m,n}) = 2n - m$.

Then, we proceed by induction on n. Assume for any integer $1 \le j \le n-1$, $\pi(F_{i,j}) = 2j - i$ are known. On the basis of (8.6.8), since

$$\begin{aligned}\pi(y_1 y_2^l F_{m-1,n-1-l}) &= 1 + 2l + 2(n-1-l) - (m-1)\\ &= 1 + 2(n-1) - (m-1)\\ &= 2n - m\end{aligned}$$

and

$$\begin{aligned}\pi(y_{i-m+2} y_2^l F_{i,n-1-l}) &= (i - m + 2) + 2l + 2(n-1-l) - i\\ &= (-m+2) + 2(n-1)\\ &= 2n - m,\end{aligned}$$

we have $\pi(F_{m,n}) = 2n - m$.

On the other hand, if there exists $\mathbf{n} \notin \mathcal{N}_{m,n}$, $|\mathbf{n}| = n$ as the power vector in $F_{m,n}$, then $\pi(F_{m,n}) = \pi(\mathbf{n}) \neq 2n - m$. A contradiction to $\pi(F_{m,n}) = 2n - m$ occurs.

Therefore, the conclusion is true. □

This lemma tells us that, for any integers $m, n \ge 1$, $F_{m,n}$ is a homogeneous polynomial of degree n with $|\mathcal{N}_{m,n}|$ terms. Let

$$\sigma_n(x) = \Big(\sum_{i\ge 1} y_i x^i\Big)^n \tag{8.7.5}$$

and let

$$\begin{aligned}\mathcal{L}_{2n-m} = \{\mathbf{n} \mid &\text{ a power vector of } \mathbf{y} \text{ in } \partial_x^{2n-m}\sigma_n(x)\\ &\text{for } \mathbf{n} = (n_1, n_2, n_3, \ldots) \text{ and } n_j = 0,\\ &j \ge i+1\}.\end{aligned} \tag{8.7.6}$$

Lemma 8.7.3. *For any integers $m, n \ge 1$, we have*

$$\mathcal{N}_{m,n} = \mathcal{L}_{2n-m} \tag{8.7.7}$$

where $n = |\mathbf{n}|$.

Proof. First, for any $\mathbf{i} \in \mathcal{L}_{2n-m}$, because of

$$\sum_{j\ge 1} j i_j = 2n - m$$

and $|\mathbf{i}| = n$, we see that $\mathbf{i} \in \mathcal{N}_{m,n}$.

Then, for any $\mathbf{n} \in \mathcal{N}_{m,n}$, because of

$$\sum_{i\ge 1} i n_i = 2n - m,$$

induced from $|\mathbf{n}| = n$, we also see that $\mathbf{n} \in \mathcal{L}_{2n-m}$. □

In fact, Lemma 8.7.3 implies that, for any integers $n \geq m \geq 1$, $F_{m,n} = \partial_x^{2n-m}\sigma_n(x)$.

Lemma 8.7.4. *For any integers $m, n \geq 1$, we have*

$$\begin{aligned} F_{m,n} &= \frac{1}{n}\partial_x^{2n-m}\sigma_n(x) \\ &= \sum_{\mathbf{i}\in\mathcal{L}_{2n-m}} \frac{(n-1)!}{(n_{m+1}+1)\mathbf{i}!}\mathbf{y}^{\mathbf{i}} \end{aligned} \tag{8.7.8}$$

where $n = |\mathbf{i}|$.

Proof. From Lemma 8.7.3, by considering the form of $\sigma_n(x)$, the conclusion is drawn. □

From this lemma, we are allowed to determine coefficients of each term in the solution of equation (8.7.1) in the form of a summation.

Theorem 8.7.5. *Let f_{gtree} be the solution of equation* (8.7.1), *then, for any integer $m \geq 0$,*

$$\partial_x^m f_{\text{gtree}} = \begin{cases} 1, & \text{when } m = 0; \\ \sum_{\substack{\mathbf{n}\in\mathcal{N}_{m,|\mathbf{n}|} \\ |\mathbf{n}|\geq m}} \frac{(|\mathbf{n}|-1)!}{(n_{m+1}+1)\mathbf{n}!}\mathbf{y}^{\mathbf{n}}, & \text{when } m \geq 1, \end{cases} \tag{8.7.9}$$

where $\mathcal{N}_{m,|\mathbf{n}|}$ is shown in Lemma 8.7.2.

Proof. From Lemmas 8.7.2–8.7.4, the conclusion is drawn. □

Example 1. On vertex partition of plane trees. Let m and $\underline{n} = (n_1, n_2, n_3, \ldots)$ be, respectively, the valency m of root-vertex and the vertex partition vector $\mathbf{n} = (n_1, n_2, n_3, \ldots)$, *i. e.*, n_i, $i \geq 1$, is the number of non-rooted vertices of valency i. Denote by $P_{m,n}$ the polynomial of $\mathbf{y}$ whose coefficient of $\mathbf{y}^{\mathbf{n}}$ is the number of root-isomorphic classes of plane trees with root-vertex valency m and vertex partition vector $\mathbf{n}$ where $n = |\underline{n}|$, the number of non-rooted vertices, $n+1$ is the order. Because of the triviality for $0 \leq m+n < 2$, it is only necessary to discuss the case $2 \leq m+n \leq 5$.

When $m+n = 2$, only $P_{1,1} = y_1$. It is shown by a in Figure 8.3.1.

When $m+n = 3$, we have $P_{2,1} = 0$ and $P_{1,2} = y_1y_2$. The latter is shown by b in Figure 8.3.1.

When $m+n = 4$, we have $P_{3,1} = 0$, $P_{2,2} = y_1^2$ and $P_{1,3} = y_1^2y_2 + y_1y_2^2$. $P_{1,3}$ is shown by c and d in Figure 8.3.1. $P_{2,2}$ is shown by a in Figure 8.7.1.

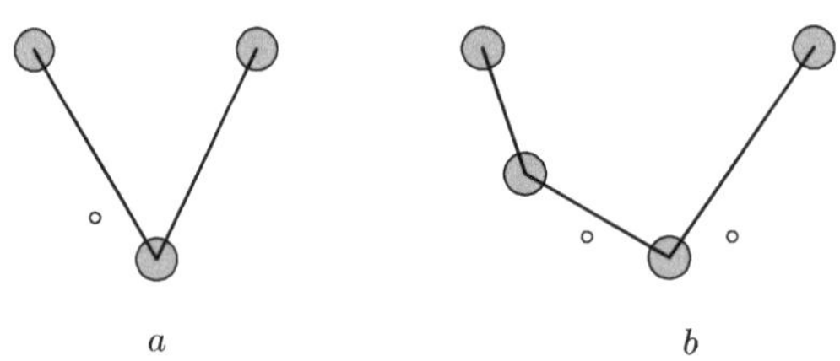

Figure 8.7.1: Classes of plane trees with vertex partition.

When $m+n=5$, we have $P_{4,1}=0$, $P_{3,2}=0$, $P_{2,3}=2y_1^2y_2$ and $P_{1,4}=y_1y_2^3+3y_1^2y_2y_3+y_4y_1^3$. $P_{1,4}$ is shown by a, b and c in Figure 8.3.2. $P_{2,3}$ is shown by b in Figure 8.7.1.

All the above show $P_{m,n}=F_{m,n}$ in (8.7.2), or (8.7.8).

8.8 Explicit expressions general

In this section, we address the explicit expression of the solution of equation (8.5.5), as well as equation (8.5.1), on the basis of the restriction for $a_0 = a_1 = a_2 = 1$ in Section 8.7.

Observation 8.8.1. *If A is the coefficient of a term with $\mathbf{y}^{\mathbf{n}}$ in the solution of equation* (8.7.1), *then $a_0a_2^{|\mathbf{n}|}A$ is the coefficient of a term with $\mathbf{y}^{\mathbf{n}}$ in the solution of equation* (8.5.1).

Proof. On the basis of 8.6.9, we proceed by induction on n. When $n=0$ and $n=1$ (hence $m=0$ and $m=1$), the conclusion is found. For general $n\geq 2$, assume when $k\leq n-1$ (hence $1\leq m\leq k-1$), the conclusion is found as we prove for $k=n$ (hence $1\leq m\leq n$). By employing (8.6.9), because of $F_{\mathbf{a}[i,n-1]}=a_0a_2^{n-1}F_{i,n-1}$ for $1\leq i\leq n$, we have

$$\begin{aligned} F_{\mathbf{a}[m,n]} &= a_2\sum_{i=m-1}^{n-1} y_{i-m+2}F_{\mathbf{a}[i,n-1]} \\ &= a_2\left(a_0a_2^{n-1}\sum_{i=m-1}^{n-1} y_{i-m+1}F_{i,n-1}\right) \\ &= a_0a_2^{n}\sum_{i=m-1}^{n-1} y_{i-m+1}F_{i,n-1}. \end{aligned}$$

This is the conclusion. □

This observation enables us to evaluate an explicit expression of the solution of equation (8.5.1) from its restriction on $a_1=a_1=a_2=1$.

Theorem 8.8.2. *Let f_{gtree} be the solution of equation* (8.5.1), *then, for any integer $m\geq 0$,*

$$\partial_x^m f_{\text{gtree}} = \begin{cases} a_0, & \text{when } m=0; \\ a_0a_2^{|\mathbf{n}|}\sum_{\substack{\mathbf{n}\in\mathcal{N}_{m,|\mathbf{n}|}\\ |\mathbf{n}|\geq m}} \frac{(|\mathbf{n}|-1)!}{(n_{m+1}+1)\mathbf{n}!}\mathbf{y}^{\mathbf{n}}, & \text{when } m\geq 1, \end{cases} \tag{8.8.1}$$

where $\mathcal{N}_{m,|\mathbf{n}|}$ is shown in Lemma 8.7.2.

Proof. On account of Theorem 8.7.5, by Observation 8.8.1, the conclusion can be found. □

On the basis of (8.5.13), the explicit expression can also be obtained via an infinite matrix analysis.

8.9 Notes

8.9.1. In Liu [61] (2016, Book 22, p. 10736, Program 76), one can find the equation

$$\begin{cases} f = ay_1 + b \int_y \dfrac{y^2 f}{1 - cyf}; \\ f|_{\mathbf{y}=\mathbf{0}} = d, \end{cases} \tag{8.9.1}$$

for seeking the condition of being well-defined. When a, b, c and d are all constants, the method employed in the first three sections of this chapter is still available. The earliest occurrences of the specific case of $a = b = c = 1$ and $d = 0$ are in Liu [18] (1985) and [20] (1986). This specific case is equivalent to equation (8.1.1).

8.9.2. In Liu [61] (2016, Book 22, p. 10737, Program 77), one can find the equation

$$\begin{cases} f = a + b \int_y y\delta_{x,y}(uf|_{x=u}); \\ f|_{x=0,\mathbf{y}=\mathbf{0}} = c. \end{cases} \tag{8.9.2}$$

When $\mathbf{a} = (a_0, a_1, a_2) = (a, b, c)$, this equation is equivalent to equation (8.5.1).

8.9.3. In Liu [61] (2016, Book 22, p. 10737, Program 78), one can find the equation

$$\begin{cases} f = \left(a - bx \int_y yf|_{x=y}\right)^{-1}; \\ f|_{x=0,\mathbf{y}=\mathbf{0}} = c. \end{cases} \tag{8.9.3}$$

The earliest occurrence of the specific case of $a = b = c = 1$ is in Wu and Liu [88] (2000).

8.9.4. In Liu [61] (2016, Book 22, p. 10737, Program 79), one might find the equation

$$\begin{cases} f = a + bxf \int_y yf|_{x=y}; \\ f|_{x=0,\mathbf{y}=\mathbf{0}} = c. \end{cases} \tag{8.9.4}$$

The earliest occurrence of the specific case of $a = b = c = 1$ is in Wu and Liu [88] (2000).

8.9.5. In Liu [61] (2016, Book 22, p. 10738, Program 80), one can find the equation

$$\begin{cases} f - ay_1 - d = bxf \int_y \dfrac{y^2(f-1)}{1 - y - cyf}; \\ f|_{x=0,\mathbf{y}=\mathbf{0}} = c. \end{cases} \tag{8.9.5}$$

Its specific case of $a = b = c = d = 1$ occurs in Liu [59] (2015, p. 219). If $d = 0$ instead of $d = 1$, it is equivalent to equation (8.1.1).

8.9.6. For the enumeration of planted and plane trees with vertex partition vector as parameter, a number of methods have been used since Tutte [83] (1964) by analysis of the functions. Others are referred to in Liu [18] (1985), [20] (1986) and Wu and Liu [88] (2000) by analysis of infinite dimensional matrices, Liu [32] (1989) and [43] (1993) by Lagrangian inversion, Liu [49] and [50] (2008) by elementary combinatorics, Liu [59] (2015) by solving a combinatorial functional equation, etc.

9 Near-tree equations first part

9.1 Unicycle model

Consider the equation

$$\begin{cases} a_2 x \int\limits_y y\partial_{x,y} f|_{x=u} = f - a_1 x^2 f_{\text{Tree}}; \\ f|_{x=0,\mathbf{y}=\mathbf{0}} = a_0 \end{cases} \tag{9.1.1}$$

where $a_0, a_1, a_2 \in \mathcal{R}_+$, $f_{\text{Tree}} \in \mathcal{R}\{x, \mathbf{y}\}$ (*i. e.*, f_{regtree} given by (8.7.9) in Chapter 8) is the solution of equation (8.7.1) in Chapter 8 and

$$\partial_{x,y} f|_{x=u} = \frac{yf(x) - xf(y)}{x - y}$$

called the *slope difference* of the function $f(u)$ between x and y.

This is equation (32) in Introduction.

Because of a solution of equation (9.1.1) for $a_0 = 0$, $a_1 = a_2 = 1$ is meaningful in unicyclic plane maps as shown by Liu [18, 20] (1985), this equation is called a *unicycle model*.

For convenience, let $F_{\mathbf{a}[m]}(\mathbf{y}) = \partial_x^m f$, or in brief, $[f]_m$, $\mathbf{a} = (a_0, a_1, a_2)$, $m \geq 0$, and $\text{id}(\mathbf{y}) = |\mathbf{n}|$, where $\mathbf{n} = (n_1, n_2, n_3, \ldots)$, *i. e.*, the power vector of $\mathbf{y}$, $|\mathbf{n}| = n_1 + n_2 + n_3 + \cdots$. For any integer $n \geq 0$, write

$$F_{\mathbf{a}[m,n]} = F_{\mathbf{a}[m]}(\mathbf{y})|_{\text{id}(\mathbf{y})=n} \left(= \sum_{|\mathbf{n}|=n} F_{\mathbf{a}[m,\mathbf{n}]} \mathbf{y}^{\mathbf{n}} \right),$$

or, as we may say, $F_{\mathbf{a}[m,n]}$ is the sum of all terms of $\mathbf{y}$ with degree n in $F_{\mathbf{a}[m]}(\mathbf{y})$. This is a homogeneous polynomial of degree n.

Because of the initial condition of equation (9.1.1),

$$F_{\mathbf{a}[0]}(\mathbf{y}) = a_0 = 0. \tag{9.1.2}$$

Therefore,

$$f = \sum_{m \geq 1} F_{\mathbf{a}[m]}(\mathbf{y}) x^m. \tag{9.1.3}$$

Since

$$\begin{aligned} \partial_{x,y}(f|_{x=u}) &= \frac{xf - yf|_{u=y}}{x - y}, \quad \text{from (9.1.3)}, \\ &= \frac{x \sum_{m \geq 1} F_{\mathbf{a}[m]}(\mathbf{y}) x^m - y \sum_{m \geq 1} F_{\mathbf{a}[m]}(\mathbf{y}) y^m}{x - y}, \end{aligned}$$

by extracting the common factor in the numerator,

https://doi.org/10.1515/9783110625837-009

$$= xy \sum_{m\geq 1} F_{\mathbf{a}[m]}(\mathbf{y}) \frac{x^{m-1} - y^{m-1}}{x - y}$$

and

$$x^{m-1} - y^{m-1} = (x - y)\left(\sum_{i=0}^{m-2} x^i y^{m-i-2} \right),$$

we have

$$\begin{aligned}
\partial_{x,y}(f|_{x=u}) &= xy \sum_{m\geq 1} F_{\mathbf{a}[m]}(\mathbf{y}) \left(\sum_{i=0}^{m-2} x^i y^{m-i-2} \right), \\
&\quad \text{by exchanging two sums,} \\
&= \sum_{i\geq 0} \sum_{m\geq i+2} F_{\mathbf{a}[m]}(\mathbf{y}) y^{m-i-1} x^{i+1}, \\
&\quad \text{by interchanging two indices } m \text{ and } i, \\
&= \sum_{m\geq 2} \left(\sum_{i\geq m+2} F_{\mathbf{a}[i]}(\mathbf{y}) y^{i-m-1} \right) x^{m+1} \\
&= \sum_{\substack{i\geq m+1 \\ m\geq 3}} F_{\mathbf{a}[i]}(\mathbf{y}) y^{i-m} x^m,
\end{aligned}$$

and hence

$$\begin{aligned}
x \int_y y\delta x, y(uf|_{x=u}) &= \sum_{\substack{i\geq m+1 \\ m\geq 3}} F_{\mathbf{a}[i]}(\mathbf{y}) y_{i-m+1} x^{m+1} \\
&= \sum_{\substack{i\geq m \\ m\geq 4}} F_{\mathbf{a}[i]}(\mathbf{y}) y_{i-m+2} x^m.
\end{aligned}$$

Because of

$$f_{\text{Tree}} = 1 + \sum_{m\geq 1} \partial_x^m f_{\text{Tree}} x^m$$

given in Theorem 8.7.5 in Chapter 8, let $T_m = \partial_x^m f_{\text{regtree}}$, $m \geq 1$, then from equation (9.1.1), we have

$$\begin{aligned}
f &= a_1 x^2 \left(1 + \sum_{m\geq 1} T_m x^m \right) + a_2 \sum_{\substack{i\geq m \\ m\geq 4}} F_{\mathbf{a}[i]}(\mathbf{y}) y_{i-m+2} x^m \\
&= a_1(x^2 + T_1 x^3) + \sum_{m\geq 4} \left(a_1 T_{m-2} + a_2 \sum_{i\geq m} F_{\mathbf{a}[i]}(\mathbf{y}) y_{i-m+2} \right) x^m \qquad (9.1.4) \\
&= \sum_{m\geq 2} a_1 T_{m-2} + a_2 \sum_{\substack{i\geq m \\ m\geq 2}} F_{\mathbf{a}[i]}(\mathbf{y}) y_{i-m+2} x^m.
\end{aligned}$$

On account of the equivalence between equation (9.1.4) and equation (9.1.1) for $a_0 = 0$ on $\mathcal{R}\{x, \mathbf{y}\}$, (9.1.2) enables us only to observe equation (9.1.4) for evaluating a solution of equation (9.1.1).

Lemma 9.1.1. *The system of equations*

$$\begin{cases} F_{\mathbf{a}[0]} = F_{\mathbf{a}[1]} = 0, & \text{when } m = 0 \text{ and } 1; \\ F_{\mathbf{a}[m]} = a_1 T_{m-2} + a_2 \sum_{i \geq m} y_{i-m+2} F_{\mathbf{a}[i]}, & \text{when } m \geq 2, \end{cases} \tag{9.1.5}$$

about $F_{\mathbf{a}[m]} = F_{\mathbf{a}[m]}(\mathbf{y})$, $m \geq 0$, *is equivalent to equation* (9.1.1) *on* $\mathcal{R}\{\mathbf{y}\}$ *where* T_m, $m \geq 0$, *are given by Theorem* 8.7.5.

Proof. From the equivalences between equation (9.1.5) and equation (9.1.4) and between equation (9.1.4) and equation (9.1.1), the conclusion is drawn. □

The lemma enables us to establish the property of being well-defined of equation (9.1.1) via equation (9.1.5).

Theorem 9.1.2. *Equation* (9.1.1) *for* $a_0 = 0$ *is well-defined on* $\mathcal{R}\{x, \mathbf{y}\}$.

Proof. Because of $F_{\mathbf{a}[0]}$ and $F_{\mathbf{a}[1]}$ in equation (9.1.5), it is only necessary to observe the vector $\mathbf{f} = (F_2, F_3, F_4, \ldots) \in \mathcal{R}\{\mathbf{y}\}^{\infty}$.

For $\mathbf{f}$, equation (9.1.5) provides the following equation:

$$(I - a_2 \mathbf{Y}_{\text{unicl}})\mathbf{f}^{\mathrm{T}} = a_1 \mathbf{t}^{\mathrm{T}} \tag{9.1.6}$$

where $\mathbf{t} = (T_0, T_1, T_2, \ldots)$ are from (8.7.9) in Chapter 8 and $\mathbf{Y}_{\text{unicl}} = (y_{i,j})_{i \geq 2, \geq 2}$,

$$y_{i,j} = \begin{cases} y_{j-i+2}, & \text{when } j \geq i; \\ 0, & \text{otherwise, } i.e., j < i. \end{cases} \tag{9.1.7}$$

Because of the existence of the inverse of the coefficient matrix,

$$(I - a_2 \mathbf{Y}_{\text{unicl}})^{-1} = \sum_{l \geq 0} (a_2 \mathbf{Y}_{\text{unicl}})^l, \tag{9.1.8}$$

we see that

$$\mathbf{f}^{\mathrm{T}} = a_1 (I - a_2 \mathbf{Y}_{\text{unicl}})^{-1} \mathbf{t}^{\mathrm{T}} \tag{9.1.9}$$

is the solution of equation (9.1.6) and hence of equation (9.1.1).

Therefore, the conclusion is drawn. □

This theorem shows the basic theoretical aspect in the systematization stage for the equation with constant coefficients of the unicycle model.

9.2 Solution unicyclic

In order to accomplish the other two stages: efficientization and intelligentization, a solution in form as a finite sum of all terms positive is particularly favorite.

Let $F_{\mathbf{a}[m,n]} = F_{\mathbf{a}[m]}|_{n=|\underline{n}|} = \langle F_{\mathbf{a}[m]} \rangle_n$, $m \geq 2$, $n \geq 0$. From (9.1.5),

$$F_{\mathbf{a}[m,n\rangle} = a_1 \langle T_{m-2} \rangle_n + a_2 \sum_{i \geq m} y_{i-m+2} \langle F_{\mathbf{a}[i]} \rangle_{n-1}. \tag{9.2.1}$$

Lemma 9.2.1. *For any integer $n \geq 0$, $F_{\mathbf{a}[m,n\rangle} = 0$, $m \geq n+3$.*

Proof. When $n = 0$, from (9.2.1), we have

$$F_{\mathbf{a}[m,0\rangle} = \begin{cases} \langle T_0 \rangle_0 = a_1, & \text{when } m = 2; \\ 0, & \text{when } m \geq 3. \end{cases}$$

This means that $F_{\mathbf{a}[m,0\rangle} = 0$ when $m \geq n+3 = 0+3 = 3$.

For any integer $n \geq 1$, assume $F_{\mathbf{a}[m,n-1\rangle} = 0$, $m \geq (n-1)+3 = n+2$. By induction, we show that $F_{\mathbf{a}[m,n\rangle} = 0$ for $m \geq n+3$. Because of $T_{m,n} = 0$ for $m \geq n+1$, we have $T_{m-2,n} = 0$ for $m-2 \geq n+1$, *i. e.*, $m \geq n+3$. If $m \geq n+3$, then, for any integer $i \geq m$, we have $i \geq n+3 > (n-1)+3 = n+2$. By the assumption, $F_{\mathbf{a}[i,n-1\rangle} = 0$ for $i \geq m$. From (9.1.6), $F_{\mathbf{a}[m,n\rangle} = 0$ for $m \geq n+3$. □

This lemma enables us to see that, for any integer $n \geq 0$ given, it is enough only to determine a finite number of $F_{\mathbf{a}[m,n\rangle}$ for $2 \leq m \leq n+2$ while (9.2.1) becomes

$$\begin{aligned} F_{\mathbf{a}[m,n\rangle} &= a_1 \langle T_{m-2} \rangle_n + a_2 \sum_{i=m}^{n+1} y_{i-m+2} \langle F_{\mathbf{a}[i]} \rangle_{n-1}, \\ &\qquad \text{by using } j = i - m, \\ &= a_1 \langle T_{m-2} \rangle_n + a_2 \sum_{j=0}^{n-m+1} y_{j+2} \langle F_{\mathbf{a}[j+m]} \rangle_{n-1}. \end{aligned} \tag{9.2.2}$$

Lemma 9.2.2. *Given an integer $m \geq 2$. For any integer $n \geq 0$, $F_{\mathbf{a}[m,n\rangle}$ is independent of y_i, $i \geq n+2$.*

Proof. First, we check that, for $n = 0$ and 1, $F_{\mathbf{a}[m,n\rangle}$ is independent of y_i, $i \geq n+2$.

Then, for general integer $n \geq 2$, by induction, assume $F_{\mathbf{a}[m,l\rangle}$, $l \leq n-1$, is independent of y_i, $i \geq l+2$. We prove that $F_{\mathbf{a}[m,n\rangle}$ is independent of y_i, $i \geq n+2$. Because of $T_{m-2,n} = \langle \partial_x^{m-2} f_{\text{rsgree}} \rangle_n$ and (8.7.9) in Chapter 8, $T_{m-2,n}$ is independent of y_i, $i \geq n+1$. From the assumption, $F_{\mathbf{a}[j+m,n-1\rangle}$, $0 \leq j \leq n-m+1$, is independent of y_i, $i \geq n+1$. Because of the occurrence of y_{n+1} in (9.2.2), $F_{\mathbf{a}[m,n\rangle}$ is independent of y_i, $i \geq n+2$. □

This lemma enables us to consider the infinite vector as a finite one of $n+1$ entries, denoted by $\mathbf{y}_{n+1}$, in determining $F_{\mathbf{a}[m,n\rangle}$ for any integers $m \geq 2$ and $n \geq 0$.

Lemma 9.2.3. *For any two integers $n \geq 1$ and $m = n+2$, $F_{\mathbf{a}[m,n\rangle} = F_{\mathbf{a}[n+2,n\rangle} = a_1 y_1^n$.*

Proof. From (9.2.2), $F_{\mathbf{a}[n+2,n\rangle} = a_1 \langle T_n \rangle_n$. From (8.7.9) in Chapter 8 and $n_{n+3} = 0$, $\langle T_n \rangle_n = \langle \partial_x^n f_{\text{regtree}} \rangle_n = y_1^n$, $n \geq 1$. The conclusion is drawn. □

In what follows, for $n = 1, 2$ and 3, $F_{\mathbf{a}[m,n]}$ $(m \geq 2)$ are evaluated.

We proceed on the basis of Lemma 9.1.1 and Lemmas 9.2.1–9.2.3, let the integer n be given, then it is only necessary to evaluate $F_{\mathbf{a}[m,n\rangle}$ for $2 \leq m \leq n+1$. As a matter of fact, they are all homomorphic polynomials of $\mathbf{y}$ with degree n.

Let $n = 1$ and then $2 \leq m \leq n+2 = 3$. Because $F_{\mathbf{a}[3,1\rangle} = a_1 y_1$ (Lemma 9.2.3), only $F_{\mathbf{a}[2,1\rangle}$ is considered. By (9.2.2),

$$\begin{aligned} F_{\mathbf{a}[2,1\rangle} &= a_1 \langle T_0 \rangle_1 + a_2 y_2 \langle F_{\mathbf{a}[2]} \rangle_0, \quad \text{by Theorem 8.7.5 and (9.2.1),} \\ &= 0 + a_1 a_2 y_2 = a_1 a_2 y_2. \end{aligned} \tag{9.2.3}$$

Let $n = 2$ and hence $2 \leq m \leq n+2 = 4$. Because $F_{\mathbf{a}[4,2\rangle} = a_1 y_1^2$ (Lemma 9.2.3), only $F_{\mathbf{a}[3,2\rangle}$ and $F_{\mathbf{a}[2,2\rangle}$ are considered. From (9.2.2),

$$\begin{aligned} F_{\mathbf{a}[3,2\rangle} &= a_1 \langle T_1 \rangle_2 + a_2 y_2 \langle F_{\mathbf{a}[3]} \rangle_1, \\ &\qquad \text{by (8.7.9) in Chapter 8 and Lemma 9.2.3,} \\ &= a_1 y_1 y_2 + a_1 a_2 y_1 y_2 = a_1(1+a_2) y_1 y_2 \end{aligned} \tag{9.2.4}$$

and

$$\begin{aligned} F_{\mathbf{a}[2,2\rangle} &= a_1 \langle T_0 \rangle_2 + a_2 (y_2 \langle F_{\mathbf{a}[2]} \rangle_1 + y_3 \langle F_{\mathbf{a}[3]} \rangle_1), \\ &\qquad \text{by (8.7.9) in Chapter 8 and Lemma 9.2.3,} \\ &= a_1 a_2 (a_2 y_2^2 + y_1 y_3). \end{aligned} \tag{9.2.5}$$

When $n = 3$ and hence $2 \leq m \leq n+2 = 5$. Because $F_{\mathbf{a}[5,3\rangle} = a_1 y_1^3$ (Lemma 9.2.3), only $F_{\mathbf{a}[4,3\rangle}$, $F_{\mathbf{a}[3,3\rangle}$ and $F_{\mathbf{a}[2,3\rangle}$ are considered. By (9.2.2),

$$\begin{aligned} F_{\mathbf{a}[4,3\rangle} &= a_1 \langle T_2 \rangle_3 + a_2 y_2 \langle F_{\mathbf{a}[4]} \rangle_2, \\ &\qquad \text{by (8.7.9) in Chapter 8 and Lemma 9.2.3,} \\ &= 2a_1 y_1^2 y_2 + a_2 y_2 a_1 y_1^2 = a_1(2+a_2) y_1^2 y_2, \end{aligned} \tag{9.2.6}$$

$$\begin{aligned} F_{\mathbf{a}[3,3\rangle} &= a_1 \langle T_1 \rangle_3 + a_2 (y_2 \langle F_{\mathbf{a}[3]} \rangle_2 + y_3 \langle F_{\mathbf{a}[4]} \rangle_2), \\ &\qquad \text{by (8.7.9) in Chapter 8, (9.2.4) and Lemma 9.2.3,} \\ &= a_1 (y_1 y_2^2 + y_1^2 y_3) + a_2 (y_2 (a_1 (1+a_2) y_1 y_2) + y_3 (a_1 y_1^2)) \\ &= a_1 ((1+a_2) y_1^2 y_3 + (1 + a_2 + a_2^2) y_1 y_2^2) \end{aligned} \tag{9.2.7}$$

and

$$\begin{aligned} F_{\mathbf{a}[2,3\rangle} &= a_1 \langle T_0 \rangle_3 + a_2 (y_2 \langle F_{\mathbf{a}[2]} \rangle_2 + y_{\mathbf{a}[3]} \langle F_3 \rangle_2 + y_4 \langle F_{\mathbf{a}[4]} \rangle_2), \\ &\qquad \text{by (9.2.4), (9.2.5) and Lemma 9.2.3,} \\ &= a_2 (y_2 (a_1 a_2 (a_2 y_2^2 + y_1 y_3)) + y_3 (a_1 (1+a_2) y_1 y_2) + y_4 (a_1 y_1^2)) \\ &= a_1 a_2 (a_2^2 y_2^3 + (1 + 2a_2) y_1 y_2 y_3 + y_1^2 y_4). \end{aligned} \tag{9.2.8}$$

Theorem 9.2.4. *Equation (9.1.1) for $a_0 = 0$ has, and is the only one to have, its solution as determined by*

$$F_{\mathbf{a}[m,n\rangle} = \begin{cases} a_1, & \text{when } m = 2, n = 0; \\ a_1 y_1, & \text{when } m = 3, n = 1; \\ a_1 a_2 y_2, & \text{when } m = 2, n = 1; \\ a_1 T_{m-2,n} + a_2 \sum_{j=0}^{n-m+1} y_{j+2} F_{\mathbf{a}[j+m,n-1\rangle}, & \text{when } 2 \le m \le n+2, n \ge 2; \\ 0, & \text{otherwise,} \end{cases} \tag{9.2.9}$$

for integers $m, n \ge 0$ on $\mathcal{R}\{x, \mathbf{y}\}$.

Proof. From the case of $n = 0$ in the proof of Lemma 9.2.1, (9.2.3) and (9.2.2), the conclusion is drawn. □

9.3 Explicit expressions unicyclic

Let $\mathbf{y}_n = (y_1, y_2, y_3, \ldots, y_n)$, then from Lemma 9.2.1 and Lemma 9.2.2, for any integer $n \ge 1$, $F_{\mathbf{a}[m,n\rangle}$ is a homogeneous polynomial of degree n in $\mathbf{y}_{n+1}$. Write $\mathcal{J}(F_{\mathbf{a}[m,n\rangle})$, or in brief,

$$\mathcal{J}_{m,n} = \{\mathbf{i}_{n+1} \mid F_{\mathbf{a}[m,n\rangle} \text{ has a term of degree } n = |\mathbf{i}_{n+1}|\}. \tag{9.3.1}$$

Lemma 9.3.1. *For any vector $\mathbf{i}_{n+1} \in \mathcal{J}_{m,n}$,*

$$\sum_{j=1}^{n+1} j i_j = 2(n+1) - m.$$

Proof. We proceed by induction on n. From (9.2.3)–(9.2.8), $F_{\mathbf{a}[m,n\rangle}$, $m \ge 2$, are checked to satisfy the conclusion for $n \le 3$.

For $n \ge 4$, assume that, for any integer $l \le n-1$, all $\mathbf{i}_{l+1} \in \mathcal{J}_{m,l}$ satisfy the conclusion. We prove the case of $l = n$. For integral vector $\mathbf{i} = (i_1, i_2, i_3, \ldots)$, let

$$\pi(\mathbf{i}) = \sum_{j \ge 1} j i_j.$$

Because of $\pi(\mathbf{i}) = 2n - m$ for any $\mathbf{i} \in \mathcal{J}(T_{m,n})$, $\pi(T_{m-2,n}) = 2n - (m-2) = 2(n+1) - m$. By the assumption, for any $\mathbf{i} \in \mathcal{J}(y_{j+2} F_{\mathbf{a}[j+m,n-1\rangle})$, $\pi(\mathbf{i}) = (j+2) + \pi(F_{\mathbf{a}[j+m,n-1\rangle}) = ((j+2) + 2n) - (j+m) = 2(n+1) - m$, $0 \le j \le n - m + 1$. Therefore, from (9.2.2), for any $\mathbf{i} \in \mathcal{J}_{m,n}$, $\pi(\mathbf{i}) = 2(n+1) - m$. This is the desired conclusion. □

Because the conclusion is independent of the choice of $\mathbf{i}_{n+1} \in \mathcal{J}_{m,n}$, we are allowed to write

$$\pi(F_{m,n}) = \sum_{j=1}^{n+1} j i_j, \text{ for } \mathbf{i}_{n+1} \in \mathcal{J}_{m,n}, \tag{9.3.2}$$

i. e., $\pi(F_{m,n}) = \pi(\underline{i}_{n+1}) = 2(n+1) - m$. Let

$$\lambda_n(z) = \left(\sum_{j=1}^{n+1} y_j x^j\right)^n \tag{9.3.3}$$

and write

$$\mathcal{I}_{2n-m+2} = \{\underline{i} \mid \text{a power vector of terms in } \partial_z^{2n-m+2}\lambda_n(z)\}. \tag{9.3.4}$$

Lemma 9.3.2. *For two integers $m \geq 2$ and $n \geq 0$,*

$$\mathcal{I}_{2n-m+2} = \{\mathbf{i} \geq \underline{0} \mid |\mathbf{i}| = n,\ \pi(\mathbf{i}) = 2n - m + 2\}. \tag{9.3.5}$$

Proof. For any $\mathbf{i} \in \mathcal{I}_{2n-m+2}$, from (9.3.5), there exists a term with power vector

$$\mathbf{i} \text{ in } \partial_z^{2n-m+2}\lambda_n(z) \quad \text{s. t.} \quad |\mathbf{i}| = n,\ \pi(\underline{i}) = 2n - m + 2.$$

This implies that the set on the left hand side of the equality is a subset of the set on the right hand side.

Conversely, for $\mathbf{i}$ in the set on the right hand side, since $x^m\mathbf{y}^\mathbf{i}$ in $F_{\mathbf{a}[m,n\rangle}$ corresponds to $\mathbf{y}^\mathbf{i}x^{\pi(\mathbf{i})}$ in $\partial_z^{2n-m+2}\lambda_n(z)$, we have $\mathbf{i} \in \mathcal{I}_{2n-m+2}$. This implies that the set on the right hand side of the equality is a subset of the set on the left hand side. □

This lemma enables us to investigate structural relations between $F_{\mathbf{a}[m,n\rangle}$ and $\partial_z^{2n-m+2}\lambda_n(z)$.

Lemma 9.3.3. *Given any integer $n \geq 3$. For $2 \leq m \leq n+2$, $F_{\mathbf{a}[m,n]}$ is independent of y_i, $i \geq n - m + 4$.*

Proof. For $n = 2$, 3 and 4, (9.2.4)–(9.2.8), the conclusions are checked.

For $n \geq 5$, by induction on n, assume that $F_{\mathbf{a}[m,n-1\rangle}$ satisfies the conclusion, *i. e.*, only depends on y_i, $1 \leq i \leq (n-1)-m+4 = n-m+2$. We prove that $F_{\mathbf{a}[m,n\rangle}$ only depends on y_i, $1 \leq i \leq n-m+3$.

Let $\delta(F) = \max\{i \mid y_i \text{ in } F\}$. On the basis of (9.2.9), it suffices to determine

$$\max\left\{\delta(T_{m-2,n}), \delta\left(\sum_{j=0}^{n-m+1} y_{j+2}F_{\mathbf{a}[j+m,n-1\rangle}\right)\right\}.$$

From (8.7.8) in Chapter 8, we have $\delta(T_{m-2,n}) \leq n - m + 3$. Because of

$$\delta(y_{n-m+3}F_{\mathbf{a}[n+1,n-1\rangle}) = \max\{\delta(y_{j+2}F_{\mathbf{a}[j+m,n-1\rangle}) \mid 0 \leq j \leq n-m+1\},$$

we have

$$\begin{aligned}\delta(F_{\mathbf{a}[m,n]}) &= \delta\left(\sum_{j=0}^{n-m+1} y_{j+2}F_{\mathbf{a}[j+m,n-1\rangle}\right)\\ &= \delta(y_{n-m+3}F_{\mathbf{a}[n+1,n-1\rangle}), \quad \text{by the assumption,}\\ &= n-m+3.\end{aligned}$$

Hence, $F_{\mathbf{a}[m,n\rangle}$ is independent of y_i, $i \geq n - m + 4$. □

This lemma enables us to construct the set $\{\mathbf{i} \geq \underline{0} \mid |\mathbf{i}| = n, \pi(\mathbf{i}) = 2n - m + 2\}$ by n recursively.

Lemma 9.3.4. *For any integers $m \geq 2$ and $n \geq 0$, $\mathcal{J}_{m,n} = \mathcal{I}_{2n-m+2}$.*

Proof. Let $\mathcal{A}$ be the set on the right hand side of (9.3.5). From Lemma 9.3.2, it is only necessary to prove $\mathcal{J}_{m,n} = \mathcal{A}$.

First, given a vector $\mathbf{i} \in \mathcal{J}_{m,n}$. We prove $\mathbf{i} \in \mathcal{A}$. From (9.3.1), $\mathbf{i}$ is a power vector of a term in $F_{\mathbf{a}[m,n\rangle}$. Since $F_{\mathbf{a}[m,n\rangle}$ is a homogeneous polynomial of degree n, $|\mathbf{i}| = n$. By Lemma 9.3.1, $\pi(\mathbf{i}) = 2(n+1) - m$. Therefore, $\mathbf{i} \in \mathcal{A}$.

Then, given a vector $\mathbf{i} \in \mathcal{A}$, we prove $\mathbf{i} \in \mathcal{J}_{m,n}$. By following the procedure appearing in the proof of Lemma 9.3.3, $\mathbf{i} \in \mathcal{J}_{m,n}$, is easily obtained. □

The conclusion of this lemma shows that, for any $\mathbf{i} \in \mathcal{J}_{m,n}$, there exists a factor $\alpha_{m,n}(\underline{i})$ so that

$$F_{\mathbf{a}[m,n\rangle} = \sum_{\underline{i} \in \mathcal{I}_{2n-m+2}} \alpha_{m,n}^{a_1,a_2}(\mathbf{i}) \frac{|\mathbf{n}|!}{\mathbf{i}!} \mathbf{y}^{\mathbf{i}}. \tag{9.3.6}$$

On the basis of this lemma, an explicision (or explicit expression) of the solution of equation (9.1.1) can be done.

Theorem 9.3.5. *The solution of equation (9.1.1) for $a_0 = 0$ is determined by the following explicision:*

$$F_{\mathbf{a}[m,n\rangle} = \begin{cases} a_1, & \text{when } m = 2, n = 0; \\ a_1 y_1, & \text{when } m = 3, n = 1; \\ a_1 a_2 y_2, & \text{when } m = 2, n = 1; \\ a_1 T_{m-2,n} + a_2 \sum_{j=0}^{n-m+1} y_{j+2} \Sigma_{\mathbf{a}[j,n-1\rangle}, & \text{when } 2 \leq m \leq n+2, n \geq 2; \\ 0, & \text{otherwise}, \end{cases} \tag{9.3.7}$$

where

$$\Sigma_{\mathbf{a}[j,n-1\rangle} = \sum_{\mathbf{i} \in \mathcal{I}_{2n-j-m}} \alpha_{j+m,n-1}^{a_1,a_2}(\mathbf{i}) \frac{(|\mathbf{n}|-1)!}{\mathbf{i}!} \mathbf{y}^{\mathbf{i}}.$$

Proof. From (9.1.5), (9.2.1) and (9.3.6), the conclusion is drawn. □

In fact, on the basis of (9.2.2), by following the procedure from (9.2.3) to (9.2.8), $\alpha_{m,n}^{a_1,a_2}(\mathbf{i})$ can be recursively determined.

On the other hand, another method for extracting an explicision of the solution of equation (9.1.1) for $a_0 = 0$ can be realized on the basis of (9.1.9).

Because of

$$(I - a_2 \mathbf{Y}_{\text{unicl}})^{-1} = \sum_{i \geq 0} (a_2 \mathbf{Y}_{\text{unicl}})^i,$$

we have

$$\mathbf{f}_{\mathbf{a}}^{\mathrm{T}} = \sum_{k\geq 0} a_1(a_2\mathbf{Y}_{\mathrm{unicl}})^k \mathbf{t}^{\mathrm{T}} \tag{9.3.8}$$

where $(a_2Y_{\mathrm{unicl}})^0 = I$, $\mathbf{f}_{\mathbf{a}} = (F_{\mathbf{a}[2]}, F_{\mathbf{a}[3]}, F_{\mathbf{a}[4]}, \ldots)$ and $\mathbf{t} = (T_0, T_1, T_2, \ldots)$, which are known from (8.7.9) in Chapter 8.

Then the only thing we have to do is to calculate $\mathbf{Y}_{\mathrm{unicl}}^k = (y_{i,j}^{[k]})_{i,j\geq 1}$ from $\mathbf{Y}_{\mathrm{unicl}} = (y_{i,j})_{i,j\geq 1}$ where

$$y_{i,j} = \begin{cases} y_{j-i+2}, & \text{when } j \geq i; \\ 0, & \text{otherwise,} \end{cases} \tag{9.3.9}$$

and

$$y_{i,j}^{[k]} = \begin{cases} y_{j-i+2}^{[k]}, & \text{when } j \geq i; \\ 0, & \text{otherwise.} \end{cases} \tag{9.3.10}$$

In fact, for $k = 1$, $y_{i,j}^{[1]} = y_{i,j}$ and for $k \geq 2$,

$$y_{j-i+2}^{[k]} = \sum_{l=2}^{j-i+2} y_l y_{j-i+2-l}^{[k-1]}. \tag{9.3.11}$$

Since all matrices $\mathbf{Y}_{\mathrm{unicl}}^{[k]}$, $k \geq 1$, are upper-triangular, we have $y_{i,j}^{[k]} = y_{1,j-i+1}^{[k]}$ for any $i, j \geq 1$. It is only necessary to determine the first row $y_{1,j}^{[k]} = y_{j+1}^{[k]}$ of $y_{j-i+2}^{[k]}$ $(j \geq 1)$ for $\mathbf{Y}_{\mathrm{unicl}}^{[k]}$.

Given an integer $n \geq 0$. We have $F_{\mathbf{a}[m,n\rangle}$ $(2 \leq m \leq n+2)$ from Lemma 9.2.1, let $\mathbf{f}_{\mathbf{a}[n+1]} = (F_{\mathbf{a}[2]}, F_{\mathbf{a}[3]}, \ldots, F_{\mathbf{a}[n+2]})$. From Lemma 9.2.2, only $\mathbf{y}_{n+1} = (y_1, y_2, \ldots, y_{n+1})$ is available for $\mathbf{y}$. Then (9.3.8) becomes

$$(I_{n+1} - a_2\mathbf{Y}_{n+1}|_{y_i=0, i\geq n+2})\mathbf{f}_{\mathbf{a}[n+1]}^{\mathrm{T}} = a_1\mathbf{t}_{n+1}^{\mathrm{T}} \tag{9.3.12}$$

where $\mathbf{t}_{n+1} = (T_0, T_1, \ldots, T_n)$ and $\mathbf{Y}_{n+1}|_{y_i=0,i\geq n+2}$ is the first principal sub-matrix of order $n+1$ restricted on $y_i = 0$, $i \geq n+2$.

Theorem 9.3.6. *For any integer $n \geq 0$, the solution of equation* (9.3.12) *has the following explicision:*

$$\mathbf{f}_{\mathbf{a}[n+1]}^{\mathrm{T}}|_{|\mathbf{n}|=n} = \sum_{k=0}^{n} a_1(a_2\mathbf{Y}_{n+1})^k|_{y_i=0,i\geq n+2}\mathbf{t}_{n+1}^{\mathrm{T}}. \tag{9.3.13}$$

Proof. Because of Lemmas 9.1.1 and 9.2.2, the solution of equation (9.1.1) and hence equation (9.1.6) from Theorem 9.1.2 is a polynomial of $\mathbf{y}$ with degree at most n. It is determined by (9.3.13). The conclusion is drawn. □

Example 1. Given $n = 1$ and hence $2 \le m \le 3$. Because of

$$\begin{pmatrix} \langle F_{\mathbf{a}[2]}\rangle_{\le 1} \\ \langle F_{\mathbf{a}[3]}\rangle_{\le 1} \end{pmatrix} = a_1 \left[\begin{pmatrix} 1 & 0 \\ 0 & 1 \end{pmatrix} + a_2 \begin{pmatrix} y_2 & y_3 \\ 0 & y_2 \end{pmatrix} \right] \begin{pmatrix} 1 \\ y_1 \end{pmatrix},$$

by $y_3 = 0$,

$$= a_1 \left[\begin{pmatrix} 1 \\ y_1 \end{pmatrix} + a_2 \left\langle \begin{matrix} y_2 \\ y_1 y_2 \end{matrix} \right\rangle_{\le 1} \right],$$

by deleting term of degree 2,

$$= a_1 \begin{pmatrix} a_1 + a_1 a_2 y_2 \\ a_1 y_1 \end{pmatrix}.$$

They are checked by (9.3.7).

9.4 Restrictions unicyclic

When $a_0 = 0$, $a_1 = a_2 = 1$, equation (9.1.1) becomes

$$\begin{cases} x \int\limits_{y} y \partial_{x,y} f|_{x=u} = f - x^2 f_{\text{regtree}}; \\ f|_{x=0,\mathbf{y}=\mathbf{0}} = 0. \end{cases} \tag{9.4.1}$$

This is the equation seen in Liu [18] (1985), [20] (1986). Its solution is the enumerating function for counting non-isomorphic unicyclic plane asymmetric maps with vertex partition vector as parameter.

Attention. The asymmetry is meant as the root edge on the circuit and on the side in the outface of the unicyclic plane map.

Theorem 9.4.1. *Equation* (9.4.1) *has, and is the only one to have, a solution as determined by*

$$F_{m,n} = \begin{cases} 1, & \text{when } m = 2, n = 0; \\ y_1, & \text{when } m = 3, n = 1; \\ y_2, & \text{when } m = 2, n = 1; \\ T_{m-2,n} + \sum_{j=0}^{n-m+1} y_{j+2} F_{j+m,n-1}, & \text{when } 2 \le m \le n+2, n \ge 2; \\ 0, & \text{otherwise}, \end{cases} \tag{9.4.2}$$

on $\mathcal{R}\{x, \mathbf{y}\}$ *for integers* $m, n \ge 0$.

Proof. Because of equation (9.4.1) as a specific case of equation (9.1.1) with $a_0 = 0$, $a_1 = a_2 = 1$, from Theorem 9.1.2, equation (9.4.1) is well-defined on $\mathcal{R}\{x, \mathbf{y}\}$. Then, by Theorem 9.2.4, the conclusion is drawn. □

This theorem enables us to evaluate all $F_{m,n}$ for $m, n \geq 0$ in the order from $n = 0$ on increasing one by one.

On the basis of Lemmas 9.2.1 and 9.2.2, for any integer $n \geq 0$, it is only necessary to consider $2 \leq m \leq n+2$ and $\mathbf{y}_{n+1}$ instead of $\mathbf{y}$ for determining $F_m (m \geq 0)$. Let $\mathbf{f}_{n+1} = \langle (F_2, F_3, \dots, F_{n+2}) \rangle_{\leq n}$, $\mathbf{t}_{n+1} = \langle (T_0, T_2, \dots, T_n) \rangle_{\leq n}$ and $\mathbf{Y}_{n+1} = \langle (y_{i,j})_{1 \leq i,j \leq n+1} \rangle_{\leq n}$ where

$$y_{i,j} = \begin{cases} y_{j-i+2}, & \text{when } 0 \leq j - i \leq n-1; \\ 0, & \text{otherwise.} \end{cases} \tag{9.4.3}$$

Lemma 9.4.2. *For any integer $n \geq 1$, $\mathbf{t}_{n+1} = \langle (T_0, T_2, \dots, T_n) \rangle_n$ satisfies the following system of equations:*

$$\mathbf{f}_{n+1}^{\mathrm{T}} = \mathbf{Y}_{n+1} \mathbf{f}_{n+1}^{\mathrm{T}} + \mathbf{t}_{n+1}^{\mathrm{T}}. \tag{9.4.4}$$

Proof. The result of (9.1.6) can be restricted to $a_1 = a_2 = 1$ and at most a term with degree n is given. □

Example 1. When $n = 1$, equation (9.4.4) becomes

$$\begin{pmatrix} \langle F_2 \rangle_{\leq 1} \\ \langle F_3 \rangle_{\leq 1} \end{pmatrix} = \begin{pmatrix} y_2 & 0 \\ 0 & y_2 \end{pmatrix} \begin{pmatrix} \langle F_2 \rangle_{\leq 1} \\ \langle F_3 \rangle_{\leq 1} \end{pmatrix} + \begin{pmatrix} \langle T_0 \rangle_{\leq 1} \\ \langle T_1 \rangle_{\leq 1} \end{pmatrix}.$$

Theorem 9.4.3. *An explicision of the solution of equation* (9.4.4) *is determined by*

$$\mathbf{f}_{n+1}^{\mathrm{T}} = \sum_{i=0}^{n} \mathbf{Y}_{n+1}^{i} \mathbf{t}_{n+1}^{\mathrm{T}}. \tag{9.4.5}$$

Proof. Since equation (9.4.4) is equivalent to

$$(I_{n+1} - \mathbf{Y}_{n+1}) \mathbf{f}_{n+1}^{\mathrm{T}} = \mathbf{t}_{n+1}^{\mathrm{T}}$$

in $\mathcal{R}\{x, \mathbf{y}\}$ and the inverse of $(I_{n+1} - \mathbf{Y}_{n+1})$ is

$$(I_{n+1} - \mathbf{Y}_{n+1})^{-1} = \sum_{i=0}^{n} \mathbf{Y}_{n+1}^{i}$$

for the restriction on n, the conclusion is drawn. □

Example 2. When $n = 1$, the solution of equation (9.4.4) is determined by

$$\begin{aligned} \begin{pmatrix} \langle F_2 \rangle_{\leq 1} \\ \langle F_3 \rangle_{\leq 1} \end{pmatrix} &= \left[\begin{pmatrix} 1 & 0 \\ 0 & 1 \end{pmatrix} + \begin{pmatrix} y_2 & 0 \\ 0 & y_2 \end{pmatrix} \right]_{\leq 1} \begin{pmatrix} 1 \\ y_1 \end{pmatrix} \\ &= \begin{pmatrix} 1 + y_2 & 0 \\ 0 & 1 + y_2 \end{pmatrix} \begin{pmatrix} 1 \\ y_1 \end{pmatrix}. \end{aligned}$$

This leads to $\langle F_2 \rangle_{\leq 1} = 1 + y_2$ and $\langle F_3 \rangle_{\leq 1} = y_1$, the case of $a_1 = a_2 = 1$ in Example 1 of Section 9.3.

Another manner for inducing an explicision of the solution of equation (9.4.1) is to employ the method described in Section 9.3.

Theorem 9.4.4. *The solution of equation* (9.4.1) *is determined by the following explicision:*

$$F_{m,n} = \begin{cases} 1, & \text{when } m = 2, n = 0; \\ y_1, & \text{when } m = 3, n = 1; \\ y_2, & \text{when } m = 2, n = 1; \\ T_{m-2,n} + \sum_{j=0}^{n-m+1} y_{j+2} \Sigma_{j,n-1}, & \text{when } 2 \le m \le n+2, n \ge 2; \\ 0, & \text{otherwise,} \end{cases} \tag{9.4.6}$$

where

$$\Sigma_{j,n-1} = \sum_{\mathbf{i} \in \mathcal{I}_{2n-j-m}} \alpha_{j+m,n-1}^{1,1}(\mathbf{i}) \frac{(|\mathbf{n}| - 1)!}{\mathbf{i}!} \mathbf{y}^{\mathbf{i}}.$$

Proof. We have the specific case of $a_1 = a_2 = 1$ in (9.3.7). □

Example 3. Root-isomorphic classes of unicyclic maps with face partition vector given. A map is called a *unicycle model* if its underlying graph has, and is the only one to have, a circuit. Because of rootedness, the root is restricted to the circuit. By considering

$$(m; \mathbf{i}) = (\text{valency of root-vertex; partition vector of non-root-vertices})$$

is given, we find all the root-isomorphic classes. In Figure 9.4.1, one might see that

$a = y_2$: $F_{2,1} = y_2$;
$b = y_1y_2$ and $c = y_1y_2$: $F_{3,2} = 2y_1y_2$;
$d = y_2^2$ and $e = y_1y_3$: $F_{2,2} = y_2 + y_1y_3$;
$f = 2y_1^2y_2$ and $g = y_1^2y_2$: $F_{4,3} = 3y_1^2y_2$;
$h = y_1y_2^2$, $i = y_1y_2^2$, $j = y_1y_2^2$, $k = y_1^2y_3$ and $l = y_1^2y_3$: $F_{3,3} = 3y_1y_2^2 + 2y_1^2y_3$;
$m = 2y_1y_2y_3$, $n = y_1y_2y_3$, $o = y_2^3$ and $p = y_1^2y_4$: $F_{2,3} = 3y_1y_2y_3 + y_2^3 + y_1^2y_4$.

9.5 Notes

9.5.1. In (9.3.6), the coefficient $\alpha_{m,n}^{a_1,a_2}(\mathbf{i})$ can be determined by the procedure shown in (9.3.7), or on the basis of (9.3.13). We proceed for the case of $\alpha_1 = \alpha_2 = 1$, by using (9.4.6), or on the basis of (9.4.5). It might be helpful to do the specific case before doing the general case.

9.5.2. We introduce new functionals from the basis $\{x, x^2, x^3, \ldots\}$ to itself to extract a simplest explicision. Let Λ be all the functionals $\lambda : \partial_{x,\mathbf{y}}^{m,\mathbf{n}} \lambda f = F_{m,\mathbf{n}}$ for $f \in \mathcal{R}\{x, \mathbf{y}\}$. Obviously, the identity functional is in Λ and the Lagrangian inversion as a functional is in Λ as well.

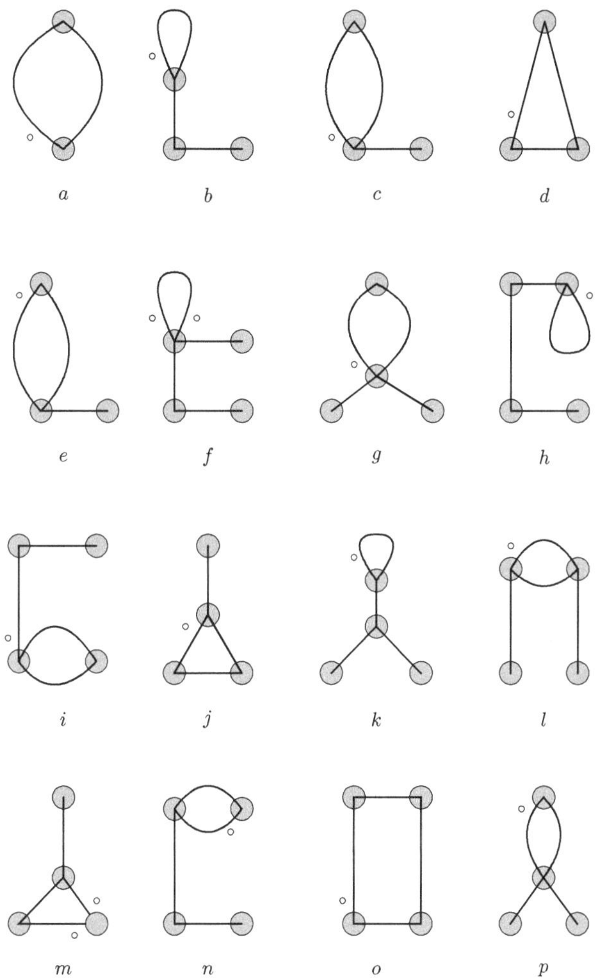

Figure 9.4.1: Classes of unicyclic maps of order not greater than 4.

9.5.3. By exploiting the constructions of Λ, we observe a way to identify if is there a functional λ such that $\partial_{x,\mathbf{y}}^{m,\mathbf{n}}\Lambda f$ is a finite sum of all terms positive or summation free for a function $f \in \mathcal{R}\{x, \mathbf{y}\}$ of certain condition when positive integers m and n are given.

9.5.4. It looks a type of problems on a null space of infinite dimensional vector where some new operations necessarily should be introduced. It looks necessary to classify the functionals shown in Section 9.5.3 in a suitable way.

9.5.5. Equation (9.1.1) can be more generalized to the case that some of a_0, a_1 and a_2 are polynomials of a variable, x, or y_i ($i \geq 1$ given). First, observe the condition, or conditions, for establishing a qualitative theory for the consistency. Then, observe how to evaluate a solution by a series of transformations, or operations, on $\mathcal{R}\{x, \mathbf{y}\}$. This might be the central part of Program 81 posed in Liu YP [61] (Book 22, 2016, p. 10738).

9.5.6. As a matter of fact, this chapter finishes, in principle, the first level of the Program 81 emphasized in the qualitative theory in Liu YP [61] (Book 22, 2016, p. 10738) for all of $a_0 = c$, $a_1 = a$ and $a_2 = b$ constant in $\mathbb{Z}_+$ and even $\mathbb{R}_+$. However, if some of a_0, a_1 and a_2 are allowed to be negative, it is definitely not straightforward unless an additional condition, or additional conditions, is or are to be further investigated because the meaning in combinatorics is not yet known.

10 Near-tree equations second part

10.1 Wintersweets model

Consider the equation

$$\begin{cases} a_2 x \int\limits_y y\delta_{x,y}(uf|_{x=u}) = \left(1 - \dfrac{a_3 x y_3}{1-y_2}\right) f - a_1; \\ f|_{x=0 \Leftrightarrow \mathbf{y}=\mathbf{0}} = a_0, \end{cases} \tag{10.1.1}$$

on $\mathcal{R}\{x,\mathbf{y}\}$ where $a_0, a_1, a_2, a_3 \in \mathcal{R}_+$.

This is equation (33) in Introduction.

Because a solution of equation (10.1.1) for $a_0 = a_1 = a_2 = a_3 = 1$ is meaningful in wintersweets as plane maps, as shown in Liu [18] (1985), and [20] (1986), this equation is called the *wintersweets model*.

For convenience, the first line of equation (10.1.1) is represented by the equivalent form as follows in $\mathcal{R}\{x,\mathbf{y}\}$:

$$f = a_1 + \frac{a_3 x y_3}{1-y_2} f + a_2 x \int\limits_y \left(y\delta_{x,y}(uf|_{x=u})\right). \tag{10.1.2}$$

Let $F_{\mathbf{a}[m]} = \partial_x^m f \ (= [f]_m)$, $m \geq 0$, and $\mathrm{id}(\mathbf{y}) = |\mathbf{i}|$, where $\mathbf{a} = (a_0, a_1, a_2, a_3)$, $\mathbf{i} = (i_1, i_2, i_3, \ldots)$, *i. e.*, the power vector of $\mathbf{y}$, $|\mathbf{i}| = i_1 + i_2 + i_3 + \cdots$. For integer $n \geq 0$, write

$$F_{\mathbf{a}[m,n\rangle} = F_{\mathbf{a}[m]}|_{\mathrm{id}(\mathbf{y})=n} \left(= \sum_{|\mathbf{i}|=n} F_{\mathbf{a}[m,\mathbf{i}]} \mathbf{y}^{\mathbf{i}} \right) = \langle F_{\mathbf{a}[m]} \rangle_n.$$

Or we may say: $F_{m,n}$ is the sum of all terms with degree n of $\mathbf{y}$ in F_m, a homogeneous polynomial of degree n for $\mathbf{y}$. Then it is seen that

$$f = \sum_{m \geq 0} [f]_m x^m = \sum_{m \geq 0} F_{\mathbf{a}[m]} x^m. \tag{10.1.3}$$

Because of

$$\begin{aligned} \delta_{x,y}(uf|_{x=u}) &= \frac{xf - yf|_{u=y}}{x-y}, \quad \text{by (10.1.3)}, \\ &= \frac{x \sum_{m \geq 0} F_{\mathbf{a}[m]} x^m - y \sum_{m \geq 0} F_{\mathbf{a}[m]} y^m}{x-y}, \\ &\quad \text{in view of the common factor from the numerator}, \\ &= \sum_{m \geq 0} F_{\mathbf{a}[m]} \frac{x^{m+1} - y^{m+1}}{x-y} \end{aligned}$$

https://doi.org/10.1515/9783110625837-010

and

$$x^{m+1} - y^{m+1} = (x-y)\left(\sum_{i=0}^{m} x^i y^{m-i}\right),$$

we have

$$\begin{aligned}\delta_{x,y}(uf|_{x=u}) &= \sum_{m\geq 0} F_{\mathbf{a}[m]}\left(\sum_{i=0}^{m} x^i y^{m-i}\right), \quad \text{by interchanging two } \Sigma,\\ &= \sum_{\mathbf{i}\geq 0}\sum_{m\geq i} F_{\mathbf{a}[m]} y^{m-i} x^i, \quad \text{by interchanging } m \text{ and } i,\\ &= \sum_{m\geq 0}\left(\sum_{i\geq m} F_{\mathbf{a}[i]} y^{i-m}\right) x^m,\end{aligned}$$

and hence

$$\int_y y\delta_{x,y}(uf|_{x=u}) = \sum_{m\geq 0}\left(\sum_{i\geq m} y_{i-m+1} F_{\mathbf{a}[i]}\right) x^m.$$

From equation (10.1.2),

$$\begin{aligned}f &= a_1 + \frac{a_3 x y_3}{1-y_2} f + a_2 \sum_{m\geq 0}\left(\sum_{i\geq m} y_{i-m+1} F_{\mathbf{a}[i]}\right) x^{m+1},\\ &\quad \text{by substituting } m+1 \text{ for } m,\\ &= a_1 + \frac{a_3 x y_3}{1-y_2} f + a_2 \sum_{m\geq 1}\left(\sum_{i\geq m-1} y_{i-m+2} F_{\mathbf{a}[i]}\right) x^m, \qquad (10.1.4)\\ &\quad \text{by (10.1.3)},\\ &= a_1 + \sum_{m\geq 1}\left(\frac{a_3 y_3}{1-y_2} F_{\mathbf{a}[m-1]} + a_2 \sum_{i\geq 0} y_{i+1} F_{\mathbf{a}[i+m-1]}\right) x^m.\end{aligned}$$

On the basis of (10.1.3) and (10.1.4), we have

$$\begin{aligned}x^0 :\ & F_{\mathbf{a}[0]} = a_1 = a_0, \quad \text{as the initiation of equation (10.1.1)},\\ & \Longrightarrow F_{\mathbf{a}}[m,0] = 0,\ m \geq 1;\ F_{\mathbf{a}[0,n]} = 0,\ n \geq 1,\end{aligned} \qquad (10.1.5)$$

$$\begin{aligned}x^1 :\ F_{\mathbf{a}[1]} &= \left(\frac{a_3 y_3}{1-y_2} + a_2 y_1\right) a_1 + a_2 \sum_{i\geq 1} y_{i+1} F_{\mathbf{a}[i]}\\ &= \left(\frac{a_3 y_3}{1-y_2} + a_2 y_1\right) a_1 + a_2 (y_2 F_{\mathbf{a}[1]} + y_3 F_{\mathbf{a}[2]}\\ &\quad + y_4 F_{\mathbf{a}[3]} + \cdots),\end{aligned} \qquad (10.1.6)$$

$$\begin{aligned}x^2 :\ F_{\mathbf{a}[2]} &= \left(\frac{a_3 y_3}{1-y_2} + a_2 y_1\right) F_{\mathbf{a}[1]} + a_2 \sum_{i\geq 1} y_i F_{\mathbf{a}[i+1]}\\ &= \left(\frac{a_3 y_3}{1-y_2} + a_2 y_1\right) F_{\mathbf{a}[1]} + a_2 (y_2 F_{\mathbf{a}[2]} + y_3 F_{\mathbf{a}[3]}\\ &\quad + y_4 F_{\mathbf{a}[4]} + \cdots),\end{aligned} \qquad (10.1.7)$$

$$
\begin{aligned}
x^3: \quad F_{\mathbf{a}[3]} &= \left(\frac{a_3 y_3}{1-y_2} + a_2 y_1\right) F_{\mathbf{a}[2]} + a_2 \sum_{i \geq 1} y_{i+1} F_{\mathbf{a}[i+2]} \\
&= \left(\frac{a_3 y_3}{1-y_2} + a_2 y_1\right) F_{\mathbf{a}[2]} + a_2 (y_2 F_{\mathbf{a}[3]} + y_3 F_{\mathbf{a}[4]} \\
&\quad + y_4 F_{\mathbf{a}[5]} + \cdots),
\end{aligned}
\tag{10.1.8}
$$

$$
\begin{aligned}
x^4: \quad F_{\mathbf{a}[4]} &= \left(\frac{a_3 y_3}{1-y_2} + a_2 y_1\right) F_{\mathbf{a}[3]} + a_2 \sum_{i \geq 1} y_{i+1} F_{\mathbf{a}[i+3]} \\
&= \left(\frac{a_3 y_3}{1-y_2} + a_2 y_1\right) F_{\mathbf{a}[3]} + a_2 (y_2 F_{\mathbf{a}[4]} + y_3 F_{\mathbf{a}[5]} \\
&\quad + y_{\mathbf{a}[4]} F_{\mathbf{a}[6]} + \cdots),
\end{aligned}
\tag{10.1.9}
$$

and, for any integer $m \geq 5$,

$$
x^m: \quad F_{\mathbf{a}[m]} = \left(\frac{a_3 y_3}{1-y_2} + a_2 y_1\right) F_{\mathbf{a}[m-1]} + a_2 \sum_{l \geq 1} y_{l+1} F_{\mathbf{a}[l+m-1]}. \tag{10.1.10}
$$

On the basis of (10.1.5)–(10.1.10), let us write $\mathbf{f_a} = (F_{\mathbf{a}[1]}, F_{\mathbf{a}[2]}, F_{\mathbf{a}[3]}, \ldots)$ and $\mathbf{x} = (x_1, x_2, x_3, \ldots)$, then $f = a_1 + \mathbf{f_a}\mathbf{x}^{\mathrm{T}}$. This enables us to only consider $\mathbf{f}$ instead of f for solving equation (10.1.1).

Lemma 10.1.1. *Equation* (10.1.1) *is equivalent to the following vector equation on* $\mathcal{R}\{x, \mathbf{y}\}$*:*

$$
\mathbf{f_a^{\mathrm{T}}} = \mathbf{Y}_{\mathbf{a}[\mathrm{wnt}]} \mathbf{f}^{\mathrm{T}} + \mathbf{c_a^{\mathrm{T}}} \tag{10.1.11}
$$

where

$$
\begin{cases}
\mathbf{c_a} = \left(\dfrac{a_3 y_3}{1-y_2} + a_2 y_1\right) a_1 \mathbf{e}_1 = c_{\mathbf{a}} a_1 \mathbf{e}_1 & (\textit{Recall } \mathbf{e}_1 = (1, 0, 0, \ldots)!); \\
c_{\mathbf{a}} = \dfrac{a_3 y_3}{1-y_2} + a_2 y_1,
\end{cases}
$$

and $\mathbf{Y}_{\mathbf{a}[\mathrm{wnt}]} = (y_{\mathbf{a}[i,j]})_{i,j \geq 1}$,

$$
y_{\mathbf{a}[i,j]} = \begin{cases}
a_2 y_{j-i+2}, & \textit{when } j - i \geq 0; \\
c_{\mathbf{a}}, & \textit{when } j - i = -1; \\
0, & \textit{when } j - i \leq -2.
\end{cases}
$$

Proof. Because of all equivalences on $\mathcal{R}\{x, \mathbf{y}\}$ from equation (10.1.1) to (10.1.11) via (10.1.5)–(10.1.10), the conclusion is drawn. □

This lemma enables us to establish the theorem of the property of being well-defined for equation (10.1.1).

Theorem 10.1.2. *Equation* (10.1.1) *has, and is the only one to have, a solution on* $\mathcal{R}\{x, \mathbf{y}\}$ *if, and only if,* $a_0 = a_1$.

Proof. The condition $a_0 = a_1$ is derived from (10.1.5).

Because of the equivalence between equation (10.1.1) and

$$(I - \mathbf{Y}_{\mathbf{a}[\text{wnt}]})\mathbf{f}^{\mathrm{T}} = c_{\mathbf{a}}\mathbf{e}_1^{\mathrm{T}}$$

on $\mathcal{R}\{x, \mathbf{y}\}$, the existence and uniqueness of the inverse of $(I - \mathbf{Y}_{\mathbf{a}[\text{winst}]})$ on $\mathcal{R}\{x, \mathbf{y}\}$ lead to the conclusion. □

Throughout, for equation (10.1.1), $a_0 = a_1$ is always considered to hold wherever it occurs.

10.2 Solution for wintersweets

In order to determine a solution of equation (10.1.1), the relevant structures of $F_{\mathbf{a}[m,n\rangle}$ for all $m, n \geq 1$ are investigated.

For any integer $m \geq 1$, the determination of $F_{\mathbf{a}[m,n\rangle}$, $n \geq 1$, is discussed. Let $\min_{\mathbf{y}}(F_{\mathbf{a}[m]})$ be the minimum of degrees among all nonzero terms of $F_{\mathbf{a}[m]}$ for $\mathbf{y}$.

Lemma 10.2.1. *Given the integer $m \geq 1$. For any integer $n \geq 1$, if $n \leq m - 1$, then $F_{\mathbf{a}[m,n\rangle} = 0$.*

Proof. From (10.1.5), $F_{\mathbf{a}[m,0\rangle} = 0$ for any integer $m \geq 1$. Because of $\min_{\mathbf{y}}(F_{\mathbf{a}[i]}) \geq i$, $i \geq 1$, from (10.1.6), $\min_{\mathbf{y}}(F_{\mathbf{a}[1]}) = \min_{\mathbf{y}}(y_3 F_{\mathbf{a}[0]}) = 1$. For $m = 2$, from (10.1.7), $\min_{\mathbf{y}}(F_{\mathbf{a}[2]}) = \min_{\mathbf{y}}(y_3 F_{\mathbf{a}[1]}) = 1 + \min_{\mathbf{y}}(F_{\mathbf{a}[1]}) = 2$. Hence, $F_{\mathbf{a}[m,j\rangle} = 0$, $0 \leq j \leq 1$. In general, assume that, for $2 \leq s \leq m - 1$, we have $F_{\mathbf{a}[s,j\rangle} = 0$, $1 \leq j \leq m - 2$. By induction on m, we prove that, for $s = m$, we have $F_{\mathbf{a}[m,j\rangle} = 0$, $0 \leq j \leq m - 1$. From the assumption, $\min_{\mathbf{y}}(F_{\mathbf{a}[m-1]}) = m - 1$. On account of (10.1.10),

$$\begin{aligned}\min_{\mathbf{y}}(F_{\mathbf{a}[m]}) &= \min_{\mathbf{y}}(y_3 F_{\mathbf{a}[m-1]}) \\ &= 1 + \min_{\mathbf{y}}(F_{\mathbf{a}[m-1]}) = 1 + (m - 1) \\ &= m.\end{aligned}$$

Therefore, $F_{\mathbf{a}[m,j\rangle} = 0$, $1 \leq j \leq m - 1$. □

This lemma enables us to omit all $F_{\mathbf{a}[m,n\rangle}$, $n \leq m - 1$, for any $m \geq 1$ given.

Lemma 10.2.2. *For any two integer m, $n \geq 1$, $F_{\mathbf{a}[m,n\rangle} = (a_2 y_1 + a_3 y_3)^m$ if, and only if, $n = m$.*

Proof. When $m = n = 1$, from (10.1.6), $F_{\mathbf{a}[1,1\rangle} = a_3 y_3 F_{\mathbf{a}[0,0\rangle} + a_2 y_1 F_{\mathbf{a}[0,0\rangle} = a_2 y_1 + a_3 y_3$. When $m = n \geq 2$, assume for any integer j, $1 \leq j \leq m - 1$, $F_{\mathbf{a}[j,j\rangle} = (a_2 y_1 + a_3 y_3)^j$. By induction on $m = n$, we prove $F_{\mathbf{a}[m,m\rangle} = (a_2y_1 + a_3y_3)^m$. From (10.1.10),

$$\begin{aligned}
F_{\mathbf{a}[m,m\rangle} &= \left\langle\left(\frac{\mathbf{a}_3y_3}{1-y_2} + a_2y_1\right)F_{\mathbf{a}[m-1]}\right\rangle_m \\
&= \left\langle\frac{\mathbf{a}_3y_3}{1-y_2} + a_2y_1\right\rangle_1 F_{\mathbf{a}[m-1,m-1\rangle} \\
&= (a_2y_1 + a_3y_3)F_{\mathbf{a}[m-1,m-1\rangle}, \\
&\quad \text{by the assumption,} \\
&= (a_2y_1 + a_3y_3)^m.
\end{aligned}$$

This is the necessity.

By considering the uniqueness of homogeneous polynomial of degree m for $\mathbf{y}$ in $F_{m,m}$, the sufficiency is proved. □

The last two lemmas provide the maximum lower bound of term degrees for $\mathbf{y}$ in $F_{\mathbf{a}[m,n]}$ whenever m is given. However, what is more important to us is to estimate the minimum upper bound of the degrees via (10.1.10).

Lemma 10.2.3. *For any two integers m and n ($n \geq m \geq 1$, $mn \neq 1$), $F_{\mathbf{a}[m,n\rangle}$ is independent of y_i, $i \geq n - m + 2$.*

Proof. From (10.1.10), for any two integers $n \geq m \geq 1$, we have

$$\begin{aligned}
F_{\mathbf{a}[m,n\rangle} &= \sum_{l\geq 0}\langle a_3y_3y_2^lF_{\mathbf{a}[m-1]}\rangle_n + a_2\sum_{i\geq m-1}\langle y_{i-m+2}F_{\mathbf{a}[i]}\rangle_n, \\
&= \sum_{l\geq 0} a_3y_3y_2^lF_{\mathbf{a}[m-1,n-l-1\rangle} + a_2\sum_{i\geq m-1} y_{i-m+2}F_{\mathbf{a}[i,n-1\rangle}, \\
&\quad \text{by Lemma 10.2.1,} \\
&= \sum_{l=0}^{n-m} a_3y_3y_2^lF_{\mathbf{a}[m-1,n-l-1\rangle} + a_2\sum_{i=m-1}^{n-1} y_{i-m+2}F_{\mathbf{a}[i,n-1\rangle}.
\end{aligned} \tag{10.2.1}$$

On this basis, for small integers m and n, it is easily checked that the $F_{\mathbf{a}[m,n\rangle}$ satisfy the conclusion. In general, assume $F_{\mathbf{a}[i,n-1\rangle}$, for $m-1 \leq i \leq n-1$ satisfy the conclusion. By induction on n, because of

$$\{y_{i-m+2} \mid m-1 \leq i \leq n-1\} \cap \{y_i \mid i \geq n-m+2\} = \emptyset,$$

and the assumption, we see that $F_{\mathbf{a}[m,n\rangle}$ is independent of y_i, $i \geq n - m + 2$. □

Based on the last three lemmas, we are allowed to determine all $F_{\mathbf{a}[m,n\rangle}$ for $m, n \geq 1$ in the order of $m + n \geq 2$ from smaller to greater and then m from greater to smaller. For instance, $m + n = 2$: $F_{\mathbf{a}[1,1\rangle}$; $m + n = 3$: $F_{\mathbf{a}[2,1\rangle}$, $F_{\mathbf{a}[1,2\rangle}$; $m + n = 4$: $F_{\mathbf{a}[3,1\rangle}$, $F_{\mathbf{a}[2,2\rangle}$, $F_{\mathbf{a}[1,3\rangle}$; $m + n = 5$: $F_{\mathbf{a}[4,1\rangle}$, $F_{\mathbf{a}[3,2\rangle}$, $F_{\mathbf{a}[2,3\rangle}$, $F_{\mathbf{a}[1,4\rangle}$; etc.

From Lemma 10.2.1, $F_{\mathbf{a}[2,1\rangle} = 0$, $F_{\mathbf{a}[3,1\rangle} = 0$ and $F_{\mathbf{a}[4,1\rangle} = F_{\mathbf{a}[3,2\rangle} = 0$. From Lemma 10.2.2, $F_{\mathbf{a}[1,1\rangle} = a_2y_1a_3f_5$, $F_{\mathbf{a}[2,2\rangle} = (a_2y_1a_3f_5)^2$ and $F_{\mathbf{a}[3,3\rangle} = (a_2y_1a_3f_5)^3$.

For convenience, for any function $g \in \mathcal{R}\{\underline{y}\}$, denote by $\langle g \rangle_n$ the homogeneous part of degree n for $\mathbf{y}$ in g.

When $m + n = 1$, it is only necessary to evaluate $F_{\mathbf{a}[1,1\rangle} = a_2 y_1 + a_3 y_3$ known from Lemma 10.2.2.

When $m + n = 3$, it is only necessary to evaluate $F_{\mathbf{a}[1,2\rangle}$. From (10.2.1),

$$\begin{aligned} F_{\mathbf{a}[1,2\rangle} &= \sum_{l=0}^{1} a_3 y_3 y_2^l F_{\mathbf{a}[0,1-l\rangle} + a_2 \sum_{i=0}^{1} y_{i+1} F_{\mathbf{a}[i,1\rangle}, \\ &\quad \text{by (10.1.5)}, \\ &= a_3 y_3 y_2 F_{\mathbf{a}[0,0\rangle} + a_2 y_2 F_{\mathbf{a}[1,1\rangle} \\ &= a_1 a_3 y_2 y_3 + a_2 y_2 (a_2 y_1 + a_3 y_3) \\ &= a_2^2 y_1 y_2 + (a_1 + a_2) a_3 y_2 y_3. \end{aligned} \tag{10.2.2}$$

When $m + n = 4$, we only study $F_{\mathbf{a}[1,3\rangle}$ and $F_{\mathbf{a}[2,2\rangle}$. From Lemma 10.2.2, $F_{\mathbf{a}[2,2\rangle} = (a_2 y_1 + a_3 y_3)^2$. For $F_{\mathbf{a}[1,3\rangle}$, from (10.2.1),

$$\begin{aligned} F_{\mathbf{a}[1,3\rangle} &= \sum_{l=0}^{2} a_3 y_3 y_2^l F_{\mathbf{a}[0,2-l\rangle} + a_2 \sum_{i=0}^{2} y_{i+1} F_{\mathbf{a}[i,2\rangle} \\ &= a_3 y_3 y_2^2 F_{\mathbf{a}[0,0\rangle} + a_2 (y_2 F_{\mathbf{a}[1,2\rangle} + y_3 F_{\mathbf{a}[2,2\rangle}) \\ &= a_1 a_3 y_2^2 y_3 + a_2 (y_2 (a_2^2 y_1 y_2 + (a_1 + a_2) a_3 y_2 y_3) \\ &\quad + y_3 (a_2 y_1 + a_3 y_3)^2) \\ &= a_1 a_3 y_2^2 y_3 + a_2^3 y_1 y_2^2 + (a_1 + a_2) a_2 a_3 y_2^2 y_3 \\ &\quad + 2 a_2^2 a_3 y_1 y_3^2 + a_2 y_3 (a_2^2 y_1^2 + a_3^2 y_3^2) \\ &= a_2^3 y_1 y_2^2 + 2 a_2^2 a_3 y_1 y_3^2 + a_2^3 y_1^2 y_3 \\ &\quad + (a_1 + a_1 a_2) a_2 a_3 y_2^2 y_3 + a_2 a_3^2 y_3^3. \end{aligned} \tag{10.2.3}$$

When $m + n = 5$, we only study $F_{\mathbf{a}[1,4\rangle}$ and $F_{\mathbf{a}[2,3\rangle}$. First, from (10.2.1),

$$\begin{aligned} F_{\mathbf{a}[2,3\rangle} &= a_3 \sum_{l=0}^{1} y_3 y_2^l F_{\mathbf{a}[1,2-l\rangle} + a_2 \sum_{i=1}^{2} y_i F_{\mathbf{a}[i,2\rangle} \\ &= a_3 (y_3 F_{\mathbf{a}[1,2\rangle} + y_2 y_3 F_{\mathbf{a}[1,1\rangle}) + a_2 (y_1 F_{\mathbf{a}[1,2\rangle} + y_2 F_{\mathbf{a}[2,2\rangle}) \\ &= a_3 y_2 y_3 F_{\mathbf{a}[1,1\rangle} + (a_3 y_3 + a_2 y_1) F_{\mathbf{a}[1,2\rangle} + a_2 y_2 F_{\mathbf{a}[2,2\rangle} \\ &= a_3 y_2 y_3 (a_2 y_1 + a_3 y_3) + (a_2 y_1 + a_3 y_3)(a_2^2 y_1 y_2 \\ &\quad + (a_1 + a_2) a_3 y_2 y_3) + a_2 y_2 (a_2 y_1 + a_3 y_3)^2 \\ &= ((a_1 + a_2 + a_3 + a_2 a_3) y_2 y_3 + 2 a_2^2 y_1 y_2)(a_2 y_1 + a_3 y_3) \\ &= (a_1 + a_2 + a_3 + a_2 a_3 + 2 a_2^2 a_3) y_1 y_2 y_3 \\ &\quad + 2 a_2^3 y_1^2 y_2 + (a_1 + a_2 + a_3 + a_2 a_3) a_3 y_2 y_3^2. \end{aligned} \tag{10.2.4}$$

Let $a = a_1 + a_2 + a_3 + a_2a_3$. From (10.2.1), we have

$$\begin{aligned}
F_{\mathbf{a}[1,4\rangle} &= a_3 \sum_{l=0}^{3} y_3 y_2^l F_{\mathbf{a}[0,3-l\rangle} + a_2 \sum_{i=0}^{3} y_{i+1} F_{\mathbf{a}[i,3\rangle} \\
&= a_3 y_3 y_2^3 F_{\mathbf{a}[0,0\rangle} + a_2(y_2 F_{\mathbf{a}[1,3\rangle} + y_3 F_{\mathbf{a}[2,3\rangle} + y_4 F_{\mathbf{a}[3,3]\mathrm{rng}}) \\
&= a_1 a_3 y_3 y_2^3 + a_2(y_2(a_2^3 y_1 y_2^2 + 2a_2^2 a_3 y_1 y_3^2 + a_2^3 y_1^2 y_3 \\
&\quad + (a_1 + a_1 a_2) a_2 a_3 y_2^2 y_3 + a_2 a_3^2 y_3^3) + y_3((a + 2a_2^2 a_3) y_1 y_2 y_3 \\
&\quad + 2a_2^3 y_1^2 y_2 + a a_3 y_2 y_3^2) + y_4(a_2 y_1 + a_3 y_3)^3) \\
&= a_2^4 y_1 y_2^3 + a_2(2a_2^2 + a a_2 + 2a_2^2 a_3) y_1 y_2 y_3^2 + 3a_2^4 y_1^2 y_2 y_3 \\
&\quad + 3a_2^4 y_1^2 y_2 y_3 + 3a_2^3 a_3 y_1^2 y_3 y_4 + a_2^4 y_1^3 y_4 + a_2 a_3 (a + a_2 a_3) y_2 y_3^3 \\
&\quad + a_1 a_2 (1 + a_2) a_3 y_2^3 y_4 + a_2 a_3^3 y_3^3 y_4.
\end{aligned} \tag{10.2.5}$$

Theorem 10.2.4. *The solution of equation* (10.1.1) *is determined by* $F_{\mathbf{a}[m,n\rangle}$ *for* $m, n \geq 0$ *all in the form of sums of finite positive terms as*

$$F_{\mathbf{a}[m,n\rangle} = \begin{cases}
a_1, & \text{when } m = n = 0,\ m + n = 0; \\
a_2 y_1 + a_3 y_3, & \text{when } m = n = 1,\ m + n = 2; \\
a_2^2 y_1 y_2 + (a_1 + a_2) a_3 y_2 y_3, & \text{when } m = 1,\ n = 2,\ m + n = 3; \\
(a_2 y_1 + a_3 y_3)^2, & \text{when } m = 2,\ n = 2,\ m + n = 4; \\
a_2^3 y_1 y_2^2 + 2a_2^2 a_3 y_1 y_3^2 + a_2^3 y_1^2 y_3 & \\
\quad + (a_1 + a_1 a_2) a_2 a_3 y_2^2 y_3 & \\
\quad + a_2 a_3^2 y_3^3, & \text{when } m = 1,\ n = 3,\ m + n = 4; \\
(a_2 y_1 + a_3 y_3)^n, & \text{when } m = n,\ m + n \geq 5; \\
a_3 y_3 \sum_{l=0}^{n-m} y_2^l F_{\mathbf{a}[m-1,n-l-1\rangle} & \\
\quad + a_2 \sum_{i=m-1}^{n-1} y_{i-m+2} F_{\mathbf{a}[i,n-1\rangle}, & \text{when } n - 1 \geq m \geq 3, ,m + n \geq 5; \\
0, & \text{otherwise.}
\end{cases} \tag{10.2.6}$$

Proof. From Lemma 10.2.1, the first case is clear. The second, the fourth and the sixth cases are from Lemma 10.2.2. The third and the fifth cases are, respectively, from (10.2.2) and (10.2.3). In general, all other cases are from Lemma 10.2.3. □

This theorem enables us to evaluate the solution of equation (10.1.1) by the parameters m and n recursively for investigating efficientization and further intelligentization.

10.3 Explicit expression for wintersweets

In mathematics, an explicision of an implicit function, particularly the solution of an equation, is very favorable for usages in mathematical reasoning. Two ways for getting an explicision are discussed in this section.

First, the direct method. For any $\mathbf{y}^{\mathbf{n}}$, write

$$\pi(\mathbf{n}) = \sum_{i=1}^{n} in_i(\text{or } \pi(\mathbf{y}^{\mathbf{n}})).$$

For a polynomial $F_{\mathbf{a}[m,n]}$ of $\underline{y}$, denote

$$\mathcal{P}(F_{m,n}) = \{\mathbf{n} \mid \mathbf{n} \text{ is a power vector of a term in } F_{m,n}\}.$$

Lemma 10.3.1. *For any integer* $m = n \geq 1$, *let* $F_{\mathbf{a}[n,n\rangle} = F_{\mathbf{a}[n]}$, *then* $\pi(F_{\mathbf{a}[n]}) = \{\pi_i(n) \mid 0 \leq i \leq n\}$, *where*

$$\pi_i(n) = \{\pi(\mathbf{n}) = 3n - 2i \mid \mathbf{n} \in \mathcal{P}(F_{\mathbf{a}[n]})\}, \quad 0 \leq i \leq n. \tag{10.3.1}$$

Proof. From Lemma 10.2.2, since for any integer $n \geq 1$,

$$F_{\mathbf{a}[n,n\rangle} = \sum_{i=0}^{n} \binom{n}{i} a_2^i a_3^{n-i} y_1^i y_3^{n-i}$$

and $\pi(y_1^i y_3^{n-i}) = i + 3(n-i) = 3n - i$, we have

$$\pi(F_{\mathbf{a}[n,n\rangle}) = \{3n, 3n-2, 3n-4, \ldots, n+2, n\}.$$

Because of the distinction among all the $n+1$ elements pairwise and exact $n+1$ terms in $F_{\mathbf{a}[n,n\rangle}$, $\pi_i(\mathbf{n}) = \pi(y_1^i y_3^{n-i})$ for $0 \leq i \leq n$. Therefore, (10.3.1) is certainly obtained. □

This lemma tells us that, for any integer $m = n \geq 1$, $F_{\mathbf{c}[n,n]}$ is a homogeneous polynomial of degree n and $n+1$ terms. Moreover,

$$\pi(F_{\mathbf{a}[n,n\rangle}) = \sum_{i=0}^{n} \pi(F_{\mathbf{a}[n]i})$$

where

$$\pi(F_{[n]i}) = \{\pi(\mathbf{n}) \mid \pi(\mathbf{n}) = \pi(y_1^{n-i} y_3^i) = 3n - 2i,\ \mathbf{n} \in \mathcal{P}(F_{[n]})\}$$

for $0 \leq i \leq n$.

Lemma 10.3.2. *For any integers* $n - 1 \geq m \geq 1$, *we have* $\pi(F_{m,n}) = \{\pi(\mathbf{n}) \mid 0 \leq i \leq n-1\}$, *where*

$$\pi_i(\mathbf{n}) = 2n - m + 2i, \quad 0 \leq i \leq n-1. \tag{10.3.2}$$

Proof. First, for $2 \le m+n \le 5$, $n \ge m \ge 1$. On the basis of (10.2.2)–(10.2.6), from Lemma 10.3.1, it is only necessary to consider $F_{\mathbf{a}[1,2\rangle}$, $F_{\mathbf{a}[1,3\rangle}$, $F_{\mathbf{a}[2,3\rangle}$ and $F_{\mathbf{a}[1,4\rangle}$.

For $F_{\mathbf{a}[1,2\rangle}$, we have

$$F_{\mathbf{a}[1,2\rangle} = y_1y_2 + 2y_2y_3 \implies \pi(F_{1,2}) = \{\pi(y_1y_2), \pi(y_2y_3)\} = \{3,5\}.$$

Because of $3 = 2\times 2 - 1 + 2\times 0 = \pi_0(F_{1,2})$ and $5 = 2\times 2 - 1 + 2\times 1 = \pi_1(F_{1,2})$, we see that $F_{1,2}$ satisfies the conclusion. Therefore,

$$\begin{aligned}\mathcal{P}(F_{1,2}) &= \mathcal{P}_0(F_{1,2}) + \mathcal{P}_1(F_{1,2}) \\ &= \{(1,1,0)\} + \{(0,1,1)\}.\end{aligned}$$

For $F_{\mathbf{a}[1,3\rangle}$, we have

$$\begin{aligned}F_{1,3} = y_1y_2^2 + 2y_1y_3^2 + y_1^2y_3 + 3y_2^2y_3 + y_3^3 &\implies \\ \pi(F_{1,3}) = \{5,7,5,7,9\} = \{5,7,9\}.&\end{aligned}$$

Because of $5 = 2\times 3 - 1 + 2\times 0 = \pi_0(F_{1,3})$, $7 = 2\times 3 - 1 + 2\times 1 = \pi_1(F_{1,3})$ and $9 = 2\times 3 - 1 + 2\times 2 = \pi_2(F_{1,3})$, it is seen that $F_{1,3}$ satisfies the conclusion. Therefore,

$$\begin{aligned}\mathcal{P}(F_{1,3}) &= \mathcal{P}_0(F_{1,3}) + \mathcal{P}_1(F_{1,3}) + \mathcal{P}_2(F_{1,3}) \\ &= \{(1,2,0),(2,0,1)\} + \{(1,0,2),(0,2,1)\} + \{(0,0,3)\}.\end{aligned}$$

Similarly, $F_{\mathbf{a}[2,3\rangle}$ and $F_{\mathbf{a}[1,4\rangle}$ can be checked.

Then, for $n \ge m \ge 3$ in general, assume all $F_{\mathbf{a}[s,t\rangle}$ satisfy the conclusion for $t \le s \le n-1$. We prove that $F_{\mathbf{a}[m,n\rangle}$ shows satisfaction.

From (10.3.2), for $n \ge m \ge 3$,

$$\pi(F_{m,n}) = \bigcup_{l=0}^{n-m} \pi(y_3y_2^lF_{m-1,n-l-1}) \cup \bigcup_{l=m-1}^{n-1} \pi(y_{l-m+2}F_{l,n-1}).$$

By the assumption,

$$\begin{aligned}\pi(y_3y_2^lF_{m-1,n-l-1}) &= \{3 + 2l + (2(n-l+1) - (m-1) + 2i) \mid 0 \le i \le n-l-2\} \\ &= \{2n - m + 2i + 2 \mid 0 \le i \le n-l-2\} \\ &= \{2n - m + 2i \mid 1 \le i \le n-l-1\}\end{aligned}$$

and

$$\begin{aligned}\pi(y_{l-m+2}F_{l,n-1}) &= \{l - m + 2 + (2(n-1) - l + 2i) \mid 0 \le i \le n-2\} \\ &= \{2n - m + 2i \mid 0 \le i \le n-2\}.\end{aligned}$$

Since when $l = 0$, the former is in the case: $i = n-1$, from the union of the two,

$$\pi(F_{m,n}) = \{2n - m + 2i \mid 0 \le i \le n-1\}.$$

This is the conclusion. □

When $m = n$, if the case $i = n$ is put in (10.3.2) in Lemma 10.3.2, then (10.3.1) in Lemma 10.3.1 is again done.

Corollary 10.3.3. *For any integers* $n \geq m \geq 1$, $\pi(F_{\mathbf{a}[m,n\rangle}) = \{\pi_i(F_{\mathbf{a}[m,n\rangle}) \mid 0 \leq i \leq n\}$, *where*

$$\pi_i(F_{\mathbf{a}[m,n\rangle}) = 2n - m + 2i, \quad 0 \leq i \leq n. \tag{10.3.3}$$

Proof. This is a direct result of Lemma 10.3.2. □

Let

$$\omega(x) = \omega(x, \mathbf{y}) = \sum_{k \geq 1} y_k x^k, \tag{10.3.4}$$

then, for any integers $n \geq m \geq 1$, $i \geq 0$, write

$$\mathcal{L}_{2n-m+2i} = \{\mathbf{n} \mid \mathbf{n} \text{ is the power vector of a tern in } \partial_x^{2n-m+2i} \omega^n(x)\}. \tag{10.3.5}$$

For any integers $n \geq m \geq 1$, $n - 1 \geq i \geq 0$, write

$$\mathcal{N}_i(m, n) = \{\mathbf{n} \geq \mathbf{0} \mid |\mathbf{n}| = n, \pi(\mathbf{n}) = 2n - m + 2i\}. \tag{10.3.6}$$

Lemma 10.3.4. *For any integers* $m, n \geq 1$, $\mathcal{N}_i(m, n) = \mathcal{L}_{2n-m+2i}$.

Proof. First, for any $\mathbf{n} \in \mathcal{L}_{2n-m+2i}$, because of

$$\sum_{j \geq 1} j n_j = 2n - m + 2i$$

and $|\mathbf{n}| = n$, we have $\mathbf{n} \in \mathcal{N}_i(m, n)$.

Then, for any $\mathbf{n} \in \mathcal{N}_i(m, n)$, since $|\mathbf{n}| = n$ leads to

$$\sum_{j \geq 1} j n_j = 2n - m + 2i,$$

we have $\mathbf{n} \in \mathcal{L}_{2n-m+2i}$.

Therefore, the conclusion is drawn. □

For a homogeneous polynomial P of $\mathbf{y}$ with degree at least 1, denote by $\hat{P}$ the polynomial obtained by taking all coefficients of terms in P as 1. As a matter of fact, Corollary 10.3.3 implies that, for any integers $n \geq m \geq 1$, $\hat{F}_{i\mathbf{a}[m,n\rangle} = \hat{\partial}_x^{2n-m+2i} \omega^n(x)$, $0 \leq i \leq n$.

Lemma 10.3.5. *Given integers* $m, n \geq 1$, *there exists a constant* $\alpha_i^{\mathbf{a}[m,n]}$, *such that*

$$\begin{aligned} F_{i\mathbf{a}[m,n\rangle} &= \alpha_i^{\mathbf{a}[m,n]} \partial_x^{2n-m} \sigma_n(x) \\ &= \sum_{\mathbf{n} \in \mathcal{L}_{2n-m+2i}} \frac{\alpha_i^{\mathbf{a}[m,n]} n!}{n\mathbf{n}!} \mathbf{y}^{\mathbf{n}} \end{aligned} \tag{10.3.7}$$

where $n = |\mathbf{n}|$.

Proof. From Corollary 10.3.3 and the form of $\omega^n(x)$, the conclusion is drawn. □

This lemma enables us to provide the solution of equation (10.1.1) such that all coefficients of its terms are summation free.

Theorem 10.3.6. *Let f_{wnt} be the solution of equation* (10.1.1), *then*

$$\langle \partial_x^m f_{\mathrm{wnt}} \rangle_n = \begin{cases} 0, & \text{when } m = 0 \text{ but } n \neq 0, \\ & \text{or } m \geq 1 \text{ and }, n = 0, \text{ or } n < m; \\ a_1, & \text{when } m = 0 \text{ and } n = 0; \\ \sum_{\substack{\mathbf{n}\in\mathcal{N}_i(m,n) \\ 0\leq i\leq n}} \frac{a_i^{\mathbf{a}[m,n]} n!}{n\mathbf{n}!} \mathbf{y}^{\mathbf{n}}, & \text{otherwise,} \end{cases} \tag{10.3.8}$$

where $\mathcal{N}_{m,n}$ is shown in (10.3.6).

Proof. From Lemma 10.3.5 and the form of equation (10.1.1), the conclusion is easily drawn. □

Second, we have the matrix method. We work on the basis of (10.1.11) in Lemma 10.1.1. Because of Lemma 10.2.1, for any integer $n \geq 1$, it is only necessary to discuss $\mathbf{f}_{\mathbf{a}[n]} = (F_{\mathbf{a}[1]}, F_{\mathbf{a}[2]}, \ldots, F_{\mathbf{a}[n]})$ instead of $\mathbf{f}$. Because of Lemma 10.2.3, $\mathbf{y}$ can be replaced by $\mathbf{y}_{n-m+1} = (y_1, y_2, \ldots, y_{n-m+1})$.

Lemma 10.3.7. *Given any integer $n = |\mathbf{n}| \geq 4$, equation* (10.1.1) *for $a_1 = a_0$ is equivalent to*

$$\mathbf{f}_{\mathbf{a}[n]}^{\mathrm{T}} = \langle \mathbf{Y}_{\mathbf{a}[\mathrm{wnt}]} \rangle_n \mathbf{f}_{\mathbf{a}[n]}^{\mathrm{T}} + c_{\mathbf{a}[n]} a_1 \mathbf{e}_{1_n}^{\mathrm{T}} \tag{10.3.9}$$

where $c_{\mathbf{a}[n]} = a_2 y_1 + a_3 y_3 (1 + y_2 + y_2^2 + \cdots + y_2^{n-1})$, $\mathbf{e}_{1_n} = (1, 0, \ldots, 0)_n$ and $\mathbf{Y}_{\mathbf{a}[\mathrm{wnt}]n} = (y_{\mathbf{a}[i,j]})_{1\leq i,j\leq n}$ such that

$$y_{\mathbf{a}[i,j]} = \begin{cases} c_{\mathbf{a}[n]}, & \text{when } j - i = -1; \\ a_2 y_{j-i+2}, & \text{when } j - i \geq 0; \\ 0, & \text{when } j - i \leq -2. \end{cases}$$

Proof. On the basis of Lemma 10.1.1, by deleting all non-relevant terms and variables from Lemma 10.2.1 and Lemma 10.2.3, the conclusion is drawn. □

Because of the equivalence between equation (10.3.9) and the equation

$$\langle I - \mathbf{Y}_{\mathbf{a}\mathrm{wnt}} \rangle_n \mathbf{f}_{\mathbf{a}[n]}^{\mathrm{T}} = c_{\mathbf{a}[n]} a_1 \mathbf{e}_{1_n}^{\mathrm{T}}, \tag{10.3.10}$$

we are allowed to only investigate equation (10.3.10) instead of equation (10.3.9).

Theorem 10.3.8. *Given an integer $n \geq 4$, the solution of equation* (10.3.10) *is of the form*

$$\mathbf{f}_{\mathbf{a}[n]}^{\mathrm{T}} = \sum_{l=0}^{n} c_{\mathbf{a}[n]} a_1 \langle \mathbf{Y}_{\mathbf{a}[\mathrm{wnt}]} \rangle_n^l \mathbf{e}_{1_n}^{\mathrm{T}}. \tag{10.3.11}$$

Proof. Because of $\langle \mathbf{Y}_{\mathbf{a}[\mathrm{wnt}]}\rangle^{n+1} = 0$,

$$\langle I - \mathbf{Y}_{\mathbf{a}[\mathrm{wnt}]}\rangle_n^{-1} = \sum_{i=0}^{n} \langle \mathbf{Y}_{\mathbf{a}[\mathrm{wnt}]}\rangle_n^i$$

From Lemma 10.3.7 and (10.3.10), the conclusion is drawn. □

Based on this theorem, it is only necessary for us to calculate the powers of matrix $\langle \mathbf{Y}_{\mathbf{a}\mathrm{wnt}}\rangle_n$ for getting an explicision of the solution of equation (10.3.9), and hence equation (10.1.1) from Lemma 10.3.7.

Let $y_{\mathbf{a}[i,j]}^{[k]}$ be the entry at position (i,j), $1 \le i,j \le n$, in $(\mathbf{Y}_{\mathbf{a}\mathrm{wnt}})_n^{[k]}$, then $\langle \mathbf{Y}_{\mathbf{a}\mathrm{wnt}}\rangle_n^k = (y_{\mathbf{a}[i,j]}^{[k]})_{1\le i,j\le n}$ where

$$y_{\mathbf{a}[i,j]}^{k} = \begin{cases} y_{\mathbf{a}[i,j]}, & \text{when } k = 1; \\ \mathbf{y}_{\mathbf{a}[i,*]}\mathbf{y}_{\mathbf{a}[*,j]}^{[k-1]}, & \text{when } n \ge k \ge 2, \end{cases}$$

in which $\mathbf{y}_{\mathbf{a}[i,*]}$ and $\mathbf{y}_{\mathbf{a}[*,j]}$ are, respectively, the row vector and the column vector of a matrix.

10.4 Restrictions for wintersweets

Now, let us to consider the equation

$$\begin{cases} x\int_y y\delta_{x,y}(uf|_{x=u}) = \left(1 - \dfrac{xy_3}{1-y_2}\right)f - 1; \\ f|_{x=0\Leftrightarrow \mathbf{y}=\mathbf{0}} = 1, \end{cases} \tag{10.4.1}$$

on $\mathcal{R}\{x,\mathbf{y}\}$.

This is the equation obtained from equation (10.1.1) in the case of $a_0 = a_1 = a_2 = a_3 = 1$, for the first time addressed in Liu [18] (1985) and then in [20] (1986), etc., for enumerating the number of non-isomorphic classes of rooted wintersweets with the vertex partition vector as parameter.

However, in those papers, the equation appeared only in the form of

$$f = 1 + \frac{xy_3}{1-y_2}f + x\int_y (y\delta_{x,y}(uf|_{x=u})), \tag{10.4.2}$$

the equivalent form of the first line of equation (10.4.1) in $\mathcal{R}\{x,\mathbf{y}\}$.

Let us write $\mathbf{f} = (F_1, F_2, F_3, \ldots)$ where ∂_x^i for $i \ge 1$, and $\mathbf{x} = (x_1, x_2, x_3, \ldots)$, then $f = 1 + \mathbf{f}\mathbf{x}^{\mathrm{T}}$. This enables us only to consider $\mathbf{f}$ instead of f for solving equation (10.4.2).

Lemma 10.4.1. *Equation* (10.4.2) *is equivalent to the following vector equation on* $\mathcal{R}\{x, \mathbf{y}\}$*:*

$$\mathbf{f}^{\mathrm{T}} = \mathbf{Y}_{\mathrm{wnt}}\mathbf{f}^{\mathrm{T}} + \mathbf{c}^{\mathrm{T}} \tag{10.4.3}$$

where

$$\begin{cases} \mathbf{c} = \left(\dfrac{y_3}{1-y_2} + y_1\right)\mathbf{e}_1 = c\mathbf{e}_1 & (\textit{recall that } \mathbf{e}_1 = (1,0,0,\ldots)!); \\ c = \dfrac{y_3}{1-y_2} + y_1, \end{cases}$$

and $\mathbf{Y}_{\mathrm{wnt}} = (y_{i,j})_{i,j\geq 1}$,

$$y_{i,j} = \begin{cases} y_{j-i+2}, & \textit{when } j-i \geq 0; \\ c, & \textit{when } j-i = -1; \\ 0, & \textit{when } j-i \leq -2. \end{cases}$$

Proof. This is Lemma 10.1.1 for $a_0 = a_1 = a_2 = a_3 = 1$. □

Similarly to Theorem 10.1.2, our theorem for the property of being well-defined of equation (10.4.1) is soon obtained.

Theorem 10.4.2. *Equation* (10.4.2)*, and hence equation* (10.4.1) *has, and is the only one to have, one solution on* $\mathcal{R}\{x, \mathbf{y}\}$.

Proof. On the basis of Lemma 10.4.1, because of the equivalence of equation (10.4.1) to

$$(I - \mathbf{Y}_{\mathrm{wnt}})\mathbf{f}^{\mathrm{T}} = c\mathbf{e}_1^{\mathrm{T}}$$

on $\mathcal{R}\{x, \mathbf{y}\}$, the existence and uniqueness of $(I - \mathbf{Y}_{\mathrm{wnt}})^{-1}$ on $\mathcal{R}\{x, \mathbf{y}\}$ lead to the conclusion. □

Because

$$(I - \mathbf{Y}_{\mathrm{wnt}})^{-1} = \sum_{i\geq 0}(\mathbf{Y}_{\mathrm{wnt}})^i, \tag{10.4.4}$$

Theorem 10.4.2 enables us to get the solution of equation (10.4.3) in the form of

$$\mathbf{f}^{\mathrm{T}} = \sum_{i\geq 0} c(\mathbf{Y}_{\mathrm{wnt}})^i\mathbf{e}_1^{\mathrm{T}}. \tag{10.4.5}$$

Although (10.4.5) presents an explicision of the solution as $f = 1 + \mathbf{f}\mathbf{x}^{\mathrm{T}}$, for computability, a parameter has to be introduced so that, for any given value of the parameter, the part of solution can be evaluated.

Let $n = |\mathbf{n}|$ be the degree of a term with $\mathbf{y}^{\mathbf{n}}$ in f. Because of Lemma 10.1.1 and Lemma 10.2.1, we are allowed to adopt $f_n = \mathbf{f}_n\mathbf{x}_n^{\mathrm{T}}$ for $n \geq 1$ instead of $f - 1$ in equation (10.4.1) where $\mathbf{f}_n = (F_{1,n}, F_{2,n}, \ldots, F_{n,n})$, $\mathbf{x}_n = (x, x^2, \ldots, x^n)$ and $F_{i,n} = \langle \partial_x^i f\rangle_n$, $1 \leq i \leq n$.

Lemma 10.4.3. *For integer $n \geq 1$ given, equation* (10.4.3) *is equivalent to*

$$(I - \langle \mathbf{Y}_{\mathrm{wnt}} \rangle_n)\mathbf{f}_n^{\mathrm{T}} = \mathbf{c}_n^{\mathrm{T}} \tag{10.4.6}$$

on $\mathcal{R}\{x, \mathbf{y}\}$.

Proof. Since all transformations from equation (10.4.3) to equation (10.4.6) and vice versa are equivalent on $\mathcal{R}\{x, \mathbf{y}\}$, the conclusion follows. □

This lemma enables us to get a solution of equation (10.4.6) in the form of

$$\mathbf{f}_n^{\mathrm{T}} = \sum_{i=0}^{n} c \langle \mathbf{Y}_{\mathrm{wnt}} \rangle_n^i \langle \mathbf{e}_1 \rangle_n^{\mathrm{T}} \tag{10.4.7}$$

where $\langle \mathbf{e}_1 \rangle_n$ is an n-dimensional vector with only the first entry 1 and all others 0. We have $y_{i,j}^{[1]} = y_{i,j}$ and

$$y_{i,1}^{[1]} = \begin{cases} y_2, & \text{when } i = 1; \\ c, & \text{when } i = 2; \\ 0, & \text{otherwise,} \end{cases}$$

and for any integer $k \geq 2$,

$$y_{i,1}^{[k]} = \begin{cases} \mathbf{y}_{i,*} \mathbf{y}_{*,1}^{[k-1]}, & \text{when } 1 \leq i \leq k; \\ c, & \text{when } i = k + 1; \\ 0, & \text{otherwise,} \end{cases} \tag{10.4.8}$$

we see that, for any integers $n \geq 5$ and $1 \leq k \leq n - 1$,

$$\mathbf{Y}_{\mathrm{wnt}}^k \langle \mathbf{e}_1 \rangle_n = \mathbf{y}_{*,1}^{[k]} \tag{10.4.9}$$

where $\mathbf{y}_{*,1}^{[k]} = (y_{1,1}^{[k]}, y_{2,1}^{[k]}, \ldots, y_{n,1}^{[k]})^{\mathrm{T}}$ as given in (10.4.8).

Theorem 10.4.4. *For an integer $n \geq 1$ and hence $1 \leq m \leq n$, the solution of equation* (10.4.6) *is determined by*

$$\mathbf{f}_n^{\mathrm{T}} = \sum_{k=0}^{n-1} \langle c \mathbf{y}_{*,1}^{[k]} \rangle_n \tag{10.4.10}$$

where $c = y_3/(1 - y_2) + y_1$ and $\mathbf{y}_{*,1}^{[k]}$ is given in (10.4.9) *and* (10.4.8).

Proof. Because $f = 1 + \mathbf{x}\mathbf{f}^{\mathrm{T}}$, for $n \geq 1$, $f_n = \mathbf{x}_n \mathbf{f}_n^{\mathrm{T}}$. On account of Lemma 10.4.3, the conclusion is drawn. □

Similarly to the discussion of Theorem 10.3.6 we extract another explicision of the solution of equation (10.4.1).

Theorem 10.4.5. *Let f_{wnst} be the solution of equation* (10.4.1), *then*

$$\langle \partial_x^m f_{\mathrm{wnst}} \rangle_n = \begin{cases} 0, & \text{when } m = 0 \text{ but } n \neq 0, \\ & \text{or } m \geq 1 \text{ and }, n = 0, \text{ or } n < m; \\ 1, & \text{when } m = 0 \text{ and } n = 0; \\ \sum_{\substack{\mathbf{n} \in \mathcal{N}_i(m,n) \\ 0 \leq i \leq n}} \frac{\alpha_i^{[m,n]} n!}{n\mathbf{n}!} \mathbf{y}^{\mathbf{n}}, & \text{otherwise}, \end{cases} \tag{10.4.11}$$

where $\mathcal{N}_{m,n}$ is shown in (10.3.6) *and $\alpha_i^{[m,n]} = \alpha_i^{\mathbf{a}[m,n]}|_{\mathbf{a}=(1,1,1,1)}$.*

Proof. The conclusion is drawn from the specific case of $a_0 = a_1 = a_2 = a_3 = 1$ in (10.3.8) of Theorem 10.3.6. □

Although two explicisions of the solution of equation (10.4.1) have been obtained, the exact result is still not easy to calculate even on computers for big parameters. This is why we concentrate our attention to finding recursive expressions particularly in the form of a sum of finite terms all of which are positive.

Theorem 10.4.6. *The solution of equation* (10.4.1) *is determined by $F_{m,n}$ for $m, n \geq 0$, all of which are in the form of polynomials with positive terms,*

$$F_{m,n} = \begin{cases} 1, & \text{when } m = n = 0, m + n = 0; \\ y_1 + y_3, & \text{when } m = n = 1, m + n = 2; \\ y_1 y_2 + 2y_2 y_3, & \text{when } m = 1, n = 2, m + n = 3; \\ (y_1 + y_3)^2, & \text{when } m = 2, n = 2, m + n = 4; \\ y_1 y_2^2 + 2y_1 y_3^2 + y_1^2 y_3 & \\ \quad + 2y_2^2 y_3 + y_3^3, & \text{when } m = 1, n = 3, m + n = 4; \\ (y_1 + y_3)^n, & \text{when } m = n, m + n \geq 5; \\ y_3 \sum_{l=0}^{n-m} y_2^l F_{m-1,n-l-1} & \\ \quad + \sum_{i=m-1}^{n-1} y_{i-m+2} F_{i,n-1}, & \text{when } n - 1 \geq m \geq 3, m + n \geq 5; \\ 0, & \text{otherwise}. \end{cases} \tag{10.4.12}$$

Proof. This is the specific case of $a_0 = a_1 = a_2 = a_3 = 1$ in (10.2.6) of Theorem 10.2.4. □

Example 1. Root-isomorphic classes of wintersweets with root-vertex valency and vertex partition vector. A map seen as a plane tree with each non-rooted end appending at most one circuit of length not greater than two is called a *wintersweet*. The root is restricted not to be on a circuit. The solution of equation (10.4.1) is the enufunction f of root-isomorphic classes of wintersweets with root-vertex valency (power of x) and the vertex partition vector (power vector of $\mathbf{y}$) as parameters. This implies that

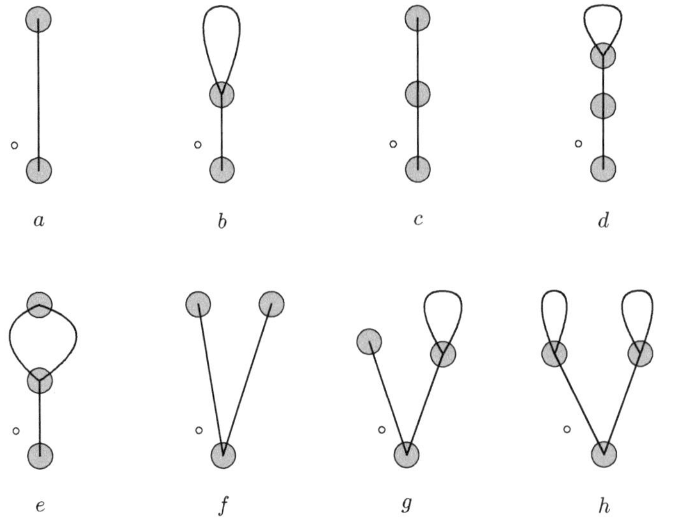

Figure 10.4.1: Classes of wintersweets with order 2–3.

$F_{m,n}|_{n=\mathbf{n}} = \partial_{x,\mathbf{y}}^{m,\mathbf{n}} f$ is the number of root-isomorphic classes of wintersweets with root-vertex valency m and order $n+1$. For example, in Figure 10.4.1,

$$\begin{aligned}(a+b)+(c+d+e) &= (y_1+y_3)+(y_1y_2+y_2y_3+y_2y_3)\\ &= (y_1+y_3)+(y_1y_2+2y_2y_3)\\ &= F_{1,1}+F_{1,2}\end{aligned}$$

and $f+g+h = y_1^2+2y_1y_3+y_3^2 = F_{2,2}$.

In Figure 10.4.2,

$$\begin{aligned}a+b+c+(d+f+g)+e &= y_1y_2^2+2y_1y_3^2+y_1^2y_3+(y_2^2y_3+y_2^2y_3+y_2^2y_3)+y_3^3\\ &= y_1y_2^2+2y_1y_3^2+y_1^2y_3+3y_2^2y_3+y_3^3\\ &= F_{1,3},\end{aligned}$$

$$\begin{aligned}h+i+j+k &= y_1^3+3y_1^2y_3+3y_1y_3^2+y_3^3\\ &= (y_1+y_2)^3\\ &= F_{3,3},\end{aligned}$$

and

$$\begin{aligned}(l+m+n)+o+(p+q) &= (2y_1y_2y_3+2y_1y_2y_3+2y_1y_2y_3)+2y_1^2y_2+(2y_2y_3^2+2y_2y_3^2)\\ &= 6y_1y_2y_3+2y_1^2y_2+4y_2y_3^2\\ &= F_{2,3}.\end{aligned}$$

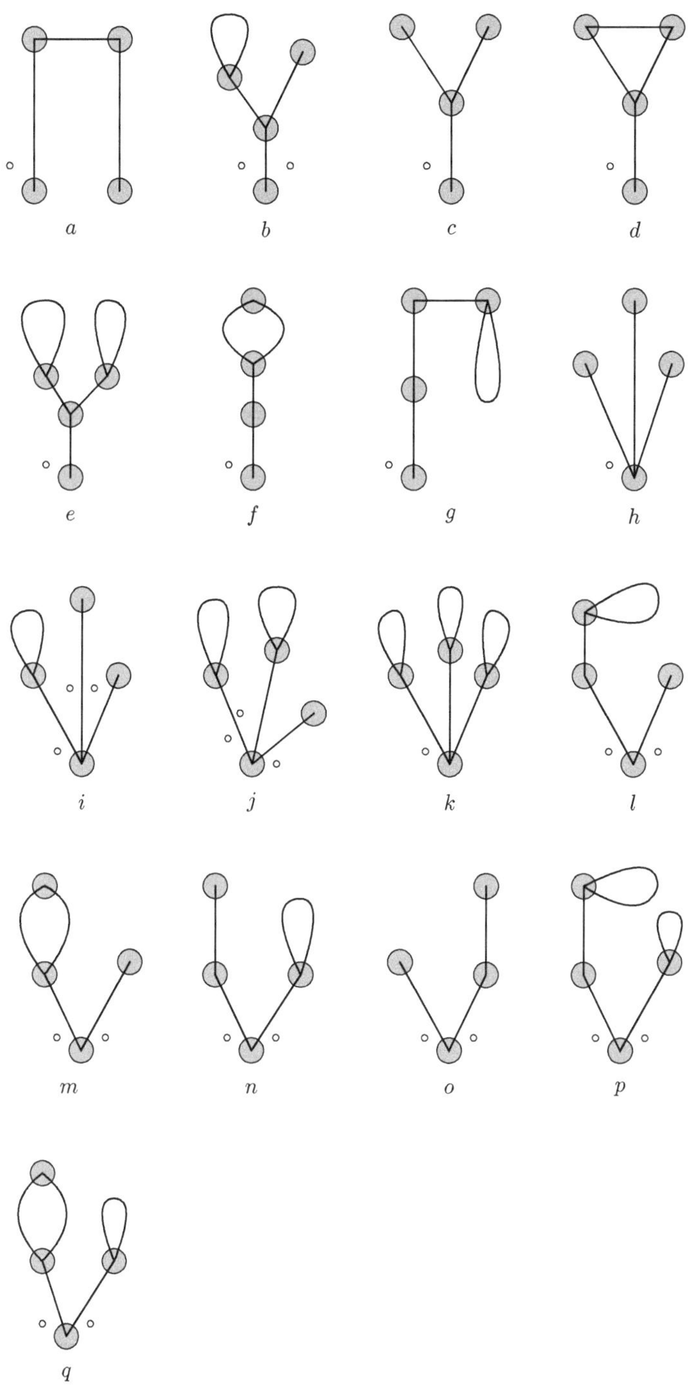

Figure 10.4.2: Classes of wintersweets with order 4.

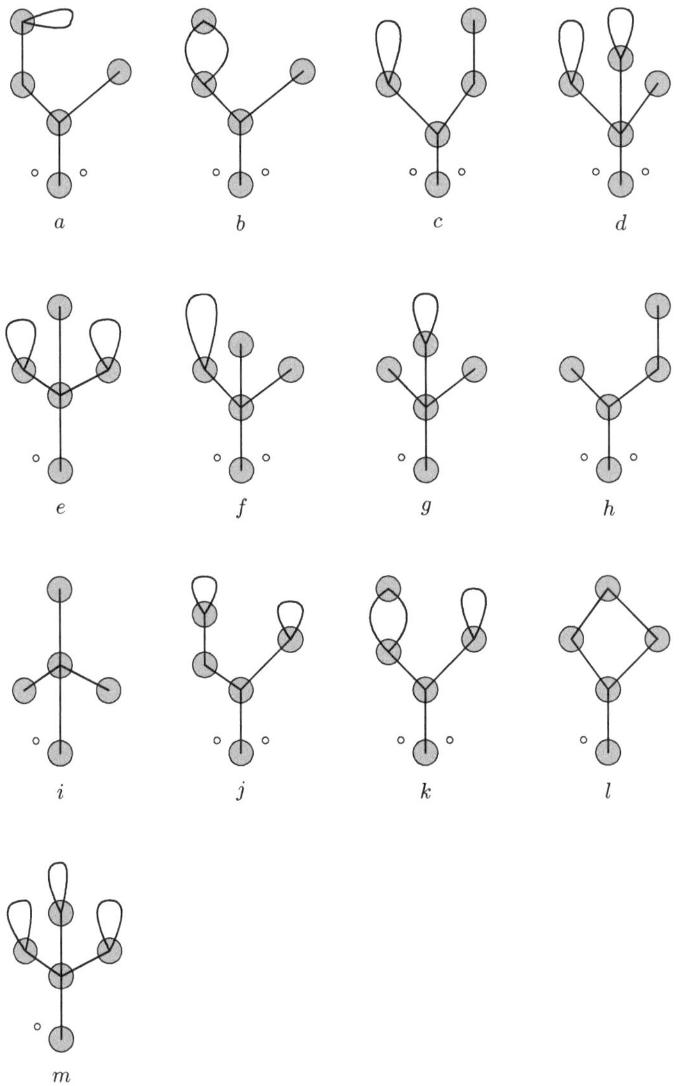

Figure 10.4.3: Classes of wintersweets with order 5.

In Figure 10.4.3,

$$\begin{aligned}(a+b+c)+(d+e)+h+(f+g)+i+(j+k)+l+m\\ &= (2+2+2)y_1y_2y_3^2+(2+1)y_1y_3^2y_4\\ &\quad +2y_1^2y_2y_3+(1+2)y_1^2y_3y_4+y_1^3y_4+(2+2)y_2y_3^2+y_2^3y_3+y_3^3y_4\\ &= 6y_1y_2y_3^2+3y_1y_3^2y_4+2y_1^2y_2y_3+3y_1^2y_3y_4+y_1^3y_4+4y_2y_3^2+y_2^3y_3+y_3^3y_4\\ &= F_{1,4}.\end{aligned}$$

10.5 Notes

10.5.1. In (10.3.7), the coefficient $\alpha^{m,n}_{\mathbf{a}[m,n]}(\mathbf{i})$ can be determined by the procedure shown in (10.2.6), or on the basis of (10.3.11). For the case of $\alpha_1 = \alpha_2 = \alpha_3 = 1$, we may work by using (10.4.10), or on the basis of (10.4.7). It might be helpful to do the specific case before the general case.

10.5.2. We may introduce new functionals from the basis $\{x, x^2, x^3, \ldots\}$ to itself to extract a simplest explicision. Let Λ stand for all the functionals $\lambda : \partial^{m,\mathbf{n}}_{x,\mathbf{y}} \lambda f = F_{m,\mathbf{n}}$ for $f \in \mathcal{R}\{x, \mathbf{y}\}$. Obviously, the identity functional is in Λ and the Lagrangian inversion as a functional is in Λ as well.

10.5.3. We may work by exploiting the constructions of Λ, to observe a way to identify if is there a functional λ such that $\partial^{m,\mathbf{n}}_{x,\mathbf{y}} \lambda f$ is a finite sum of all terms positive or summation free for a function $f \in \mathcal{R}\{x, \mathbf{y}\}$ of a certain condition when positive integers m and n are given.

10.5.4. It looks a type of problems on a null space of infinite dimensional vector with some new operations necessarily should be introduced. It looks necessary to classify the functionals shown in 10.5.3 in a suitable way.

10.5.5. Equation (10.1.1) can be generalized to the case that some of a_0, a_1, a_2 and a_3 are polynomials of a variable, x, or y_i ($i \geq 1$ given). First, observe the condition, or conditions, for establishing a qualitative theory for the consistency. Then we observe how to evaluate a solution by a series of transformations, or operations, on $\mathcal{R}\{x, \mathbf{y}\}$. This might be the central part of Program 84 posed in Liu YP [61] (Book 22, 2016, p. 10739).

10.5.6. As a matter of fact, this chapter finishes, in principle, the first level of the Program 84 in Liu YP [61] (Book 22, 2016, p. 10739) for all of $a_0 = d$, $a_1 = b$, $a_2 = c$ and $a_3 = a$ constants in $\mathbb{Z}_+$ and $\mathbb{R}_+$. However, if some of a_0, a_1, a_2 and a_3 are allowed to be negative, it is definitely not straightforward unless additional condition, or conditions, are to be further investigated because the meaning in combinatorics is not yet known.

Bibliography

[1] Bender EA, Wormald NC, The number of loopless planar maps, *Discrete Math.*, 54 (1985), 235–237.
[2] Birkhoff G, *Lattice Theory* (3rd edition), AMS, Providence, 1967.
[3] Blissard J, Theory of generic equations, *Quad. J. Pure Appl. Math.*, 4 (1861), 279–305.
[4] Blissard J, Theory of generic equations, *Quad. J. Pure Appl. Math.*, 5 (1862), 58–75.
[5] Brown WG, Enumeration of quadrangular dissections of the disc, *Can. J. Math.*, 17 (1965), 302–317.
[6] Cai JL, Liu YP, The number of nearly 4-regular planar maps, *Util. Math.*, 58 (2000), 243–254.
[7] Cai JL, Liu YP, The enumeration of general rooted planar maps, *Acta Math. Sin. Engl. Ser.* 21 (2005), 215–224.
[8] Dong FM, Liu YP, Counting rooted near-triangulations on the sphere, Preprint, Institute of Applied Math. Academia Sinica, 1990. Also in *Discrete Math.*, 123 (1993), 35–45.
[9] Hao RX, Liu YP, Enumeration of general rooted maps on the torus, *OR Trans.*, 6(1) (2002), 19–28.
[10] Li ZX, Liu YP, Enumeration of loopless maps on the projective plane, *J. Appl. Math. Comput.*, 10 (2002), 145–155.
[11] Liu WZ, Liu, YP, Enumeration of three kinds of rooted maps on the Klein bottle, *J. Appl. Math. Comput.*, 24 (2007), 411–419.
[12] Liu YP, On the number of rooted c-nets, Combin. Optim. CORR83-35, Univ. Waterloo, 1983; Also in *J. Comb. Theory*, B36 (1984), 118–123.
[13] Liu YP, Enumeration of rooted separable planar maps, Combin. Optim CORR82-4, Univ. Waterloo, 1982; Also in *Util. Math.*, 25 (1984), 77–94.
[14] Liu YP, Counting rooted planar maps, Combin. Optim. CORR83-26, Univ. Waterloo, 1983; Also in *J. Math. Res. Exposition*, 3 (1984), 37–46.
[15] Liu YP, Chromatic sum equations for rooted planar maps, Contributed to the 15th S-E Conf. Combin. Graph Theory and Comp., Baton Rouge, 1984. Also in *Congr. Numer.*, 45 (1984), 275–280.
[16] Liu YP, Enumerating rooted loopless planar maps, Combin. Optim. CORR83-4, Univ. Waterloo, 1983; Also in *Acta Math. Appl. Sin. Engl. Ser.*, 2 (1985), 14–26.
[17] Liu YP, An enumerating equation of nonseparable planar maps with vertex partition (in Chinese), *Kexue Tongbao*, 30 (1985), 646–649.
[18] Liu YP, Some enumerating problems with vertex partition (in Chinese), *Kexue Tongbao*, 30 (1985), 1521–1524.
[19] Liu YP, A functional equation for enumerating nonseparable planar maps with vertex partition, *Sci. Bull.*, 31 (1986), 73–77.
[20] Liu YP, Some emumerating problems of maps with vertex partition, *Sci. Bull.*, 31 (1986), 1009–1014.
[21] Liu YP, An equation for enumerating nonseparable Eulerian planar maps with vertex partition (in Chinese), *Kexue Tongbao*, 31 (1986), 81–84.
[22] Liu YP, An equation for enumerating outer planar rooted maps with vertex partition (in Chinese), *Kexue Tongbao*, 31 (1986), 1045–1048.
[23] Liu YP, An equation for enumerating nonseparable outer planar rooted maps with vertex partition (in Chinese), *Kexue Tongbao*, 31 (1986), 1681–1684.
[24] Liu YP, A functional equation of enumerating nonseparable Eulerian planar maps with vertex partition, *Sci. Bull.*, 31 (1986), 941–945.

https://doi.org/10.1515/9783110625837-011

[25] Liu YP, Enumeration of rooted outerplanar maps with vertex partition, *Sci. Bull.*, 32 (1987), 295–299.

[26] Liu YP, Enumeration of non-separable outerplanar maps with vertex partition, *Sci. Bull.*, 32 (1987), 1664–1668.

[27] Liu YP, An Enumerating Equation of Simple Planar Maps with Face Partition, RUTCOR Research Report RRR36-87, Rutgers Univ., 1987. Main part also in On enumerating equation of simple planar maps with face partition (in Chinese), *Acta Math. Appl. Sin.*, 12 (1989), 292–295.

[28] Liu YP, An Enumerating Equation of Rooted Loopless Planar Maps with Vertex Partition, RUTCOR Research Report RRR 36-87, Rutgers Univ., 1987. Main part also in Enumerating equation loopless planar rooted maps with vertex partition (in Chinese), *Acta Math. Appl. Sin.*, 12 (1989), 213–217.

[29] Liu YP, On the enumeration of simple planar maps (in Chinese), *Acta Math. Sin.*, 31 (1988), 279–282.

[30] Liu YP, Enumeration of nonseparable outer planar rooted maps (in Chinese), *Kexue Tongbao*, 33 (1988), 473.

[31] Liu YP, Enumeration of simple planar maps, RUTCOR Research Report RRR 36-87, Rutgers Univ., 1987. Also in *Util. Math.*, 34 (1988), 97–104.

[32] Liu YP, On enumerating equations of planar maps (in Chinese), *Adv. Math. China*, 18 (1989), 446–460.

[33] Liu YP, Enumeration of non-separable outerplanar maps, *Acta Math. Appl. Sin. Engl. Ser.* 5 (1989), 169–175.

[34] Liu YP, Vertex partition of loopless planar rooted maps (in Chinese), *Acta Math. Sin.*, 12 (1989), 213–217.

[35] Liu YP, On face partition equation for enumerating simple planar maps (in Chinese), *Acta Math. Sin.*, 12 (1989), 292–295.

[36] Liu YP, On the vertex partition equation of rooted loopless planar maps, *Acta Math. Sci.*, 10 (1990), 167–172.

[37] Liu YP, On loopless Eulerian vertex partition equation (in Chinese), *Kexue Tongbao*, 35 (1990), 1041–1044.

[38] Liu YP, On the loopless Eulerian vertex partition equation, *Chin. Sci. Bull.*, 36 (1991), 1585–1589.

[39] Liu YP, Enumeration of planar rooted maps with vertex partition (in Chinese), *Acta Appl. Math. Cumput. Math.*, 5(1991), 1–20.

[40] Liu YP, On the number of Eulerian planar maps, Research Report, Series A, No. 24, Dept. Statistics, Univ. Rome "La Sapienza", 1989; Also in *Acta Math. Sci.*, 12 (1992), 418–423.

[41] Liu YP, On the vertex partition equation of loopless Eulerian planar maps, *Acta Math. Appl. Sin. Engl. Ser.*, 8 (1992), 45–58.

[42] Liu YP, A note on the number of bipartite planar maps, *Acta Math. Sci.*, 12 (1992), 85–88.

[43] Liu YP, On functional equations arising from map enumerations, *Discrete Math.*. 123 (1993), 93–109.

[44] Liu YP, *Enumerative Theory of Maps*, Kluwer (Springer) Academic Publishers/Science Press, Dordrecht, Boston, London/Beijing, New York, 1999.

[45] Liu YP, *Theory of Counting Combinatorial Maps* (in Chinese), Science Press, Beijing, 2001.

[46] Liu YP, *Advances in Combinatorial Maps* (in Chinese), Northern Jiaotong Univ. Press, Beijing, 2003.

[47] Liu YP, Census of rooted Petal Bundles (in Chinese), *J. Xinyang Norm. Univ., N. S.*, 17 (2004).

[48] Liu YP, *Theory of Counting Combinatorial Maps* (in Chinese, 2nd edition), Science Press, Beijing, 2008.

[49] Liu YP, New method for counting trees with vertex partition (in Chinese), *Sci. China*, A38 (2008), 477–480.
[50] Liu YP, A new method for counting trees with vertex partition, *Sci. China Math.*, A5111 (2008), 2000–2004.
[51] Liu YP, *General Theory of Map Census*, Science Press, Beijing, 2009.
[52] Liu YP, *Theory of Graphic Embeddability* (in Chinese, 2nd edition), Science Press, Beijing, 2010.
[53] Liu YP, Partial differential equations of quadrangulations on surfaces (in Chinese), *J. Kunming Univ. Technol., N. S.*, 37(5) (2012), 79–87, 94.
[54] Liu YP, Partial differential equations on surfaces (in Chinese), *J. Yulin Norm. Univ., N. S.*, 33(2) (2012), 2–9.
[55] Liu YP, Non-orientable maximum genus of graphs (in Chinese), *Sci. Sin.*, Special Issue on Math. I (1979), 191–201.
[56] Liu YP, *Introductory Map Theory*, Kapa & Omega, Glendale, 2010.
[57] Liu YP, On a meson equation of surface type, *J. Kunming Univ. Sci. Tech., N. S.*, 37(1) (2012), 82–89.
[58] Liu YP, A meson equation of surface type, *J. Jishou Univ., N. S.*, 33(5) (2012), 56–62.
[59] Liu YP, *Theory of Combinatorial Functional Equations* (in Chinese), USTC Press, Hefei, 2015.
[60] Liu YP, *Selected Publications of Y. P. Liu*, Times Literature Press, Beijing, Hongkang, 2015.
[61] Liu YP, *Semi-Empty Collections in Mathematics*, Times Culture Press, Beijing, Hongkang, 2016.
[62] Liu YP, *Algebraic Elements of Graphs*, Walter de Gruyter GmbH, Berlin, Munich, Boston, 2017.
[63] Mullin RC, Rota GC, On the foundations of combinatorial theory III, theory of binomial enumeration, in *Graph Theory and Its Applications*, Ed. by Harris B, Academic Press, New York, 1970, 167–213.
[64] Pan LY, Liu YP, General investigation of loopless maps on surface, *J. Xinyang Teach. Coll., N. S.*, 26(1) (2013), 1–6.
[65] Ren H, Liu YP, On the number of regular maps on the projective plane, *Util. Math.*, 56 (1999), 201–209.
[66] Ren H, Liu YP, Counting rooted Eulerian maps on the projective plane, *Acta Math. Sci.*, 20 (2000), 169–174.
[67] Ren H, Liu YP, Counting rooted near-triangulations on the cylinder, *J. Math. Res. Exposition*, 20 (2000), 529–533.
[68] Ren H, Liu YP, Enumerating near 4-regular maps on the sphere and the torus, *Discrete Appl. Math.*, 110 (2001), 273–288.
[69] Ren H, Liu YP, 4 regular maps on the Klein bottle, *J. Comb. Theory*, B82 (2001), 118–155.
[70] Ren H, Liu YP, Enumeration of simple bipartite maps on the sphere and the projective plane, *Discrete Math.*, 242 (2002), 187–200.
[71] Ren H, Liu YP, The number of loopless 4-regular maps on the projective plane, *J. Comb. Theory*, B84 (2002), 84–99.
[72] Ren H, Liu YP, Enumeration of Eulerian maps on the torus, *J. Math. Res. Exposition*, 22(1) (2002), 42–48.
[73] Ren H, Deng M, Liu YP, Enumeration of nonseparable near triangulations in thr projective plane, *Acta Math. Sci.*, 35A (2005), 839–845.
[74] Roman SM, Rota GC, The umbral culculas, *Adv. Math.*, 27 (1978), 95–188.
[75] Rota GC, The number of partitions of a set, *Am. Math. Mon.*, 71 (1964), 498–504.
[76] Rota GC, Kahaner D, Odlyzko A, On the foundations of combinatorial theory VIII, Finite operator calculus, *J. Math. Anal. Appl.*, 42 (1973), 684–760.
[77] Rota GC, Taylor BD, The classical umbral calculus, *SIAM J. Math. Anal.*, 25 (1994), 694–711.

[78] Rota GC, Shen J, Taylor BD, All polynomial of binomial type are represented by Abel polynomials, *Ann. Sc. Norm. Super. Pisa, Cl. Sci.*, 25(4) (1997), 731–738.
[79] Shafarevich IR, *Basic Notions of Algebra*, Springer, Berlin, 2005.
[80] Tutte WT, A Census of planar triangulations, *Can. J. Math.*, 14 (1962), 21–38.
[81] Tutte WT, A census of Hamiltonian polygons, *Can. J. Math. Soc.*, 68 (1962), 402–417.
[82] Tutte WT, A census of slicings, *Can. J. Math.*, 14 (1962), 708–722.
[83] Tutte WT, A new branch of enumerative graph theory, *Bull. Am. Math. Soc.*, 68 (1962), 500–504.
[84] Tutte WT, A census of planar maps, *Can. J. Math.*, 15 (1963), 249–271.
[85] Tutte WT, The number of planted plane trees with a given partition, *Am. Math. Mon.*, 71 (1964), 272–277.
[86] Tutte WT, Chromatic solutions II, *Can. J. Math.*, 34 (1982), 952–960.
[87] Tutte WT, Dichromatic sums for rooted maps, *Proc. Symp. Pure Math.*, 19 (1971), 235–245.
[88] Wu FE, Liu YP, Enumeration of one vertex maps with the face partition, *Acta Math. Sci.*, 20 (2000), 229–234.
[89] Xu Y, Liu YP, The number of pan–fan maps on the projective plane, *Util. Math.*, 72 (2007), 279–286.
[90] Zhang YL, Liu YP, Cai JL, Enumeration of unicursal planar near-triangulations, *Ars Comb.*, 97 (2010), 383–388.

Index

Bei Fragen zur Produktsicherheit wenden Sie sich bitte an:
If you have any questions regarding product safety,
please contact:

Walter de Gruyter GmbH
Genthiner Straße 13
10785 Berlin
productsafety@degruyterbrill.com